Birkhäuser

Trends in Mathematics

Trends in Mathematics is a series devoted to the publication of volumes arising from conferences and lecture series focusing on a particular topic from any area of mathematics. Its aim is to make current developments available to the community as rapidly as possible without compromise to quality and to archive these for reference.

Proposals for volumes can be submitted using the Online Book Project Submission Form at our website www.birkhauser-science.com.

Material submitted for publication must be screened and prepared as follows:

All contributions should undergo a reviewing process similar to that carried out by journals and be checked for correct use of language which, as a rule, is English. Articles without proofs, or which do not contain any significantly new results, should be rejected. High quality survey papers, however, are welcome.

We expect the organizers to deliver manuscripts in a form that is essentially ready for direct reproduction. Any version of $\TeX$ is acceptable, but the entire collection of files must be in one particular dialect of $\TeX$ and unified according to simple instructions available from Birkhäuser.

Furthermore, in order to guarantee the timely appearance of the proceedings it is essential that the final version of the entire material be submitted no later than one year after the conference.

More information about this series at http://www.springer.com/series/4961

Shahla Molahajloo • M. W. Wong
Editors

Analysis of Pseudo-Differential Operators

Birkhäuser

Editors
Shahla Molahajloo
Department of Mathematics
Institute for Advanced Studies in Basic Sciences
Zanjan, Iran

M. W. Wong
Department of Mathematics and Statistics
York University
Toronto, ON, Canada

ISSN 2297-0215 ISSN 2297-024X (electronic)
Trends in Mathematics
ISBN 978-3-030-05167-9 ISBN 978-3-030-05168-6 (eBook)
https://doi.org/10.1007/978-3-030-05168-6

Library of Congress Control Number: 2019935501

Mathematics Subject Classification (2010): 47-XX, 46-XX

This book is published under the imprint Birkhäuser, www.birkhauser-science.com by the registered company Springer Nature Switzerland AG.
The registered company address is: Gewerbestrasse 11, 6330 Cham, Switzerland

Preface

Since the 2003 ISAAC Congress at York University, it has become a tradition that a volume based on the special session on pseudo-differential operators be published. It is not only intended to document the event, but also to provide guidance for future research on pseudo-differential operators and related topics.

The 11th ISAAC Congress was held at Linnæus University in Sweden on August 14–18, 2018. This volume, as a sequel to its predecessors, is based on talks given at the congress and invited articles by experts in the field.

There are ten chapters in this volume, titled "Analysis of Pseudo-Differential Operators." The first four chapters address the functional analysis of pseudo-differential operators in a broad range of settings, from $\mathbb{Z}$ to $\mathbb{R}^n$, to compact and Hausdorff groups. Chapters 5 and 6 focus on operators on Lie groups and manifolds with edge. The next two chapters discuss topics in probability, while the last two chapters cover topics in differential equations.

It is hoped that these volumes on pseudo-differential operators published by Birkhäuser in Basel over a span of fifteen years have served and will continue to serve as useful reference guides for young mathematicians aspiring to explore new directions in pseudo-differential operators. It is also our firm belief that these volumes on pseudo-differential operators will continue to grow and develop in unforeseen directions, thanks to the input of new generations of mathematicians.

Zanjan, Iran — Shahla Molahajloo
Toronto, ON, Canada — M. W. Wong

Contents

Discrete Analogs of Wigner Transforms and Weyl Transforms

Shahla Molahajloo and M. W. Wong

Abstract We first introduce the discrete Fourier–Wigner transform and the discrete Wigner transform acting on functions in $L^2(\mathbb{Z})$. We prove that properties of the standard Wigner transform of functions in $L^2(\mathbb{R}^n)$ such as the Moyal identity, the inversion formula, time-frequency marginal conditions, and the resolution formula hold for the Wigner transforms of functions in $L^2(\mathbb{Z})$. Using the discrete Wigner transform, we define the discrete Weyl transform corresponding to a suitable symbol on $\mathbb{Z} \times \mathbb{S}^1$. We give a necessary and sufficient condition for the self-adjointness of the discrete Weyl transform. Moreover, we give a necessary and sufficient condition for a discrete Weyl transform to be a Hilbert–Schmidt operator. Then we show how we can reconstruct the symbol from its corresponding Weyl transform. We prove that the product of two Weyl transforms is again a Weyl transform and an explicit formula for the symbol of the product of two Weyl transforms is given. This result gives a necessary and sufficient condition for the Weyl transform to be in the trace class.

Keywords Fourier–Wigner transform · Wigner transform · Weyl transform · Moyal identity · Time-frequency marginal conditions · Wigner inversion formula · Weyl inversion formula · Kernels · Hilbert–Schmidt operators · Trace class operators · Twisted convolution · Weyl calculus

Mathematics Subject Classification (2000) 47F05, 47G30

This research has been supported by the Natural Sciences and Engineering Research Council of Canada under Discovery Grant 0008562.

S. Molahajloo
Department of Mathematics, Institute for Advanced Studies in Basic Sciences, Zanjan, Iran

M. W. Wong (✉)
Department of Mathematics and Statistics, York University, Toronto, ON, Canada
e-mail: mwwong@mathstat.yorku.ca

S. Molahajloo, M. W. Wong (eds.), *Analysis of Pseudo-Differential Operators*, Trends in Mathematics, https://doi.org/10.1007/978-3-030-05168-6_1

1 Introduction

To put this paper in perspective, we first recall the Wigner transform and the Weyl transform mapping functions in $L^2(\mathbb{R}^n)$ into functions on, respectively, $\mathbb{R}^n \times \mathbb{R}^n$ and $\mathbb{R}^n$.

Let $\sigma \in L^2(\mathbb{R}^n \times \mathbb{R}^n)$. Then the Weyl transform $W_\sigma : L^2(\mathbb{R}^n) \to L^2(\mathbb{R}^n)$ corresponding to the symbol σ is defined by

$$(W_\sigma f, g)_{L^2(\mathbb{R}^n)} = (2\pi)^{-n/2} \int_{\mathbb{R}^n} \int_{\mathbb{R}^n} \sigma(x, \xi) W(f, g)(x, \xi)\, dx\, d\xi$$

for all f and g in $L^2(\mathbb{R}^n)$, where $W(f, g)$ is the Wigner transform of f and g defined by

$$W(f, g)(x, \xi) = (2\pi)^{-n/2} \int_{\mathbb{R}^n} e^{-i\xi \cdot p} f\left(x + \frac{p}{2}\right) \overline{g\left(x - \frac{p}{2}\right)} dp, \quad x, \xi \in \mathbb{R}^n.$$

Closely related to the Wigner transform $W(f, g)$ of f and g in $L^2(\mathbb{R}^n)$ is the Fourier–Wigner transform $V(f, g)$ given by

$$V(f, g)(q, p) = (2\pi)^{-n/2} \int_{\mathbb{R}^n} e^{iq \cdot y} f\left(y + \frac{p}{2}\right) \overline{g\left(y - \frac{p}{2}\right)} dy, \quad q, p \in \mathbb{R}^n.$$

Weyl transforms and Wigner transforms on $\mathbb{R}^n$ have been extensively studied in [5, 13] among others.

Weyl transforms on groups such as the Heisenberg group, the upper half plane, and the Poincaré unit disk are investigated in [8, 10–12]. Closely related to Weyl transforms are pseudo-differential operators on groups. See, for instance, [4, 7, 9, 15].

The strategy that we use to develop the Weyl transform on $\mathbb{Z}$ is to have a look at the case of $\mathbb{R}^n$, where the symbol σ is a function on $\mathbb{R}^n \times \mathbb{R}^n$. Recent works in pseudo-differential operators and Weyl transforms on topological groups G suggest that the correct phase space to work in is $G \times \widehat{G}$, where $\widehat{G}$ is the dual group of G. That the dual group of $\mathbb{R}^n$ is the same as $\mathbb{R}^n$ is the reason why the phase space on which symbols are defined is $\mathbb{R}^n \times \mathbb{R}^n$.

In the case of the group $\mathbb{Z}$ in this paper, the dual group is the unit circle $\mathbb{S}^1$ centered at the origin and the phase space $G \times \widehat{G}$ is then $\mathbb{Z} \times \mathbb{S}^1$.

For $1 \le p < \infty$, the set of all measurable functions F on $\mathbb{Z}$ such that

$$\|F\|_{L^p(\mathbb{Z})}^p = \sum_{n \in \mathbb{Z}} |F(n)|^p < \infty$$

is denoted by $L^p(\mathbb{Z})$. We define $L^p(\mathbb{S}^1)$ to be the set of all measurable functions f on the unit circle $\mathbb{S}^1$ with center at the origin for which

$$\|f\|_{L^p(\mathbb{S}^1)}^p = \frac{1}{2\pi}\int_{-\pi}^{\pi} |f(\theta)|^p\, d\theta < \infty.$$

We define the Fourier transform $\mathcal{F}_{\mathbb{Z}}F$ of $F \in L^1(\mathbb{Z})$ to be the function on $\mathbb{S}^1$ by

$$(\mathcal{F}_{\mathbb{Z}}F)(\theta) = \sum_{n\in\mathbb{Z}} e^{in\theta} F(n), \quad \theta \in [-\pi, \pi].$$

If f is a suitable function on $\mathbb{S}^1$, then we define the Fourier transform $\mathcal{F}_{\mathbb{S}^1} f$ of f to be the function on $\mathbb{Z}$ by

$$\left(\mathcal{F}_{\mathbb{S}^1} f\right)(n) = \frac{1}{2\pi}\int_{-\pi}^{\pi} e^{-in\theta} f(\theta)\, d\theta, \quad n \in \mathbb{Z}.$$

Note that $\mathcal{F}_{\mathbb{Z}} : L^2(\mathbb{Z}) \to L^2(\mathbb{S}^1)$ is a surjective isomorphism. In fact,

$$\mathcal{F}_{\mathbb{Z}} = \mathcal{F}_{\mathbb{S}^1}^{-1} = \mathcal{F}_{\mathbb{S}^1}^{*}$$

and

$$\|\mathcal{F}_{\mathbb{Z}}F\|_{L^2(\mathbb{S}^1)} = \|F\|_{L^2(\mathbb{Z})}, \quad F \in L^2(\mathbb{Z}).$$

Let H be a suitable function on $\mathbb{S}^1 \times \mathbb{Z}$. Then we define the Fourier transform $\mathcal{F}_{\mathbb{S}^1\times\mathbb{Z}}H$ of H to be the function on $\mathbb{Z} \times \mathbb{S}^1$ by

$$\left(\mathcal{F}_{\mathbb{S}^1\times\mathbb{Z}}H\right)(m,\theta) = \frac{1}{2\pi}\int_{-\pi}^{\pi} \sum_{n\in\mathbb{Z}} e^{-im\phi+in\theta} H(\phi, n)\, d\phi, \quad (m,\theta) \in \mathbb{Z} \times \mathbb{S}^1.$$

Similarly, for all suitable functions K on $\mathbb{Z} \times \mathbb{S}^1$, we define the Fourier transform $\mathcal{F}_{\mathbb{Z}\times\mathbb{S}^1}K$ of K to be the function on $\mathbb{S}^1 \times \mathbb{Z}$ by

$$(\mathcal{F}_{\mathbb{Z}\times\mathbb{S}^1}K)(\theta, m) = \frac{1}{2\pi}\int_{-\pi}^{\pi} \sum_{n\in\mathbb{Z}} e^{-im\phi+in\theta} K(n, \phi)\, d\phi, \quad (\theta, m) \in \mathbb{S}^1 \times \mathbb{Z}.$$

For $1 \le p < \infty$, we define $L^p(\mathbb{Z}\times\mathbb{S}^1)$ to be the space of all measurable functions h on $\mathbb{Z} \times \mathbb{S}^1$ such that

$$\|h\|_{L^p(\mathbb{Z}\times\mathbb{S}^1)}^p = \frac{1}{2\pi}\sum_{n\in\mathbb{Z}}\int_{-\pi}^{\pi} |h(n,\theta)|^p\, d\theta < \infty.$$

In Sect. 2, we define the Fourier–Wigner transform and the Wigner transform as mappings from $L^2(\mathbb{Z})$ into, respectively, $L^2(\mathbb{Z} \times \mathbb{S}^1)$ and $L^2(\mathbb{S}^1 \times \mathbb{Z})$. Then we show that the discrete Fourier–Wigner transform and the discrete Wigner transform satisfy the Moyal identity. We give an inversion formula to reconstruct a function from its discrete Wigner transform up to a constant factor. Then we give the time and frequency marginal conditions and a convolution theorem for the discrete Wigner transform. The results in this section are analogs of the results for the Wigner transforms on $\mathbb{R}^n$ given in [1, 13]. In Sect. 3, we use the discrete Wigner transform to define the Weyl transform on $\mathbb{Z}$. A characterization of Hilbert–Schmidt discrete Weyl transforms is also given. The Weyl inversion formula recovering a symbol from the corresponding discrete Weyl transform is given. In Sect. 4, we present the Weyl calculus giving the symbol of the adjoint of a discrete Weyl transform on $L^2(\mathbb{Z})$ and the symbol of the product of two discrete Weyl transforms. The adjoint formula gives a characterization of self-adjoint discrete Weyl transforms and the product formula gives a characterization of trace class discrete Weyl transforms.

We use $\mathbb{Z}_e$ and $\mathbb{Z}_o$ to denote, respectively, the set of all even integers and the set of all odd integers.

2 Discrete Fourier–Wigner Transforms and Discrete Wigner Transforms

Let $F \in L^2(\mathbb{Z})$. Then for all $(n, \theta) \in \mathbb{Z} \times \mathbb{S}^1$, we define $\rho(n, \theta)F$ to be the function on $\mathbb{Z}$ by

$$(\rho(n,\theta)F)(k) = \begin{cases} e^{i(k+\frac{n}{2})\theta} F(k+n), & n \in \mathbb{Z}_e, \\ e^{i(k+\frac{n-1}{2})\theta} F(k+n), & n \in \mathbb{Z}_o, \end{cases}$$

for all $k \in \mathbb{Z}$. Note that for all $(n, \theta) \in \mathbb{Z} \times \mathbb{S}^1$, $\rho(n, \theta) : L^2(\mathbb{Z}) \to L^2(\mathbb{Z})$ is a unitary operator and

$$\rho(n, \theta)^* = \rho(-n, -\theta).$$

For all functions F and G in $L^2(\mathbb{Z})$, we define the Fourier–Wigner transform $V(F, G)$ of F and G to be the function on $\mathbb{Z} \times \mathbb{S}^1$ by

$$V(F, G)(n, \theta) = (\rho(n, \theta)F, G)_{L^2(\mathbb{Z})}, \quad (n, \theta) \in \mathbb{Z} \times \mathbb{S}^1.$$

Therefore for all $(n, \theta) \in \mathbb{Z} \times \mathbb{S}^1$,

$$V(F,G)(n,\theta) = \begin{cases} \sum_{k\in\mathbb{Z}} e^{i(k+\frac{n}{2})\theta} F(k+n)\overline{G(k)}, & n \in \mathbb{Z}_e, \\ \sum_{k\in\mathbb{Z}} e^{i(k+\frac{n-1}{2})\theta} F(k+n)\overline{G(k)}, & n \in \mathbb{Z}_o. \end{cases}$$

By the change of variables from k to m using

$$\begin{cases} m = k + \frac{n}{2}, & n \in \mathbb{Z}_e, \\ m = k + \frac{n-1}{2}, & n \in \mathbb{Z}_o, \end{cases}$$

we get

$$V(F,G)(n,\theta) = \begin{cases} \sum_{m\in\mathbb{Z}} e^{im\theta} F(m+\frac{n}{2})\overline{G(m-\frac{n}{2})}, & n \in \mathbb{Z}_e, \\ \sum_{m\in\mathbb{Z}} e^{im\theta} F(m+\frac{n+1}{2})\overline{G(m-\frac{n-1}{2})}, & n \in \mathbb{Z}_o. \end{cases}$$

In fact, if we let

$$H_n(m) = \begin{cases} F(m+\frac{n}{2})\overline{G(m-\frac{n}{2})}, & n \in \mathbb{Z}_e, \\ F(m+\frac{n+1}{2})\overline{G(m-\frac{n-1}{2})}, & n \in \mathbb{Z}_o. \end{cases}$$

Then

$$V(F,G)(n,\theta) = (\mathcal{F}_{\mathbb{Z}} H_n)(\theta). \tag{2.1}$$

We have the following Moyal identity for the discrete Fourier–Wigner transform.

Theorem 2.1 *Let F_1, F_2, G_1, and G_2 be functions in $L^2(\mathbb{Z})$. Then*

$$\big(V(F_1,G_1), V(F_2,G_2)\big)_{L^2(\mathbb{Z}\times\mathbb{S}^1)} = \big(F_1,F_2\big)_{L^2(\mathbb{Z})}\overline{\big(G_1,G_2\big)_{L^2(\mathbb{Z})}}.$$

Proof For $j = 1, 2$, we let

$$H_{j,n}(m) = \begin{cases} F_j(m+\frac{n}{2})\overline{G_j(m-\frac{n}{2})}, & n \in \mathbb{Z}_e, \\ F_j(m+\frac{n+1}{2})\overline{G_j(m-\frac{n-1}{2})}, & n \in \mathbb{Z}_o. \end{cases}$$

Then by (2.1) and the Parseval identity,

$$\begin{aligned} &\frac{1}{2\pi}\int_{-\pi}^{\pi} V(F_1,G_1)(n,\theta)\overline{V(F_2,G_2)(n,\theta)}\,d\theta \\ &= \frac{1}{2\pi}\int_{-\pi}^{\pi} \big(\mathcal{F}_{\mathbb{Z}} H_{1,n}\big)(\theta)\overline{\big(\mathcal{F}_{\mathbb{Z}} H_{2,n}\big)(\theta)}\,d\theta \\ &= \sum_{m\in\mathbb{Z}} H_{1,n}(m)\overline{H_{2,n}(m)}. \end{aligned}$$

Therefore

$$(V(F_1,G_1), V(F_2,G_2))_{L^2(\mathbb{Z}\times\mathbb{S}^1)} = \sum_{n\in\mathbb{Z}}\sum_{m\in\mathbb{Z}} H_{1,n}(m)\overline{H_{2,n}(m)}.$$

If $n \in \mathbb{Z}_e$, then we make the change of variables from (m, n) to (k_1, l_1) by $k_1 = m + \frac{n}{2}$ and $l_1 = m - \frac{n}{2}$. If $n \in \mathbb{Z}_o$, then the change of variables from (m, n) to (k_2, l_2) is given by $k_2 = m + \frac{n+1}{2}$ and $l_2 = m - \frac{n-1}{2}$. We get

$$
\begin{aligned}
&(V(F_1, G_1), V(F_2, G_2))_{L^2(\mathbb{S}^1\times\mathbb{Z})}\\
&= \sum_{l_1\in\mathbb{Z}} \sum_{\substack{k_1\in\mathbb{Z}\\ k_1+l_1\in\mathbb{Z}_e}} F_1(k_1)\overline{G_1(l_1)F_2(k_1)}G_2(l_1)\\
&+ \sum_{l_2\in\mathbb{Z}} \sum_{\substack{k_2\in\mathbb{Z}\\ k_2+l_2\in\mathbb{Z}_o}} F_1(k_2)\overline{G_1(l_2)F_2(k_2)}G_2(l_2)\\
&= \sum_{k\in\mathbb{Z}} \sum_{l\in\mathbb{Z}} F_1(k)\overline{G_1(l)F_2(k)}G_2(l)\\
&= (F_1, F_2)_{L^2(\mathbb{Z})}\overline{(G_1, G_2)_{L^2(\mathbb{Z})}}.
\end{aligned}
$$

□

Let F and G be functions in $L^2(\mathbb{Z})$. Then we define the Wigner transform $W(F, G)$ of F and G to be the function on $\mathbb{S}^1 \times \mathbb{Z}$ by

$$W(F, G) = \mathcal{F}_{\mathbb{Z}\times\mathbb{S}^1} V(F, G).$$

Theorem 2.2 *For all* $(\phi, m) \in \mathbb{S}^1 \times \mathbb{Z}$,

$$
\begin{aligned}
&W(F, G)(\phi, m)\\
&= \sum_{n\in\mathbb{Z}_e} e^{in\phi} F\left(m + \frac{n}{2}\right)\overline{G\left(m - \frac{n}{2}\right)}\\
&+ \sum_{n\in\mathbb{Z}_o} e^{in\phi} F\left(m + \frac{n+1}{2}\right)\overline{G\left(m - \frac{n-1}{2}\right)}.
\end{aligned}
\tag{2.2}
$$

Proof We begin with the definition of the discrete Wigner transform to the effect that

$$
\begin{aligned}
&W(F, G)(\theta, m)\\
&= (\mathcal{F}_{\mathbb{Z}\times\mathbb{S}^1} V(F, G))(\theta, m)\\
&= \frac{1}{2\pi} \int_{-\pi}^{\pi} \sum_{n\in\mathbb{Z}} e^{-im\phi+in\theta}\, V(F, G)(n, \phi)\, d\phi.
\end{aligned}
$$

We carry out the sum over $n \in \mathbb{Z}$ by first performing the sum over $n \in \mathbb{Z}_e$ and then over $n \in \mathbb{Z}_o$. Summing over all even integers gives

$$\begin{aligned}
&\frac{1}{2\pi}\int_{-\pi}^{\pi}\left(\sum_{n\in\mathbb{Z}_e} e^{-im\phi+in\theta}\sum_{k\in\mathbb{Z}} e^{ik\phi}F\left(k+\frac{n}{2}\right)\overline{G\left(k-\frac{n}{2}\right)}\right)d\phi\\
&=\sum_{n\in\mathbb{Z}_e}\sum_{k\in\mathbb{Z}} e^{in\theta}\left(\frac{1}{2\pi}\int_{-\pi}^{\pi} e^{-i(m-k)\phi}d\phi\right)F\left(k+\frac{n}{2}\right)\overline{G\left(k-\frac{n}{2}\right)}\\
&=\sum_{n\in\mathbb{Z}_e} e^{in\theta}F\left(m+\frac{n}{2}\right)\overline{G\left(m-\frac{n}{2}\right)}
\end{aligned}$$

for all $(\theta, m) \in \mathbb{S}^1 \times \mathbb{Z}$. The sum over $n \in \mathbb{Z}_o$ can be calculated similarly. □

Similarly, we have the Moyal identity for the Wigner transform.

Theorem 2.3 *Let F_1, F_2, G_1, and G_2 be functions in $L^2(\mathbb{Z})$. Then*

$$\left(W(F_1, G_1), W(F_2, G_2)\right)_{L^2(\mathbb{S}^1\times\mathbb{Z})} = \left(F_1, G_1\right)_{L^2(\mathbb{Z})}\overline{\left(F_2, G_2\right)_{L^2(\mathbb{Z})}}.$$

As in the case of Wigner transforms on $\mathbb{R}^n$, the following proposition guarantees that for all $F \in L^2(\mathbb{Z})$, $W(F, F)$ is real.

Proposition 2.4 *Let F and G be functions in $L^2(\mathbb{Z})$. Then*

$$W(F, G) = \overline{W(G, F)}.$$

In particular, $W(F, F)$ is a real-valued function on $\mathbb{S}^1 \times \mathbb{Z}$.

Proof For all $(\phi, m) \in \mathbb{S}^1 \times \mathbb{Z}$, we get by (2.2)

$$\begin{aligned}
&\overline{W(F, G)(\phi, m)}\\
&=\sum_{n\in\mathbb{Z}_e} e^{-in\phi}G\left(m-\frac{n}{2}\right)\overline{F\left(m+\frac{n}{2}\right)}\\
&+\sum_{n\in\mathbb{Z}_o} e^{-in\phi}G\left(m-\frac{n-1}{2}\right)\overline{F\left(m+\frac{n+1}{2}\right)}.
\end{aligned}$$

If we change the index of summation from n to k by $n = -k$, then for all $(\phi, m) \in \mathbb{S}^1 \times \mathbb{Z}$,

$$\begin{aligned}
&\overline{W(F, G)(\phi, m)}\\
&=\sum_{k\in\mathbb{Z}_e} e^{ik\phi}G\left(m+\frac{k}{2}\right)\overline{F\left(m-\frac{k}{2}\right)}
\end{aligned}$$

$$+\sum_{k\in\mathbb{Z}_o} e^{ik\phi} G\left(m+\frac{k+1}{2}\right)\overline{F\left(m-\frac{k-1}{2}\right)}$$
$$= W(G,F)(\phi,m).$$

This completes the proof. □

For simplicity, we denote $W(F,F)$ by $W(F)$ for all functions $F \in L^2(\mathbb{Z})$. The following theorem states that we can reconstruct the original function F from its Wigner transform $W(F)$ up to a constant factor.

Theorem 2.5 *Let $F \in L^2(\mathbb{Z})$. Then for all $n \in \mathbb{Z}$,*

$$F(n)\overline{F(0)} = \begin{cases} \frac{1}{2\pi}\int_{-\pi}^{\pi} e^{-in\phi} W(F)(\phi, \frac{n}{2})\, d\phi, & n \in \mathbb{Z}_e, \\ \frac{1}{2\pi}\int_{-\pi}^{\pi} e^{-in\phi} W(F)(\phi, \frac{n-1}{2})\, d\phi, & n \in \mathbb{Z}_o. \end{cases}$$

Proof By (2.1) and the definition of the Wigner transform, for all m and n in $\mathbb{Z}$, we have

$$H_n(m) = (\mathcal{F}_{\mathbb{S}^1}(W(F)(\cdot, m)))(n).$$

First, we assume that $n \in \mathbb{Z}_e$. Then for all $m \in \mathbb{Z}$, we get

$$F\left(m+\frac{n}{2}\right)\overline{F\left(m-\frac{n}{2}\right)} = \frac{1}{2\pi}\int_{-\pi}^{\pi} e^{-in\phi} W(F)(\phi, m)\, d\phi. \tag{2.3}$$

Now, let $m = \frac{n}{2}$. Then

$$F(n)\overline{F(0)} = \frac{1}{2\pi}\int_{-\pi}^{\pi} e^{-in\phi} W(F)\left(\phi, \frac{n}{2}\right) d\phi.$$

Similarly, we obtain $F(n)\overline{F(0)}$, for $n \in \mathbb{Z}_o$ by letting $m = \frac{n-1}{2}$. □

We have the time and frequency marginal conditions for the discrete Wigner transform.

Proposition 2.6 *Let $F \in L^2(\mathbb{Z})$. Then*

(i) *For all $m \in \mathbb{Z}$,*

$$\int_{-\pi}^{\pi} W(F)(\phi, m)\, d\phi = 2\pi |F(m)|^2.$$

(ii) *For all $\phi \in [-\pi, \pi]$,*

$$\sum_{m\in\mathbb{Z}} W(F)(\phi, m) = |(\mathcal{F}_{\mathbb{Z}} F)(\phi)|^2.$$

Proof Let $n = 0$ in (2.3). Then we get part (i). To prove part (ii), we have for all $\phi \in [-\pi, \pi]$,

$$\begin{aligned}
&\sum_{m\in\mathbb{Z}} W(F)(\phi, m)\\
&= (\mathcal{F}_{\mathbb{Z}}(W(F)(\phi, \cdot)))(0)\\
&= \sum_{m\in\mathbb{Z}}\sum_{n\in\mathbb{Z}_e} e^{in\phi} F\left(m+\frac{n}{2}\right)\overline{F\left(m-\frac{n}{2}\right)}\\
&+ \sum_{m\in\mathbb{Z}}\sum_{n\in\mathbb{Z}_o} e^{in\phi} F\left(m+\frac{n+1}{2}\right)\overline{F\left(m-\frac{n-1}{2}\right)}.
\end{aligned}$$

For all $n \in \mathbb{Z}_e$, we make the change of variables from (m, n) to (k_1, l_1) by $k_1 = m + \frac{n}{2}$ and $l_1 = m - \frac{n}{2}$. Then we get

$$\begin{cases} m = \frac{k_1+l_1}{2}, \\ n = k_1 - l_1, \end{cases} \tag{2.4}$$

and for all $n \in \mathbb{Z}_o$, using the change of variables from (m, n) to (k_2, l_2) given by $k_2 = m + \frac{n+1}{2}$ and $l_2 = m - \frac{n-1}{2}$, we get

$$\begin{cases} m = \frac{k_2+l_2-1}{2}, \\ n = k_2 - l_2. \end{cases} \tag{2.5}$$

Therefore we get

$$\begin{aligned}
&\sum_{m\in\mathbb{Z}} W(F)(\phi, m)\, d\phi\\
&= \sum_{\substack{k_1\in\mathbb{Z}\\ k_1+l_1\in\mathbb{Z}_e}}\sum_{l_1\in\mathbb{Z}} e^{i(k_1-l_1)\phi} F(k_1)\overline{F(l_1)} + \sum_{\substack{k_2\in\mathbb{Z}\\ k_2+l_2\in\mathbb{Z}_o}}\sum_{l_2\in\mathbb{Z}} e^{i(k_2-l_2)\phi} F(k_2)\overline{F(l_2)}\\
&= \sum_{k\in\mathbb{Z}}\sum_{l\in\mathbb{Z}} e^{i(k-l)\phi} F(k)\overline{F(l)}\\
&= |\,(\mathcal{F}_{\mathbb{Z}}F)\,(\phi)|^2
\end{aligned}$$

and the proof is complete. □

Let $T : L^2(\mathbb{Z}\times\mathbb{Z}) \to L^2(\mathbb{Z}\times\mathbb{Z})$ be the twisting operator defined by

$$(TF)(n, m) = \begin{cases} F(m + \frac{n}{2}, m - \frac{n}{2}), & n \in \mathbb{Z}_e, \\ F(m + \frac{n+1}{2}, m - \frac{n-1}{2}), & n \in \mathbb{Z}_o, \end{cases}$$

for all $F \in L^2(\mathbb{Z}\times\mathbb{Z})$ and all $(n,m) \in \mathbb{Z}\times\mathbb{Z}$. In fact, $T : L^2(\mathbb{Z}\times\mathbb{Z}) \to L^2(\mathbb{Z}\times\mathbb{Z})$ is a unitary operator and its inverse T^{-1} is given by

$$(T^{-1}F)(n,m) = \begin{cases} F(n-m, \frac{m+n}{2}), & n+m \in \mathbb{Z}_e, \\ F(n-m, \frac{m+n-1}{2}), & n+m \in \mathbb{Z}_o. \end{cases}$$

Moreover, for all F and G in $L^2(\mathbb{Z})$,

$$W(F,G)(\phi,m) = \left(\mathcal{F}_{1,\mathbb{Z}}T(F\otimes\overline{G})\right)(\phi,m), \quad (\phi,m) \in \mathbb{S}^1 \times \mathbb{Z}, \tag{2.6}$$

where $F \otimes \overline{G}$ is the tensor product of F and $\overline{G}$ given by

$$(F\otimes\overline{G})(n,m) = F(n)\overline{G}(m), \quad (n,m) \in \mathbb{Z}\times\mathbb{Z},$$

and $\mathcal{F}_{1,\mathbb{Z}}T(F\otimes\overline{G})$ is the partial Fourier transform of $T(F\otimes\overline{G})$ with respect to the first variable. The following proposition gives the shift-invariance of the Wigner transform and the proof is straightforward.

Proposition 2.7 *Let $F \in L^2(\mathbb{Z})$. For $\theta \in [-\pi,\pi]$ and $k \in \mathbb{Z}$, we define the function G on $\mathbb{Z}$ by*

$$G(n) = e^{in\theta}F(n-k), \quad n \in \mathbb{Z}.$$

Then

$$W(G)(\phi,m) = W(F)(\phi+\theta, m-k), \quad (\phi,m) \in \mathbb{S}^1\times\mathbb{Z}.$$

We can now give a result on the Wigner transform of the product of two functions on $\mathbb{Z}$.

Proposition 2.8 *Let F and G be functions in $L^2(\mathbb{Z})$. Then for all (ϕ,m) in $\mathbb{S}^1\times\mathbb{Z}$,*

$$W(FG)(\phi,m) = \Big(W(F)(\cdot,m) * W(G)(\cdot,m)\Big)(\phi),$$

where $*$ is the convolution on $\mathbb{S}^1$ defined by

$$(f*g)(\phi) = \frac{1}{2\pi}\int_{-\pi}^{\pi} f(\phi-\theta)g(\theta)\,d\theta$$

for all f and g in $L^2(\mathbb{S}^1)$.

Proof Let $(\phi,m) \in \mathbb{S}^1\times\mathbb{Z}$. Then

$$\begin{aligned}&(W(F)(\cdot,m) * W(G)(\cdot,m))(\phi)\\ &= \frac{1}{2\pi}\int_{-\pi}^{\pi} W(F)(\phi-\theta,m)W(G)(\theta,m)\,d\theta\end{aligned}$$

$$= \frac{1}{2\pi}\int_{-\pi}^{\pi}\Big\{\sum_{n_1\in\mathbb{Z}_e} e^{in_1(\phi-\theta)}F\left(m+\frac{n_1}{2}\right)\overline{F\left(m-\frac{n_1}{2}\right)}$$

$$+\sum_{n_1\in\mathbb{Z}_o} e^{in_1(\phi-\theta)}F\left(m+\frac{n_1+1}{2}\right)\overline{F\left(m-\frac{n_1-1}{2}\right)}\Big\}\times$$

$$\Big\{\sum_{n_2\in\mathbb{Z}_e} e^{in_2\theta}G\left(m+\frac{n_2}{2}\right)\overline{G\left(m-\frac{n_2}{2}\right)}$$

$$+\sum_{n_2\in\mathbb{Z}_o} e^{in_2\theta}G\left(m+\frac{n_2+1}{2}\right)\overline{G\left(m-\frac{n_2-1}{2}\right)}\Big\}\,d\theta.$$

Since

$$\int_{-\pi}^{\pi} e^{-i\theta(n_1-n_2)}\,d\theta = \begin{cases} 0, & n_1\neq n_2,\\ 2\pi, & n_1=n_2,\end{cases}$$

it follows that

$$(W(F)(\cdot,m)*W(G)(\cdot,m))(\phi)$$

$$=\sum_{n\in\mathbb{Z}_e} e^{in\phi}F\left(m+\frac{n}{2}\right)G\left(m+\frac{n}{2}\right)\overline{F\left(m-\frac{n}{2}\right)G\left(m-\frac{n}{2}\right)}$$

$$+\sum_{n\in\mathbb{Z}_o} e^{in\phi}F\left(m+\frac{n+1}{2}\right)G\left(m+\frac{n+1}{2}\right)\times$$

$$\overline{F\left(m-\frac{n-1}{2}\right)G\left(m-\frac{n-1}{2}\right)}$$

$$= W(FG)(\phi,m)$$

for all $(\phi, m)\in\mathbb{S}^1\times\mathbb{Z}$. □

3 Discrete Weyl Transforms

Let $\sigma\in L^2(\mathbb{Z}\times\mathbb{S}^1)$. Then for all functions F in $L^2(\mathbb{Z})$, we define the Weyl transform $W_\sigma F$ corresponding to the symbol σ by

$$(W_\sigma F, G)_{L^2(\mathbb{Z})} = \frac{1}{2\pi}\sum_{m\in\mathbb{Z}}\int_{-\pi}^{\pi}\sigma(m,\phi)W(F,G)(\phi,m)\,d\phi$$

for all $G \in L^2(\mathbb{Z})$. In fact,

$$
\begin{aligned}
&(W_\sigma F, G)_{L^2(\mathbb{Z})} \\
&= \frac{1}{2\pi} \sum_{m\in\mathbb{Z}} \int_{-\pi}^{\pi} \sum_{n\in\mathbb{Z}_e} \sigma(m,\phi) e^{in\phi} F\left(m+\frac{n}{2}\right) \overline{G\left(m-\frac{n}{2}\right)}\, d\phi \\
&+ \frac{1}{2\pi} \sum_{m\in\mathbb{Z}} \int_{-\pi}^{\pi} \sum_{n\in\mathbb{Z}_o} \sigma(m,\phi) e^{in\phi} F\left(m+\frac{n+1}{2}\right) \overline{G\left(m-\frac{n-1}{2}\right)}\, d\phi.
\end{aligned}
$$

If $n \in \mathbb{Z}_e$, then we make the change of variables from (m, n) to (k_1, l_1) by $k_1 = m + \frac{n}{2}$ and $l_1 = m - \frac{n}{2}$. If $n \in \mathbb{Z}_o$, the change from (m, n) to (k_2, l_2) is effected by $k_2 = m + \frac{n+1}{2}$ and $l_2 = m - \frac{n-1}{2}$. (See (2.4) and (2.5) in this connection.) Therefore we obtain

$$
\begin{aligned}
&2\pi\, (W_\sigma F, G)_{L^2(\mathbb{Z})} \\
&= \int_{-\pi}^{\pi} \sum_{k_1\in\mathbb{Z}} \sum_{\substack{l_1\in\mathbb{Z} \\ k_1+l_1\in\mathbb{Z}_e}} e^{i(k_1-l_1)\phi} \sigma\left(\frac{k_1+l_1}{2}, \phi\right) F(k_1)\overline{G(l_1)}\, d\phi \\
&+ \int_{-\pi}^{\pi} \sum_{k_2\in\mathbb{Z}} \sum_{\substack{l_2\in\mathbb{Z} \\ k_2+l_2\in\mathbb{Z}_o}} e^{i(k_2-l_2)\phi} \sigma\left(\frac{k_2+l_2-1}{2}, \phi\right) F(k_2)\overline{G(l_2)}\, d\phi \\
&= \int_{-\pi}^{\pi} \sum_{l\in\mathbb{Z}} \sum_{k\in\mathbb{Z}} e^{2i(k-l)\phi} \sigma(k,\phi) F(2k-l)\overline{G(l)}\, d\phi \\
&+ \int_{-\pi}^{\pi} \sum_{l\in\mathbb{Z}} \sum_{k\in\mathbb{Z}} e^{i(2(k-l)+1)\phi} \sigma(k,\phi) F(2k+1-l)\overline{G(l)}\, d\phi.
\end{aligned}
$$

Therefore for all $l \in \mathbb{Z}$,

$$
\begin{aligned}
(W_\sigma F)(l) &= \frac{1}{2\pi} \int_{-\pi}^{\pi} \sum_{k\in\mathbb{Z}} e^{2i(k-l)\phi} \sigma(k,\phi) F(2k-l)\, d\phi \\
&+ \frac{1}{2\pi} \int_{-\pi}^{\pi} \sum_{k\in\mathbb{Z}} e^{i(2(k-l)+1)\phi} \sigma(k,\phi) F(2k+1-l)\, d\phi.
\end{aligned}
$$

By another change of variables, we get

$$
(W_\sigma F)(l) = \sum_{m\in\mathbb{Z}} k_\sigma(l, m) F(m),
$$

where k_σ is the kernel of W_σ given by

$$k_\sigma(l,m) = \begin{cases} \frac{1}{2\pi}\int_{-\pi}^{\pi} e^{i(m-l)\phi}\sigma(\frac{m+l}{2},\phi)\,d\phi, & m+l \in \mathbb{Z}_e, \\ \frac{1}{2\pi}\int_{-\pi}^{\pi} e^{i(m-l)\phi}\sigma(\frac{m+l-1}{2},\phi)\,d\phi, & m+l \in \mathbb{Z}_o. \end{cases}$$

Therefore

$$k_\sigma(l,m) = \begin{cases} \left(\mathcal{F}_{2,\mathbb{S}^1}\sigma\right)(\frac{m+l}{2}, l-m), & m+l \in \mathbb{Z}_e, \\ \left(\mathcal{F}_{2,\mathbb{S}^1}\sigma\right)(\frac{m+l-1}{2}, l-m), & m+l \in \mathbb{Z}_o. \end{cases} \tag{3.1}$$

where $\mathcal{F}_{2,\mathbb{S}^1}\sigma$ is the Fourier transform on $\mathbb{S}^1$ of σ with respect to the second variable.

The following theorem is an inversion formula for discrete Weyl transforms. It shows how we can reconstruct the symbol from the corresponding Weyl transform. For Weyl transforms on $\mathbb{R}^n$, the corresponding formula and other related formulas can be found in [2, 3, 6].

Theorem 3.1 *Let $\sigma \in L^2(\mathbb{Z}\times\mathbb{S}^1)$ be such that $\rho(n,\theta)W_\sigma$ is a trace class operator for all $(n,\theta) \in \mathbb{Z}\times\mathbb{S}^1$. Then for all $(\theta,n) \in \mathbb{S}^1 \times \mathbb{Z}$, we have*

$$(\mathcal{F}_{\mathbb{Z}\times\mathbb{S}^1}\sigma)(\theta,n) = \operatorname{tr}(\rho(n,\theta)W_\sigma).$$

Proof Let $F \in L^2(\mathbb{Z})$. First we assume that $n \in \mathbb{Z}_e$. Then for all $l \in \mathbb{Z}$,

$$\begin{aligned} &(\rho(n,\theta)W_\sigma F)(l) \\ &= e^{i(l+\frac{n}{2})\theta}\,(W_\sigma F)(l+n) \\ &= e^{i(l+\frac{n}{2})\theta}\sum_{m\in\mathbb{Z}} k_\sigma(l+n,m)F(m), \end{aligned}$$

where k_σ is the kernel of W_σ. So, for all $l \in \mathbb{Z}$,

$$(\rho(n,\theta)W_\sigma F)(l) = \sum_{m\in\mathbb{Z}} k^\theta(l,m)F(m),$$

where

$$k^\theta(l,m) = e^{i(l+\frac{n}{2})\theta}k_\sigma(l+n,m).$$

Hence

$$\begin{aligned} &\operatorname{tr}(\rho(n,\theta)W_\sigma) \\ &= \sum_{l\in\mathbb{Z}} k^\theta(l,l) \end{aligned}$$

$$= \sum_{l\in\mathbb{Z}} e^{i(l+\frac{n}{2})\theta} k_\sigma(l+n,l)$$

$$= \frac{1}{2\pi}\sum_{l\in\mathbb{Z}} e^{i(l+\frac{n}{2})\theta}\int_{-\pi}^{\pi} e^{-in\phi}\sigma\left(l+\frac{n}{2},\phi\right)d\phi.$$

By the change of variables from l to k by $k = l + \frac{n}{2}$, we get for all $(\theta, n) \in \mathbb{S}^1 \times \mathbb{Z}$,

$$\operatorname{tr}(\rho(n,\theta)W_\sigma) = \left(\mathcal{F}_{\mathbb{Z}\times\mathbb{S}^1}\sigma\right)(\theta,n).$$

Similarly, the above formula holds for all $n \in \mathbb{Z}_o$. □

4 Hilbert–Schmidt Discrete Weyl Transforms

The following proposition gives a class of bounded and Hilbert–Schmidt Weyl transforms on $L^2(\mathbb{Z})$.

Proposition 4.1 *Let $\sigma \in L^2(\mathbb{Z}\times\mathbb{S}^1)$. Then $W_\sigma : L^2(\mathbb{Z}) \to L^2(\mathbb{Z})$ is a bounded linear operator and*

$$\|W_\sigma\|_* \le \|\sigma\|_{L^2(\mathbb{Z}\times\mathbb{S}^1)},$$

where $\|\cdot\|_*$ is the norm in the C^*-algebra of all bounded linear operators on $L^2(\mathbb{Z})$. Moreover, W_σ is a Hilbert–Schmidt operator on $L^2(\mathbb{Z})$ and

$$\|W_\sigma\|_{HS(L^2(\mathbb{Z}))} = \|\sigma\|_{L^2(\mathbb{Z}\times\mathbb{S}^1)}.$$

Proof If we define the function $\tilde{\sigma}$ on $\mathbb{S}^1 \times \mathbb{Z}$ by

$$\tilde{\sigma}(\theta,n) = \sigma(n,\theta), \quad (\theta,n)\in\mathbb{S}^1\times\mathbb{Z},$$

then $\tilde{\sigma} \in L^2(\mathbb{S}^1 \times \mathbb{Z})$ and

$$\|\tilde{\sigma}\|_{L^2(\mathbb{S}^1\times\mathbb{Z})} = \|\sigma\|_{L^2(\mathbb{Z}\times\mathbb{S}^1)}.$$

Let F and G be functions in $L^2(\mathbb{Z})$. Then by Schwarz's inequality and the Moyal identity for the Wigner transform, we have

$$\begin{aligned}|(W_\sigma F, G)_{L^2(\mathbb{Z})}| &= |(\tilde{\sigma}, W(G,F))_{L^2(\mathbb{S}^1\times\mathbb{Z})}| \\ &\le \|\tilde{\sigma}\|_{L^2(\mathbb{S}^1\times\mathbb{Z})}\|F\|_{L^2(\mathbb{Z})}\|G\|_{L^2(\mathbb{Z})} \\ &= \|\sigma\|_{L^2(\mathbb{Z}\times\mathbb{S}^1)}\|F\|_{L^2(\mathbb{Z})}\|G\|_{L^2(\mathbb{Z})}.\end{aligned}$$

Hence

$$\|W_\sigma F\|_{L^2(\mathbb{Z})} \le \|\sigma\|_{L^2(\mathbb{Z}\times\mathbb{S}^1)} \|F\|_{L^2(\mathbb{Z})}.$$

Therefore

$$\|W_\sigma\|_* \le \|\sigma\|_{L^2(\mathbb{Z}\times\mathbb{S}^1)}.$$

By (3.1),

$$\begin{aligned}
\|W_\sigma\|^2_{HS(L^2(\mathbb{Z}))} &= \sum_{m\in\mathbb{Z}} \sum_{l\in\mathbb{Z}} |k_\sigma(m,l)|^2 \\
&= \sum_{\substack{m\in\mathbb{Z} \\ m+l\in\mathbb{Z}_e}} \sum_{l\in\mathbb{Z}} \left| (\mathcal{F}_{2,\mathbb{S}^1}\sigma)\left(\frac{m+l}{2}, l-m\right)\right|^2 \\
&+ \sum_{\substack{m\in\mathbb{Z} \\ m+l\in\mathbb{Z}_o}} \sum_{l\in\mathbb{Z}} \left| (\mathcal{F}_{1,\mathbb{S}^1}\sigma)\left(\frac{m+l-1}{2}, l-m\right)\right|^2.
\end{aligned}$$

By the change of variables and the Parseval identity, we get

$$\begin{aligned}
\|W_\sigma\|^2_{HS(L^2(\mathbb{Z}))} &= \sum_{k\in\mathbb{Z}} \sum_{n\in\mathbb{Z}} \left| (\mathcal{F}_{2,\mathbb{S}^1}\sigma)(n,k)\right|^2 \\
&= \frac{1}{2\pi} \int_{-\pi}^{\pi} \sum_{n\in\mathbb{Z}} |\sigma(n,\phi)|^2 \, d\phi \\
&= \|\sigma\|^2_{L^2(\mathbb{Z}\times\mathbb{S}^1)}.
\end{aligned}$$

□

A Hilbert–Schmidt operator on $A : L^2(\mathbb{Z}) \to L^2(\mathbb{Z})$ is of the form

$$(AF)(n) = \sum_{m\in\mathbb{Z}} h(n,m)F(m), \quad F \in L^2(\mathbb{Z}),$$

where h is a function in $L^2(\mathbb{Z}\times\mathbb{Z})$. The function h is called the kernel of the Hilbert–Schmidt operator A on $L^2(\mathbb{Z})$. The following theorem states that every Hilbert–Schmidt operator on $L^2(\mathbb{Z})$ is a Weyl transform with symbol in $L^2(\mathbb{Z} \times \mathbb{S}^1)$.

Theorem 4.2 *Let $A : L^2(\mathbb{Z}) \to L^2(\mathbb{Z})$ be a Hilbert–Schmidt operator. Then there exists a unique symbol σ in $L^2(\mathbb{Z} \times \mathbb{S}^1)$ such that $A = W_\sigma$.*

Proof Let $h \in L^2(\mathbb{Z} \times \mathbb{Z})$ be such that

$$(AF)(n) = \sum_{m\in\mathbb{Z}} h(n,m)F(m), \quad F \in L^2(\mathbb{Z}).$$

Then for all $F, G \in L^2(\mathbb{Z})$,

$$\begin{aligned}(AF, G)_{L^2(\mathbb{Z})} &= \sum_{n\in\mathbb{Z}}\sum_{m\in\mathbb{Z}} h(n,m)F(m)\overline{G(n)} \\ &= \sum_{n\in\mathbb{Z}}\sum_{m\in\mathbb{Z}} \tilde{h}(m,n)\left(F \otimes \overline{G}\right)(m,n) \\ &= (F \otimes \overline{G}, \overline{\tilde{h}})_{L^2(\mathbb{Z}\times\mathbb{Z})},\end{aligned}$$

where $\tilde{h}$ is the function in $L^2(\mathbb{Z} \times \mathbb{Z})$ such that

$$\tilde{h}(m,n) = h(n,m), \quad (m,n) \in \mathbb{Z} \times \mathbb{Z}.$$

We define the function σ on $\mathbb{Z} \times \mathbb{S}^1$ by

$$\overline{\sigma} = \mathcal{F}_{1,\mathbb{Z}} T \overline{\tilde{h}}^{\sim},$$

where $T : L^2(\mathbb{Z} \times \mathbb{Z}) \to L^2(\mathbb{Z} \times \mathbb{Z})$ is the twisting operator defined earlier. Then $\overline{\tilde{h}} = T^{-1}\mathcal{F}_{1,\mathbb{S}^1}\overline{\tilde{\sigma}}$. Hence we have

$$\begin{aligned}(AF, G)_{L^2(\mathbb{Z})} &= \left(F \otimes \overline{G}, T^{-1}\mathcal{F}_{1,\mathbb{S}^1}\overline{\tilde{\sigma}}\right)_{L^2(\mathbb{Z}\times\mathbb{Z})} \\ &= \left(\mathcal{F}_{1,\mathbb{Z}} T(F \otimes \overline{G}), \overline{\tilde{\sigma}}\right)_{L^2(\mathbb{S}^1\times\mathbb{Z})} \\ &= \left(W(F,G), \overline{\tilde{\sigma}}\right)_{L^2(\mathbb{S}^1\times\mathbb{Z})} \\ &= (W_\sigma F, G)_{L^2(\mathbb{Z})}.\end{aligned}$$

□

5 The Weyl Calculus

The following proposition on the adjoint of a discrete Weyl transform on $L^2(\mathbb{Z})$ follows directly from the definition of the Weyl transform and Proposition 2.4.

Proposition 5.1 *Let $\sigma \in L^2(\mathbb{Z} \times \mathbb{S}^1)$. Then $W_\sigma^* = W_{\overline{\sigma}}$, where $W_\sigma^* : L^2(\mathbb{Z}) \to L^2(\mathbb{Z})$ is the adjoint of $W_\sigma : L^2(\mathbb{Z}) \to L^2(\mathbb{Z})$. In particular, W_σ is self-adjoint if and only if σ is real-valued.*

Proof Let $\sigma \in L^2(\mathbb{Z} \times \mathbb{S}^1)$. Then

$$(W_\sigma^* F, G)_{L^2(\mathbb{Z})} = (F, W_\sigma G)_{L^2(\mathbb{Z})} = \overline{(W_\sigma G, F)_{L^2(\mathbb{Z})}}$$
$$= \frac{1}{2\pi} \sum_{m\in\mathbb{Z}} \int_{-\pi}^{\pi} \overline{\sigma(m, \phi) W(G, F)(\phi, m)}\, d\phi. \tag{5.1}$$

By Proposition 2.4,

$$\overline{W(G, F)} = W(F, G),$$

and hence by (5.1),

$$(W_\sigma^* F, G)_{L^2(\mathbb{Z})} = \frac{1}{2\pi} \sum_{m\in\mathbb{Z}} \int_{-\pi}^{\pi} \overline{\sigma}(m, \phi) W(F, G)(\phi, m)\, d\phi.$$

So,

$$(W_\sigma^* F, G)_{L^2(\mathbb{Z})} = (W_{\overline{\sigma}} F, G)_{L^2(\mathbb{Z})}$$

and the proof is complete. □

Let $\sigma \in L^2(\mathbb{Z} \times \mathbb{S}^1)$. For simplicity, we denote $\mathcal{F}_{\mathbb{Z}\times\mathbb{S}^1}\sigma$ by $\widehat{\sigma}$.

Lemma 5.2 *Let $\sigma \in L^2(\mathbb{Z} \times \mathbb{S}^1)$. Then for all $F \in L^2(\mathbb{Z})$,*

$$(W_\sigma F)(m) = \frac{1}{2\pi} \int_{-\pi}^{\pi} \sum_{n\in\mathbb{Z}} \widehat{\sigma}(\theta, n)\, (\rho(n, \theta) F)(m)\, d\theta, \quad m \in \mathbb{Z}.$$

Proof Let F and G be in $L^2(\mathbb{Z})$. Then using the adjoint formula,

$$(W_\sigma F, G)_{L^2(\mathbb{Z})} = \frac{1}{2\pi} \sum_{m\in\mathbb{Z}} \int_{-\pi}^{\pi} \sigma(m, \phi) W(F, G)(\phi, m)\, d\phi$$
$$= \frac{1}{2\pi} \int_{-\pi}^{\pi} \sum_{n\in\mathbb{Z}} \hat{\sigma}(\theta, n) V(F, G)(n, \theta)\, d\theta$$
$$= \frac{1}{2\pi} \int_{-\pi}^{\pi} \sum_{n\in\mathbb{Z}} \hat{\sigma}(\theta, n)\, (\rho(n, \theta) F, G)_{L^2(\mathbb{Z})}\, d\theta$$
$$= \frac{1}{2\pi} \int_{-\pi}^{\pi} \sum_{n\in\mathbb{Z}} \sum_{m\in\mathbb{Z}} \hat{\sigma}(\theta, n)\, (\rho(n, \theta) F)(m) \overline{G(m)}\, d\theta.$$

Hence

$$(W_\sigma F)(m) = \frac{1}{2\pi}\int_{-\pi}^{\pi}\sum_{n\in\mathbb{Z}}\widehat{\sigma}(\theta,n)\,(\rho(n,\theta)F)(m)\,d\theta, \quad m\in\mathbb{Z}.$$

□

Lemma 5.3 *For all (n,θ) and (k,ϕ) in $\mathbb{Z}\times\mathbb{S}^1$, we have*

$$\rho(n,\theta)\rho(k,\phi) = e^{i[(n,\theta);(k,\phi)]}\rho(n+k,\theta+\phi),$$

where

$$[(n,\theta);(k,\phi)] = \begin{cases} \frac{n}{2}\phi - \frac{k}{2}\theta, & n\in\mathbb{Z}_e,\ k\in\mathbb{Z}_e,\\ \frac{n-1}{2}\phi - \frac{k+1}{2}\theta, & n\in\mathbb{Z}_o,\ k\in\mathbb{Z}_o,\\ \frac{n}{2}\phi - \frac{k-1}{2}\theta, & n\in\mathbb{Z}_e,\ k\in\mathbb{Z}_o,\\ \frac{n+1}{2}\phi - \frac{k}{2}\theta, & n\in\mathbb{Z}_o,\ k\in\mathbb{Z}_e.\end{cases}$$

The proof of the lemma is straightforward. Let F and G be suitable functions on $\mathbb{S}^1\times\mathbb{Z}$. Then we define the twisted convolution $F\circledast G$ of F and G to be the function on $\mathbb{S}^1\times\mathbb{Z}$ by

$$\begin{aligned}&(F\circledast G)(\gamma,l)\\ &= \frac{1}{2\pi}\int_{-\pi}^{\pi}\sum_{k\in\mathbb{Z}} e^{i[(l-k,\gamma-\phi);(k,\phi)]}F(\gamma-\phi,l-k)G(\phi,k)\,d\phi, \quad (\gamma,l)\in\mathbb{S}^1\times\mathbb{Z}.\end{aligned}$$

Let $\sigma\in L^2(\mathbb{Z}\times\mathbb{S}^1)$. The following theorem guarantees that the product of two Weyl transforms is still a Weyl transform.

Theorem 5.4 *Let σ and τ be symbols in $L^2(\mathbb{Z}\times\mathbb{S}^1)$. Then*

$$W_\sigma W_\tau = W_\omega,$$

where

$$\hat{\omega} = \hat{\sigma}\circledast\hat{\tau}.$$

Proof Let $F\in L^2(\mathbb{Z})$. Then for all $m\in\mathbb{Z}$, we get by Lemma 5.2

$$\begin{aligned}&(W_\sigma W_\tau F)(m)\\ &= \frac{1}{2\pi}\int_{-\pi}^{\pi}\sum_{n\in\mathbb{Z}}\widehat{\sigma}(\theta,n)\,(\rho(n,\theta)W_\tau F)(m)\,d\theta\\ &= \frac{1}{(2\pi)^2}\int_{-\pi}^{\pi}\sum_{n\in\mathbb{Z}}\int_{-\pi}^{\pi}\sum_{k\in\mathbb{Z}}\widehat{\sigma}(\theta,n)\widehat{\tau}(\phi,k)\,(\rho(n,\theta)\rho(k,\phi)F)(m)\,d\phi\,d\theta.\end{aligned}$$

Now, by Lemma 5.3, we have

$$
\begin{aligned}
&(W_\sigma W_\tau F)(m) \\
&= \frac{1}{(2\pi)^2} \int_{-\pi}^{\pi} \sum_{n\in\mathbb{Z}} \int_{-\pi}^{\pi} \sum_{k\in\mathbb{Z}} \widehat{\sigma}(\theta, n)\widehat{\tau}(\phi, k) e^{i[(n,\theta);(k,\phi)]} \times \\
&\qquad\qquad (\rho(n+k, \theta+\phi)F)(m)\, d\phi\, d\theta.
\end{aligned}
$$

Let $l = n + k$ and $\gamma = \theta + \phi$. Then we get

$$
\begin{aligned}
&(W_\sigma W_\tau F)(m) \\
&= \frac{1}{(2\pi)^2} \int_{-\pi}^{\pi} \sum_{l\in\mathbb{Z}} \int_{-\pi}^{\pi} \sum_{k\in\mathbb{Z}} \widehat{\sigma}(\gamma-\phi, l-k)\widehat{\tau}(\phi, k)\, e^{i[(l-k,\gamma-\phi);(k,\phi)]} \times \\
&\qquad\qquad (\rho(l, \gamma)F)(m)\, d\phi\, d\gamma.
\end{aligned}
$$

Let $\omega \in L^2(\mathbb{Z} \times \mathbb{S}^1)$ be such that

$$
\hat{\omega} = \hat{\sigma} \circledast \hat{\tau}.
$$

Then

$$
(W_\sigma W_\tau F)(m) = \frac{1}{2\pi} \int_{-\pi}^{\pi} \sum_{l\in\mathbb{Z}} \hat{\omega}(\gamma, l)\, (\rho(l, \gamma)F)(m)\, d\gamma = (W_\omega F)(m)
$$

for all $m \in \mathbb{Z}$. □

As an application of the product formula, we give a characterization of trace class discrete Weyl transforms. It is an analog for the discrete Weyl transform of the characterization of trace class Weyl transforms on $\mathbb{R}^n$ in [14]. Let W be the set defined by

$$
W = \left\{ \mathcal{F}_{\mathbb{S}^1\times\mathbb{Z}}(\hat{\sigma} \circledast \hat{\tau}) : \sigma, \tau \in L^2(\mathbb{Z} \times \mathbb{S}^1) \right\}.
$$

Let $S_1(L^2(\mathbb{Z}))$ be the space of all trace class operators on $L^2(\mathbb{Z})$. The following theorem gives a characterization of trace class discrete Weyl transforms on $L^2(\mathbb{Z})$.

Theorem 5.5 *Let $\sigma \in L^2(\mathbb{Z} \times \mathbb{S}^1)$. Then $W_\sigma : L^2(\mathbb{Z}) \to L^2(\mathbb{Z})$ is in $S_1(L^2(\mathbb{Z}))$ if and only if $\sigma \in W$. Moreover, if $\sigma = \mathcal{F}_{\mathbb{S}^1\times\mathbb{Z}}(\hat{\alpha} \circledast \hat{\beta})$ with α and β in $L^2(\mathbb{Z} \times \mathbb{S}^1)$, then*

$$
\|W_\sigma\|_{S_1(L^2(\mathbb{Z}))} \le \|\alpha\|_{L^2(\mathbb{Z}\times\mathbb{S}^1)} \|\beta\|_{L^2(\mathbb{Z}\times\mathbb{S}^1)}.
$$

Proof First we assume that $\sigma \in W$. Then

$$\sigma = \mathcal{F}_{\mathbb{S}^1 \times \mathbb{Z}}(\hat{\alpha} \circledast \hat{\beta})$$

for some α and β in $L^2(\mathbb{Z} \times \mathbb{S}^1)$. By Theorem 5.4,

$$W_\sigma = W_\alpha W_\beta.$$

Moreover, by Proposition 4.1, W_α and W_β are Hilbert–Schmidt operators. Hence W_σ is the product of two Hilbert–Schmidt operators on $L^2(\mathbb{Z})$ and therefore is in $S_1(L^2(\mathbb{Z}))$. Conversely, assume that $W_\sigma \in S_1(L^2(\mathbb{Z}))$. Then W_σ is the product of two Hilbert–Schmidt operators on $L^2(\mathbb{Z})$. Hence by Theorem 4.2, there exist symbols α and β in $L^2(\mathbb{Z} \times \mathbb{S}^1)$ such that

$$W_\sigma = W_\alpha W_\beta.$$

So, by Theorem 5.4, $\sigma \in W$. □

References

1. L. Cohen, *Time-Frequency Analysis* (Prentice-Hall PTR, Upper Saddle River, 1995)
2. L. Cohen, *The Weyl Operator and Its Generalization* (Birkhäuser, Basel, 2013)
3. L. Cohen, Inverse Weyl transform/operator. J. Pseudo-Differ. Oper. Appl. **8**, 661–678 (2017)
4. A. Dasgupta, M.W. Wong, Pseudo-differential operators on the affine group, in *Pseudo-Differential Operators: Groups, Geometry and Applications*. Trends in Mathematics (Birkhäuser, Basel, 2017), pp. 1–14
5. M. de Gosson, *The Wigner Transform* (World Scientific, Singapore, 2017)
6. X. Duan, M.W. Wong, Pseudo-differential operators for Weyl transforms. Politehn. Univ. Bucharest Sci. Bull. Ser. A: Appl. Math. Phys. **75**, 3–12 (2013)
7. S. Molahajloo, Pseudo-differential operators on $\mathbb{Z}$, in *New Developments in Pseudo-Differential Operators*. Operator Theory: Advances and Applications, vol. 205 (Birkhäuser, Basel, 2009), pp. 213–221
8. S. Molahajloo, Pseudo-differential operators, Wigner transforms and Weyl transforms on the Poincaré unit disk. Compl. Anal. Oper. Theory **12**, 811–833 (2018)
9. S. Molahajloo, M.W. Wong, Ellipticity, Fredholmness and spectral invariance of pseudo-differential operators on $\mathbb{S}^1$. J. Pseudo-Differ. Oper. Appl. **1**, 183–205 (2010)
10. L. Peng, J. Zhao, Weyl transforms associated with the Heisenberg group. Bull. Sci. Math. **132**, 78–86 (2008)
11. L. Peng, J. Zhao, Weyl transforms on the upper half plane. Rev. Mat. Complut. **23**, 77–95 (2010)
12. T. Tate, Weyl pseudo-differential operator and Wigner transform on the Poincaré disk. Ann. Global Anal. Geom **22**, 29–48 (2002)
13. M.W. Wong, *Weyl Transforms* (Springer, New York, 1998)
14. M.W. Wong, Trace class Weyl transforms, in *Recent Advances in Operator Theory and its Applications*. Operator Theory: Advances and Applications, vol. 160 (Birkhäuser, Basel, 2005), pp. 469–478
15. M.W. Wong, *Discrete Fourier Analysis* (Birkhäuser, Basel, 2011)

Characterization and Spectral Invariance of Non-Smooth Pseudodifferential Operators with Hölder Continuous Coefficients

Helmut Abels and Christine Pfeuffer

Abstract Smooth pseudodifferential operators on $\mathbb{R}^n$ can be characterized by their mapping properties between L^p-Sobolev spaces due to Beals and Ueberberg. In applications such a characterization would also be useful in the non-smooth case, for example to show the regularity of solutions of a partial differential equation. Therefore, we will show that every linear operator P, which satisfies some specific continuity assumptions, is a non-smooth pseudodifferential operator of the symbol-class $C^\tau S^m_{1,0}(\mathbb{R}^n \times \mathbb{R}^n)$. The main new difficulties are the limited mapping properties of pseudodifferential operators with non-smooth symbols.

Keywords Non-smooth pseudodifferential operators · Characterization by mapping properties

Mathematics Subject Classification (2000) 35S05, 47G30

In this chapter we study pseudodifferential operators of the form

$$OP(p)(x) := p(x, D_x)u(x) := \int_{\mathbb{R}^n} e^{ix\cdot\xi} p(x,\xi)\hat{u}(\xi) đ\xi \qquad \text{for all } u \in \mathcal{S}(\mathbb{R}^n), x \in \mathbb{R}^n,$$

where $\mathcal{S}(\mathbb{R}^n)$ denotes the Schwartz space of all rapidly decreasing smooth functions, $\hat{u} = \mathscr{F}[u]$ is the Fourier transformation of u, $đ\xi = (2\pi)^{-n}d\xi$, and $p\colon \mathbb{R}^n \times \mathbb{R}^n \to \mathbb{C}$ is a given function, called the symbol of the pseudodifferential operator. In the case that p is a smooth function contained in a suitable symbol class, the pseudodifferential operators have a lot of nice and important algebraic properties, e.g. they are closed under arbitrary compositions and adjoints and have natural mapping properties between Sobolev spaces of arbitrary high order. Moreover, suitable classes of pseudodifferential operators with smooth symbols can

H. Abels · C. Pfeuffer (✉)
Fakultät für Mathematik, Universität Regensburg, Regensburg, Germany
e-mail: Helmut.Abels@mathematik.uni-regensburg.de;
Christine.Pfeuffer@mathematik.uni-halle.de

S. Molahajloo, M. W. Wong (eds.), *Analysis of Pseudo-Differential Operators*,
Trends in Mathematics, https://doi.org/10.1007/978-3-030-05168-6_2

be characterized in terms of their mapping properties and the mapping properties of certain iterated commutators. But, unfortunately, all this in general breaks down, when working with symbols of limited smoothness, e.g., with limited smoothness in the "space" variable x. Such operators arise naturally in the studies of nonlinear partial differential equations, where the symbol depends on the solution, which has a priori only limited smoothness. In the case of limited smoothness in the spatial variable x, the order of the Sobolev spaces, in which such a pseudodifferential operator maps continuously into, is limited by the smoothness of the symbol. As a consequence the composition of two non-smooth pseudodifferential operators is only well defined under some restrictions. Moreover, the composition is in general not a pseudodifferential operator anymore (at least not of the same class). But there are results on compositions up to certain operators of lower order in terms of their mapping properties. All this makes the characterization of pseudodifferential operators much more difficult in the non-smooth than in the smooth case. In this chapter we will present and review some first results in this direction and discuss an application of them to the spectral invariance of these operators.

Now let us comment on the known results in the case of smooth symbols. First characterizations of pseudodifferential operators by their mapping properties between L^2-Sobolev spaces were proven by Beals [5]. These results include a characterization of the Hörmander classes $S^m_{\rho,\delta}(\mathbb{R}^n \times \mathbb{R}^n)$ with $0 \leq \delta \leq \rho \leq 1$ and $\delta < 1$. Here a smooth $p\colon \mathbb{R}^n \times \mathbb{R}^n \to \mathbb{C}$ belongs to $S^m_{\rho,\delta}(\mathbb{R}^n \times \mathbb{R}^n)$ for $m \in \mathbb{R}$ and $0 \leq \delta \leq \rho \leq 1$ if and only if

$$|p|_k^{(m)} := \max_{|\alpha|,|\beta|\leq k} \sup_{x,\xi\in\mathbb{R}^n} |\partial_\xi^\alpha \partial_x^\beta p(x,\xi)| \langle \xi \rangle^{-(m-\rho|\alpha|+\delta|\beta|)} < \infty$$

for all $k \in \mathbb{N}_0$. Moreover, $OPS^m_{\rho,\delta}(\mathbb{R}^n \times \mathbb{R}^n)$ is the set of all pseudodifferential operators with symbols in $S^m_{\rho,\delta}(\mathbb{R}^n \times \mathbb{R}^n)$. A characterization of the latter classes by mapping properties between L^p-Sobolev spaces was first proved by Ueberberg [22]. Further characterizations were obtained by Kryakvin [14], Leopold and Schrohe [16] and Schrohe [19].

In the following we will use the approach by Ueberberg [22] in order to obtain a characterization of non-smooth pseudodifferential operators. Let us note that it is based on the method for characterizing algebras of pseudodifferential operators by Beals [5, 6], Coifman and Meyer [10] and Cordes [8, 9].

A first characterization of pseudodifferential operators with non-smooth symbols was obtained in [3], where symbols in $C^\tau_* S^m_{\rho,0}(\mathbb{R}^n \times \mathbb{R}^n; M)$, $\rho \in \{0, 1\}$, are considered. This characterization was extended and refined in [2] and applied to obtain results on spectral invariance. These results are based on the PhD-thesis of the second author. It is the goal of this contribution to present these results on characterization and spectral invariance in the case of symbols in $C^\tau_* S^m_{\rho,0}(\mathbb{R}^n \times \mathbb{R}^n; M)$ in a self-contained way. We will follow [2, 3, 18], respectively, closely, leaving out some proofs of technical results, generalizing some results and adding some details for a more convenient presentation. To this end let us recall some basic definitions. The Hölder-Zygmund space of order $\tau > 0$ is defined by

$$C_*^\tau(\mathbb{R}^n) := \left\{ f \in \mathcal{S}'(\mathbb{R}^n) : \|f\|_{C_*^\tau} := \sup_{j\in\mathbb{N}_0} 2^{j\tau} \|\mathscr{F}^{-1}[\varphi_j \hat{f}]\|_{L^\infty} < \infty \right\},$$

where $\mathscr{F}^{-1}[u]$ is the inverse Fourier transformation of $u \in \mathcal{S}'(\mathbb{R}^n)$, the dual space of $\mathcal{S}(\mathbb{R}^n)$. Here $(\varphi_j)_{j\in\mathbb{N}_0}$ is a standard dyadic partition of unity on $\mathbb{R}^n$, cf. e.g. [7]. Moreover, $C_*^\tau S^m_{\rho,0}(\mathbb{R}^n \times \mathbb{R}^n; M)$ for $m \in \mathbb{R}$, $\tau > 0$, $0 \le \rho \le 1$, and $M \in [0, \infty]$ is the set of all $p : \mathbb{R}^n \times \mathbb{R}^n \to \mathbb{C}$ such that

i) $\partial_x^\beta p(x, .) \in C^M(\mathbb{R}^n)$ for all $x \in \mathbb{R}^n$,
ii) $\partial_x^\beta \partial_\xi^\alpha p \in C^0(\mathbb{R}^n_x \times \mathbb{R}^n_\xi)$,
iii) $\|\partial_\xi^\alpha p(., \xi)\|_{C_*^\tau(\mathbb{R}^n)} \le C_\alpha \langle\xi\rangle^{m-\rho|\alpha|}$ for all $\xi \in \mathbb{R}^n$

holds for all $\alpha, \beta \in \mathbb{N}_0^n$ with $|\alpha| \le M$ and $|\beta| \le \tau$. If $M \notin \mathbb{N}_0 \cup \{\infty\}$, we additionally assume that for all $\alpha \in \mathbb{N}_0^n$ with $|\alpha| \le M$

iv) $\left\| \partial_\xi^\alpha p(\cdot, \xi + \eta) - \partial_\xi^\alpha p(\cdot, \xi) \right\|_{C_*^\tau(\mathbb{R}^n)} \le C_\alpha |\eta|^{M-\lfloor M \rfloor} \langle\xi\rangle^{m-\rho|\alpha|}$ for all $\eta \in \mathbb{R}^n$

holds. The associated pseudodifferential operator $OP(p) = p(x, D_x)$ is defined as above. The set of all pseudodifferential operators with symbols in the symbol-class $C_*^\tau S^m_{\rho,0}(\mathbb{R}^n \times \mathbb{R}^n; M)$ is denoted by $OPC_*^\tau S^m_{\rho,0}(\mathbb{R}^n \times \mathbb{R}^n; M)$.

The following set of operators plays the central role for the characterization and is defined similarly as in the smooth case, cf. [22] and [5].

Definition 0.1 Let $m \in \mathbb{R}$, $M \in \mathbb{N}_0 \cup \{\infty\}$ and $0 \le \rho \le 1$. Additionally let $\tilde{m} \in \mathbb{N}_0 \cup \{\infty\}$ and $1 < q < \infty$. Then we define $\mathcal{A}^{m,M}_{\rho,0}(\tilde{m}, q)$ as the set of all linear and bounded operators $P : H_q^m(\mathbb{R}^n) \to L^q(\mathbb{R}^n)$, such that for all $l \in \mathbb{N}$, $\alpha_1, \ldots, \alpha_l \in \mathbb{N}_0^n$ and $\beta_1, \ldots, \beta_l \in \mathbb{N}_0^n$ with $|\alpha| \le M$, $|\beta| \le \tilde{m}$ and $|\alpha_1| + |\beta_1| = \ldots = |\alpha_l| + |\beta_l| = 1$ the iterated commutator of P

$$\mathrm{ad}(-ix)^{\alpha_1}\, \mathrm{ad}(D_x)^{\beta_1} \ldots \mathrm{ad}(-ix)^{\alpha_l}\, \mathrm{ad}(D_x)^{\beta_l} P : H_q^{m-\rho|\alpha|}(\mathbb{R}^n) \to L^q(\mathbb{R}^n)$$

is continuous. Here $\alpha := \alpha_1 + \ldots + \alpha_l$ and $\beta := \beta_1 + \ldots + \beta_l$.

We refer to Definition 1.3 below for the definition of the iterated commutators. In the case $M = \infty$ we write $\mathcal{A}^m_{\rho,0}(\tilde{m}, q)$ instead of $\mathcal{A}^{m,\infty}_{\rho,0}(\tilde{m}, q)$. Because of [22, Satz 1.8], one has in the case $M = \tilde{m} = \infty$

$$\mathcal{A}^m_{\rho,0}(\infty, q) \subseteq OPS^m_{\rho,0}(\mathbb{R}^n \times \mathbb{R}^n)$$

with equality if $q = 2$ or $\rho = 1$.

In the case $\tilde{m} \neq \infty$ we have the following results:

Lemma 0.2 *Let* $\tau > 0$, $\tau \notin \mathbb{N}$, $m \in \mathbb{R}$ *and* $\rho \in \{0,1\}$. *Considering* $p \in C^{\tau}_{*} S^{m}_{\rho,0}(\mathbb{R}^n \times \mathbb{R}^n)$ *we get for* $\tilde{m} := \max\{k \in \mathbb{N}_0 : \tau - k > n/2\}$ *and* $1 < q < \infty$:

i) $p(x, D_x) \in \mathcal{A}^{m+n/2}_{0,0}(\lfloor \tau \rfloor, 2)$ *if* $\rho = 0$,
ii) $p(x, D_x) \in \mathcal{A}^{m}_{0,0}(\tilde{m}, 2)$ *if* $\rho = 0$,
iii) $p(x, D_x) \in \mathcal{A}^{m}_{1,0}(\lfloor \tau \rfloor, q)$ *if* $\rho = 1$.

Proof This follows from Remark 1.5 and Theorem 2.5 below. □

Conversely, one obtains for $m \in \mathbb{R}$, $\rho \in \{0,1\}$, $1 < q < \infty$, $\tilde{m} \in \mathbb{N}_0$ with $\tilde{m} > n/q$ and $M \in \mathbb{N}_0 \cup \{\infty\}$ with $M > n+1$ the inclusions

$$\mathcal{A}^{m,M}_{\rho,0}(\tilde{m}, q) \subseteq C^{s}_{*} S^{m}_{\rho,0}(\mathbb{R}^n \times \mathbb{R}^n; \tilde{M} - 1) \cap \mathscr{L}(H^m_q(\mathbb{R}^n), L^q(\mathbb{R}^n)),$$

where $\tilde{M} := M - (n+1) \geq 1$ and $s \in (0, \tilde{m} - n/q]$ with $s \notin \mathbb{N}_0$ is assumed, cf. Lemma 3.11 and Theorem 3.12 below. This is one of the main results of [3]. We note that, as in the smooth case, the characterization of non-smooth pseudodifferential operators of the symbol-class $C^{\tau}_{*} S^{m}_{1,0}(\mathbb{R}^n \times \mathbb{R}^n; M)$ is reduced to the characterization of those ones of the symbol-class $C^{\tau}_{*} S^{m}_{0,0}(\mathbb{R}^n \times \mathbb{R}^n; M)$. To this end we need:

Lemma 0.3 *Let* $m \in \mathbb{R}$, $M \in \mathbb{N}_0 \cup \{\infty\}$ *and* $0 \leq \rho_1 < \rho_2 \leq 1$. *Furthermore, let* $\tilde{m} \in \mathbb{N}_0$ *and* $1 < q < \infty$. *Then*

$$\mathcal{A}^{m,M}_{\rho_2,0}(\tilde{m}, q) \subseteq \mathcal{A}^{m,M}_{\rho_1,0}(\tilde{m}, q).$$

Proof This follows immediately from the embeddings

$$H^{m-\rho_1|\alpha|}_q(\mathbb{R}^n) \hookrightarrow H^{m-\rho_2|\alpha|}_q(\mathbb{R}^n).$$

□

Furthermore, using the characterization of non-smooth pseudodifferential operators, we will show that the inverse of a non-smooth pseudodifferential operator is again a non-smooth pseudodifferential operator (in a slightly larger symbol class) under certain assumptions. This yields immediately results on spectral invariance of non-smooth pseudodifferential operators. These results are presented in Sect. 4.1 and are based on [2].

The structure of this contribution is as follows: Sect. 2 serves to introduce all notations and mathematical basics needed for this paper. Section 2 is devoted to some properties of pseudodifferential operators with single symbols, cf. Sect. 2.1, and pseudodifferential operators with double symbols, cf. Sect. 2.3. In the first two subsections of Sect. 3 we present some auxiliary tools needed for the proof of the characterization in the case $m = 0$. In Sect. 3.1 we show that a bounded sequence in $C^{\tilde{m},s}_{*} S^{0}_{0,0}(\mathbb{R}^n \times \mathbb{R}^n; M)$ is relatively compact in $C^{\tilde{m},s}_{*} S^{0}_{0,0}(\mathbb{R}^n \times \mathbb{R}^n; \lceil M \rceil - 1)$. In Sect. 3.2 a smoothed family of operators $(T_\varepsilon)_{\varepsilon \in (0,1]}$ associated to T is considered. It is shown that $T_\varepsilon : \mathcal{S}'(\mathbb{R}^n) \to \mathcal{S}(\mathbb{R}^n)$ is continuous for all $\varepsilon \in (0, 1]$ and converges pointwise if $\varepsilon \to 0$. Moreover, all iterated commutators of T_ε are uniformly bounded with respect to ε as maps from $L^q(\mathbb{R}^n)$ to $L^q(\mathbb{R}^n)$. Sections 3.3 and 3.4 are dedicated to verify the characterization of non-smooth pseudodifferential operators. Finally, in Sect. 4 the results on spectral invariance are presented.

1 Preliminaries

For the convenience of the reader this section is dedicated to the introduction of the notation and the mathematical preliminaries for this chapter. $\mathbb{N}$ denotes the set of all natural numbers without zero and $\mathbb{N}_0 = \mathbb{N} \cup \{0\}$. We consider $n \in \mathbb{N}$ except when otherwise agreed during the whole chapter. In particular this means $n \neq 0$. For $x \in \mathbb{R}$ we set

$$x^+ := \max\{0; x\} \qquad \text{and} \qquad \lfloor x \rfloor := \max\{k \in \mathbb{Z} : k \leq x\}.$$

Additionally we define

$$\langle x \rangle := (1 + |x|^2)^{1/2} \quad \text{for all } x \in \mathbb{R}^n \qquad \text{and} \qquad đ\xi := (2\pi)^{-n} d\xi.$$

An element $\alpha = (\alpha_1, \ldots, \alpha_n) \in \mathbb{N}_0^n$ is called a *multi-index*. Partial derivatives with respect to a variable $x \in \mathbb{R}^n$ scaled with the factor $-i$ are denoted by

$$D_x^\alpha := (-i)^{|\alpha|} \partial_x^\alpha := (-i)^{|\alpha|} \partial_{x_1}^{\alpha_1} \ldots \partial_{x_n}^{\alpha_n} \qquad \text{for each } \alpha \in \mathbb{N}_0^n.$$

Moreover we denote for $j \in \{1, \ldots, n\}$ the j-th canonical unit vector by $e_j \in \mathbb{N}_0^n$. Hence $(e_j)_k = 1$ if $j = k$ and $(e_j)_k = 0$ else.

For two Banach spaces X, Y the set $\mathscr{L}(X, Y)$ consists of all linear and bounded operators $A : X \to Y$. Instead of $\mathscr{L}(X, X)$ we also write $\mathscr{L}(X)$.

Finally the dual space of a topological vector space V is denoted by V'. In view of a Banach space V we write $\langle ., . \rangle_{V;V'}$ for the duality product of V.

1.1 Functions on $\mathbb{R}^n$ and Function Spaces

In this subsection we fix the convention for all function spaces needed during this chapter. The *Hölder space* of the order $m \in \mathbb{N}_0$ with Hölder continuity $s \in (0, 1]$ is denoted by

$$C^{m,s}(\mathbb{R}^n) := \left\{ f \in C_b^m(\mathbb{R}^n) : \sup_{|\alpha| \leq m} | \sup_{x \neq y} \frac{\partial_x^\alpha f(x) - \partial_y^\alpha f(y)|}{|x - y|^s} < \infty \right\}.$$

In case $s \neq 1$ we also write $C^{m+s}(\mathbb{R}^n)$ instead of $C^{m,s}(\mathbb{R}^n)$. Note that $C^s(\mathbb{R}^n) = C^s_*(\mathbb{R}^n)$ if $s \notin \mathbb{N}_0$. For $s \in \mathbb{R}$ and $1 < p < \infty$ the *Bessel potential space* $H^s_p(\mathbb{R}^n)$ is defined by

$$H^s_p(\mathbb{R}^n) := \{f \in \mathcal{S}'(\mathbb{R}^n) : \langle D_x \rangle^s f \in L^p(\mathbb{R}^n)\},$$

where $\langle D_x \rangle^s := OP(\langle \xi \rangle^s)$.

Elements of a Bessel potential space can be characterized as follows, see, e.g., [3, Lemma 2.1]:

Lemma 1.1 *Let* $1 < p < \infty$, $s < 0$ *and* $m := -\lfloor s \rfloor$. *Then for each* $f \in H^s_p(\mathbb{R}^n)$ *there are functions* $g_\alpha \in H_p^{s-\lfloor s \rfloor}(\mathbb{R}^n)$, *where* $\alpha \in \mathbb{N}_0^n$ *with* $|\alpha| \leq m$, *such that*

- $f = \sum\limits_{|\alpha| \leq m} \partial_x^\alpha g_\alpha$,
- $\sum\limits_{|\alpha| \leq m} \|g_\alpha\|_{H_p^{s-\lfloor s \rfloor}} \leq C\|f\|_{H^s_p}$,

where C *is independent of* f *and* g_α.

Moreover we define for $y \in \mathbb{R}^n$ the translation function $\tau_y(g) : \mathbb{R}^n \to \mathbb{C}$ of $g \in L^1(\mathbb{R}^n)$ as $\tau_y(g)(x) := g(x - y)$ for all $x \in \mathbb{R}^n$.

1.2 Kernel Theorem

In this subsection we focus on an important ingredient of the characterization, namely the next kernel theorem and its applications. All results are taken from [3, Sect. 2.2].

Theorem 1.2 *Every continuous linear operator* $T : \mathcal{S}'(\mathbb{R}^n) \to \mathcal{S}(\mathbb{R}^n)$ *has a Schwartz kernel* $t(x, y)$ *which is an element of* $\mathcal{S}(\mathbb{R}^n \times \mathbb{R}^n)$. *Thus for every* $u \in \mathcal{S}(\mathbb{R}^n)$ *we have*

$$Tu(x) = \int_{\mathbb{R}^n} t(x, y)u(y)dy \qquad \text{for all } x \in \mathbb{R}^n.$$

Proof This claim is a consequence of [21, Theorem 51.6] and [4, Theorem 1.48] if one uses that $\mathcal{S}(\mathbb{R}^n)$ and $\mathcal{S}'(\mathbb{R}^n)$ are nuclear and conuclear, see, e.g., [21, p. 530]. For more details, we refer to [18, Theorem 2.62]. □

We want to apply the previous kernel theorem on some iterated commutators of a linear and bounded operator $P : \mathcal{S}'(\mathbb{R}^n) \to \mathcal{S}(\mathbb{R}^n)$ defined by:

Definition 1.3 Let $X, Y \in \{\mathcal{S}(\mathbb{R}^n), \mathcal{S}'(\mathbb{R}^n)\}$ and $T : X \to Y$ be linear. We define the linear operators $\mathrm{ad}(-ix_j)T : X \to Y$ and $\mathrm{ad}(D_{x_j})T : X \to Y$ for all $j \in \{1, \ldots, n\}$ and $u \in X$ by

$$\mathrm{ad}(-ix_j)Tu := -ix_jTu + T\left(ix_ju\right) \quad \text{and} \quad \mathrm{ad}(D_{x_j})Tu := D_{x_j}(Tu) - T\left(D_{x_j}u\right).$$

For arbitrary multi-indices $\alpha, \beta \in \mathbb{N}_0^n$ we denote the *iterated commutator* of T as

$$\mathrm{ad}(-ix)^\alpha \, \mathrm{ad}(D_x)^\beta T := [\mathrm{ad}(-ix_1)]^{\alpha_1} \ldots [\mathrm{ad}(-ix_n)]^{\alpha_n} [\mathrm{ad}(D_{x_1})]^{\beta_1} \ldots [\mathrm{ad}(D_{x_n})]^{\beta_n} T.$$

Since all iterated commutators of $T : \mathcal{S}'(\mathbb{R}^n) \to \mathcal{S}(\mathbb{R}^n)$ map $\mathcal{S}'(\mathbb{R}^n)$ to $\mathcal{S}(\mathbb{R}^n)$, Theorem 1.2 provides:

Corollary 1.4 *Let $\alpha, \beta \in \mathbb{N}_0^n$ and $T : \mathcal{S}'(\mathbb{R}^n) \to \mathcal{S}(\mathbb{R}^n)$ be a linear operator. Then the operator $\mathrm{ad}(-ix)^\alpha \, \mathrm{ad}(D_x)^\beta T : \mathcal{S}'(\mathbb{R}^n) \to \mathcal{S}(\mathbb{R}^n)$ has a Schwartz kernel $f^{\alpha,\beta} \in \mathcal{S}(\mathbb{R}^n \times \mathbb{R}^n)$, i.e., for all $u \in \mathcal{S}(\mathbb{R}^n)$*

$$\mathrm{ad}(-ix)^\alpha \, \mathrm{ad}(D_x)^\beta Tu(x) = \int_{\mathbb{R}^n} f^{\alpha,\beta}(x, y)u(y)dy \qquad \textit{for all } x \in \mathbb{R}^n. \tag{1}$$

Moreover let us mention that the iterated commutators of a non-smooth pseudodifferential operator are pseudodifferential operators if suitable conditions are fulfilled.

Remark 1.5 Let $\tilde{m} \in \mathbb{N}_0$, $M \in \mathbb{N}_0 \cup \{\infty\}$, $0 < \tau \leq 1$, $m \in \mathbb{R}$ and $0 \leq \rho \leq 1$. We assume that $p \in C^{\tilde{m},\tau} S^m_{\rho,0}(\mathbb{R}^n \times \mathbb{R}^n; M)$. We define $P := p(x, D_x)$. Using integration by parts and some properties of the Fourier transformation, one can calculate for each $u \in \mathcal{S}(\mathbb{R}^n)$ at once that

$$\mathrm{ad}(-ix_j)Pu(x) = -ix_jPu(x) + P[ix_ju(x)] = (\partial_{\xi_j} p)(x, D_x)u(x),$$
$$\mathrm{ad}(D_{x_j})Pu(x) = D_{x_j}\{Pu(x)\} - P[D_{x_j}u(x)] = (D_{x_j} p)(x, D_x)u(x)$$

for all $x \in \mathbb{R}^n$. Applying $p(x, \xi) \in C^{\tilde{m},\tau} S^m_{\rho,0}(\mathbb{R}^n_x \times \mathbb{R}^n_\xi; M)$, we obtain

$$(\partial_{\xi_j} p)(x, \xi) \in C^{\tilde{m},\tau} S^{m-\rho}_{\rho,0}(\mathbb{R}^n_x \times \mathbb{R}^n_\xi; M - 1)$$

and

$$(D_{x_j} p)(x, \xi) \in C^{\tilde{m}-1,\tau} S^m_{\rho,0}(\mathbb{R}^n_x \times \mathbb{R}^n_\xi; M).$$

Now let $l \in \mathbb{N}$, $\alpha_1, \ldots, \alpha_l \in \mathbb{N}_0^n$ and $\beta_1, \ldots, \beta_l \in \mathbb{N}_0^n$ with $|\alpha_j + \beta_j| = 1$ for all $j \in \{1, \ldots, l\}$, $|\alpha| \leq M$ and $|\beta| \leq \tilde{m}$. Here α and β are defined by $\alpha := \alpha_1 + \ldots + \alpha_l$ and $\beta := \beta_1 + \ldots + \beta_l$. By induction with respect to l we can prove that the operator

$$\mathrm{ad}(-ix)^{\alpha_1} \, \mathrm{ad}(D_x)^{\beta_1} \ldots \mathrm{ad}(-ix)^{\alpha_l} \, \mathrm{ad}(D_x)^{\beta_l} \, p(x, D_x)$$

is a pseudodifferential operator with the symbol

$$\partial_\xi^\alpha D_x^\beta p(x,\xi) \in C^{\tilde{m}-|\beta|,\tau} S^{m-\rho|\alpha|}_{\rho,0}(\mathbb{R}^n_x \times \mathbb{R}^n_\xi; M-|\alpha|).$$

If we even have $p \in S^m_{\rho,0}(\mathbb{R}^n \times \mathbb{R}^n)$, then $\partial_\xi^\alpha D_x^\beta p(x,\xi) \in S^{m-\rho|\alpha|}_{\rho,0}(\mathbb{R}^n \times \mathbb{R}^n)$.

Now we have all ingredients at hand in order to verify the application of the kernel theorem needed later on to prove the characterization of non-smooth pseudodifferential operators:

Lemma 1.6 *Let $g \in \mathcal{S}(\mathbb{R}^n)$. For all $y \in \mathbb{R}^n$ we denote $g_y := \tau_y(g) := g(.-y)$. Moreover, let $P : \mathcal{S}'(\mathbb{R}^n) \to \mathcal{S}(\mathbb{R}^n)$ be linear and continuous. We define $p : \mathbb{R}^n \times \mathbb{R}^n \times \mathbb{R}^n \to \mathbb{C}$ by*

$$p(x,\xi,y) := e^{-ix\cdot\xi} P\left(e_\xi g_y\right)(x) \qquad \text{for all } x,\xi,y \in \mathbb{R}^n.$$

Here $e_\xi(x) := e^{ix\cdot\xi}$ for all $x \in \mathbb{R}^n$. Then we have for all $\alpha,\beta,\gamma \in \mathbb{N}_0^n$:

$$\begin{aligned} &\partial_\xi^\alpha D_x^\beta D_y^\gamma p(x,\xi,y) \\ &= (-1)^\gamma \sum_{\beta_1+\beta_2=\beta} \binom{\beta}{\beta_1} e^{-ix\cdot\xi} \left(\mathrm{ad}(-ix)^\alpha \,\mathrm{ad}(D_x)^{\beta_1} P\right)\left(e_\xi D_x^{\beta_2+\gamma} g_y\right)(x). \end{aligned}$$

Proof Theorem 1.2 provides the existence of a Schwartz kernel $f \in \mathcal{S}(\mathbb{R}^n \times \mathbb{R}^n)$ of P. Due to $g \in \mathcal{S}(\mathbb{R}^n)$ and $f \in \mathcal{S}(\mathbb{R}^n \times \mathbb{R}^n)$ we get for all $x \in \mathbb{R}^n$:

$$\begin{aligned} D_y^\gamma \left\{e^{-ix\cdot\xi} P(e_\xi g_y)(x)\right\} &= e^{-ix\cdot\xi} D_y^\gamma \int f(x,z) e^{iz\cdot\xi} g_y(z)dz \\ &= e^{-ix\cdot\xi} \int f(x,z) e^{iz\cdot\xi} D_y^\gamma g_y(z)dz \\ &= (-1)^{|\gamma|} e^{-ix\cdot\xi} P(e_\xi D_x^\gamma g_y)(x). \end{aligned}$$

Inductively with respect to $|\beta|$ we can show for all $\beta,\gamma \in \mathbb{N}_0^n$ and each $x \in \mathbb{R}^n$:

$$\begin{aligned} &D_x^\beta D_y^\gamma \left\{e^{-ix\cdot\xi} P(e_\xi g_y)(x)\right\} \\ &= (-1)^{|\gamma|} D_x^\beta \left\{e^{-ix\cdot\xi} P(e_\xi D_x^\gamma g_y)(x)\right\} \\ &= (-1)^{|\gamma|} \sum_{\beta_1+\beta_2=\beta} \binom{\beta}{\beta_1} e^{-ix\cdot\xi}\, \mathrm{ad}(D_x)^{\beta_1} P\left(e_\xi D_x^{\beta_2+\gamma} g_y\right)(x). \qquad (2) \end{aligned}$$

With Corollary 1.4 at hand, the operator $\mathrm{ad}(D_x)^{\beta_1} P$ has a Schwartz kernel $f^{\beta_1} \in \mathcal{S}(\mathbb{R}^n \times \mathbb{R}^n)$. Due to $e_\xi D_x^{\beta_2+\gamma} g_y \in \mathcal{S}(\mathbb{R}^n)$ and $(ix)^{\alpha_2} e_\xi D_x^{\beta_2+\gamma} g_y(x) \in \mathcal{S}(\mathbb{R}^n_x)$ an application of the Leibniz rule and interchanging the derivatives with the integral yields for all $x \in \mathbb{R}^n$:

$$\partial_\xi^\alpha \left\{ e^{-ix\cdot\xi} \operatorname{ad}(D_x)^{\beta_1} P \left(e_\xi D_x^{\beta_2+\gamma} g_y \right)(x) \right\}$$
$$= \sum_{\alpha_1+\alpha_2=\alpha} \binom{\alpha}{\alpha_1} (-ix)^{\alpha_1} e^{-ix\cdot\xi} \int f^{\beta_1}(x,z)(iz)^{\alpha_2} e^{iz\cdot\xi} D_z^{\beta_2+\gamma} g_y(z) dz$$
$$= e^{-ix\cdot\xi} (\operatorname{ad}(-ix)^\alpha \operatorname{ad}(D_x)^{\beta_1} P) \left(e_\xi D_x^{\beta_2+\gamma} g_y \right)(x). \tag{3}$$

Finally, the combination of (2) and (3) finishes the proof:

$$\partial_\xi^\alpha D_x^\beta D_y^\gamma p(x,\xi,y)$$
$$= \partial_\xi^\alpha D_x^\beta D_y^\gamma \left\{ e^{-ix\cdot\xi} P\left(e_\xi g_y\right)(x) \right\}$$
$$= (-1)^{|\gamma|} \sum_{\beta_1+\beta_2=\beta} \binom{\beta}{\beta_1} e^{-ix\cdot\xi} (\operatorname{ad}(-ix)^\alpha \operatorname{ad}(D_x)^{\beta_1} P) \left(e_\xi D_x^{\beta_2+\gamma} g_y \right)(x)$$

for all $x, \xi, y \in \mathbb{R}^n$. □

1.3 Extension of the Space of Amplitudes

The *space of amplitudes* $\mathscr{A}_\tau^{m,N}(\mathbb{R}^n \times \mathbb{R}^n)$ $(m, \tau \in \mathbb{R}, N \in \mathbb{N}_0 \cup \{\infty\})$ is the set of all functions $a : \mathbb{R}^n \times \mathbb{R}^n \to \mathbb{C}$ with the following properties: For all $\alpha, \beta \in \mathbb{N}_0^n$ with $|\alpha| \leq N$ we have

i) $\partial_\eta^\alpha \partial_y^\beta a(y,\eta) \in C^0(\mathbb{R}_y^n \times \mathbb{R}_\eta^n)$,
ii) $\left|\partial_\eta^\alpha \partial_y^\beta a(y,\eta)\right| \leq C_{\alpha,\beta}(1+|\eta|)^m (1+|y|)^\tau$ for all $y, \eta \in \mathbb{R}^n$.

If $N = \infty$ we also write $\mathscr{A}_\tau^m(\mathbb{R}^n \times \mathbb{R}^n)$ instead of $\mathscr{A}_\tau^{m,\infty}(\mathbb{R}^n \times \mathbb{R}^n)$. The **oscillatory integrals** defined via

$$\text{Os-}\iint e^{-iy\cdot\eta} a(y,\eta) dy đ\eta := \lim_{\varepsilon \to 0} \iint \chi(\varepsilon y, \varepsilon\eta) e^{-iy\cdot\eta} a(y,\eta) dy đ\eta, \tag{4}$$

where $\chi \in \mathcal{S}(\mathbb{R}^n \times \mathbb{R}^n)$ with $\chi(0,0) = 1$, are well-defined for each function a of the space of amplitudes. They constitute an important technique for verifying statements in the theory of pseudodifferential operators.

This subsection serves to present the properties of oscillatory integrals verified in [3, Subsection 2.3]. For this we need:

Remark 1.7 For arbitrary $m = 2k$, $k \in \mathbb{N}$ we can show $e^{ix\cdot\xi} = \langle\xi\rangle^{-m}\langle D_x\rangle^m e^{ix\cdot\xi}$, if we write $\langle D_x\rangle^{2k} = \sum_{|\gamma|\le k} a_{\gamma,k} D_x^{2\gamma}$. Now let $m = 2k+1$, $k \in \mathbb{N}_0$. Using

$$\langle\xi\rangle^m = \langle\xi\rangle^{2k}\frac{\langle\xi\rangle^2}{\langle\xi\rangle} = \langle\xi\rangle^{2k}\left\{\frac{1}{\langle\xi\rangle} + \sum_{j=1}^n \frac{\xi_j^2}{\langle\xi\rangle}\right\}$$

we get:

$$e^{ix\cdot\xi} = \langle\xi\rangle^{-m}\langle\xi\rangle^m e^{ix\cdot\xi} = \langle\xi\rangle^{-m-1}\langle D_x\rangle^{m-1}e^{ix\cdot\xi} + \sum_{j=1}^n \langle\xi\rangle^{-m}\frac{\xi_j}{\langle\xi\rangle}\langle D_x\rangle^{m-1}D_{x_j}e^{ix\cdot\xi}.$$

By means of the previous remark, we define for all $m \in \mathbb{N}$

$$A^m(D_x,\xi) := \langle\xi\rangle^{-m}\langle D_x\rangle^m \qquad \text{if } m \text{ is even,}$$

$$A^m(D_x,\xi) := \langle\xi\rangle^{-m-1}\langle D_x\rangle^{m-1} - \sum_{j=1}^n \langle\xi\rangle^{-m}\frac{\xi_j}{\langle\xi\rangle}\langle D_x\rangle^{m-1}D_{x_j} \qquad \text{else.}$$

Remark 1.8 Let $m, \tau \in \mathbb{R}$ and $N \in \mathbb{N}_0 \cup \{\infty\}$. Moreover let $\mathscr{B} \subseteq \mathscr{A}_\tau^{m,N}(\mathbb{R}^n \times \mathbb{R}^n)$ be bounded, i.e., for all $\alpha, \beta \in \mathbb{N}_0^n$ with $|\beta| \le N$ we have

$$|\partial_x^\alpha \partial_\xi^\beta a(x,\xi)| \le C_{\alpha,\beta}\langle\xi\rangle^m\langle x\rangle^\tau \qquad \text{for all } a \in \mathscr{B}. \tag{5}$$

If we use the Leibniz rule and write $\langle D_x\rangle^{(l-1)} = \sum_{|\gamma|\le(l-1)/2} a_{\gamma,l} D_x^{2\gamma}$ for odd $l \in \mathbb{N}$ we obtain due to (5), $\xi_j\langle\xi\rangle^{-1} \in S^0_{1,0}(\mathbb{R}^n \times \mathbb{R}^n)$ and $\langle\xi\rangle^s \in S^s_{1,0}(\mathbb{R}^n \times \mathbb{R}^n)$ for all $\alpha, \beta \in \mathbb{N}_0^n$ with $|\beta| \le N$ and $l \in \mathbb{N}$:

$$\left|\partial_x^\alpha\partial_\xi^\beta\left\{\langle\xi\rangle^{-l-1}\langle D_x\rangle^{l-1}a(x,\xi) - \sum_{j=1}^n \langle\xi\rangle^{-l}\frac{\xi_j}{\langle\xi\rangle}\langle D_x\rangle^{l-1}D_{x_j}a(x,\xi)\right\}\right|$$
$$\le C_\alpha\langle\xi\rangle^{-l+m}\langle x\rangle^\tau$$

for all $a \in \mathscr{B}$.

The properties of oscillatory integrals needed later on are listed up in the next Theorems and Corollaries. For the proof of all these results, we refer to [2, Sect. 2.3].

Theorem 1.9 *Let $m, \tau \in \mathbb{R}$ and $N \in \mathbb{N}_0 \cup \{\infty\}$ with $N > n + \tau$. Moreover, let $\chi \in \mathcal{S}(\mathbb{R}^n \times \mathbb{R}^n)$ with $\chi(0,0) = 1$ be arbitrary. Then the* **oscillatory integral**

$$Os\text{-}\iint e^{-iy\cdot\eta}a(y,\eta)dy\,đ\eta := \lim_{\varepsilon\to 0}\iint \chi(\varepsilon y,\varepsilon\eta)e^{-iy\cdot\eta}a(y,\eta)dy\,đ\eta$$

exists for each $a \in \mathscr{A}_{\tau}^{m,N}(\mathbb{R}^n \times \mathbb{R}^n)$. *Additionally for all* $l, l' \in \mathbb{N}_0$ *with* $l > n + m$ *and* $N \geq l' > n + \tau$ *we have*

$$Os\text{-}\iint e^{-iy\cdot\eta} a(y,\eta)dy đ\eta = \iint e^{-iy\cdot\eta} A^{l'}(D_\eta, y)[A^l(D_y,\eta)a(y,\eta)]dy đ\eta.$$

Therefore the definition does not depend on the choice of χ.

Theorem 1.10 *Let* $m, \tau \in \mathbb{R}$ *and* $k \in \mathbb{N}$. *We define* $\tilde{\tau} := \tau$ *if* $\tau \geq -k$, $\tilde{\tau} := -k - 0.5$ *if* $\tau \in \mathbb{Z}$ *and* $\tau < -k$ *and* $\tilde{\tau} := -k - (|\tau| - \lfloor -\tau \rfloor)/2$ *else. Moreover, we define* $\hat{\tau} := \tau_+$ *if* $\tau \geq -k$ *and* $\hat{\tau} := \tau - \tilde{\tau}$ *else. Additionally let* $N \in \mathbb{N}_0 \cup \{\infty\}$ *and* $M := \max\{m \in \mathbb{N}_0 : N - m \geq l > k + \tilde{\tau}$ *for one* $l \in \mathbb{N}_0\}$. *Assuming an* $a \in \mathscr{A}_{\tau}^{m,N}(\mathbb{R}^{n+k} \times \mathbb{R}^{n+k})$ *we define* $b : \mathbb{R}^n \times \mathbb{R}^n \to \mathbb{C}$ *via*

$$b(y,\eta) := Os\text{-}\iint e^{-iy'\cdot\eta'} a(y,y',\eta,\eta')dy'đ\eta' \qquad \text{for all } y,\eta \in \mathbb{R}^n.$$

If there is an $\tilde{l} \in \mathbb{N}_0$ *with* $M \geq \tilde{l} > n + \hat{\tau}$, *we obtain*

$$\begin{aligned} Os\text{-}&\iiiint e^{-iy\cdot\eta - iy'\cdot\eta'} a(y,y',\eta,\eta')dydy'đ\eta đ\eta' \\ &= Os\text{-}\iint e^{-iy\cdot\eta}\left[Os\text{-}\iint e^{-iy'\cdot\eta'} a(y,y',\eta,\eta')dy'đ\eta'\right]dy đ\eta. \end{aligned} \tag{6}$$

If there is an $\tilde{l} \in \mathbb{N}_0$ *with* $N \geq \tilde{l} > k + \tau$, *we have* $b \in \mathscr{A}_{\hat{\tau}}^{m_+,M}(\mathbb{R}^n \times \mathbb{R}^n)$ *and*

$$\partial_y^\alpha \partial_\eta^\beta b(y,\eta) = Os\text{-}\iint e^{-iy'\cdot\eta'} \partial_y^\alpha \partial_\eta^\beta a(y,y',\eta,\eta')dy'đ\eta' \qquad \text{for all } y,\eta \in \mathbb{R}^n \tag{7}$$

for each $\alpha, \beta \in \mathbb{N}_0^n$ *with* $|\beta| \leq M$.

Theorem 1.11 *Let* $m, \tau \in \mathbb{R}$ *and* $N \in \mathbb{N}_0 \cup \{\infty\}$ *with* $N > n + \tau$. *Moreover, let* $l_0, \tilde{l}_0 \in \mathbb{N}_0$ *with* $\tilde{l}_0 \leq N$. *Then*

$$Os\text{-}\iint e^{-iy\cdot\eta} a(y,\eta)dyđ\eta = Os\text{-}\iint e^{-iy\cdot\eta} A^{\tilde{l}_0}(D_\eta, y)A^{l_0}(D_y,\eta)a(y,\eta)dyđ\eta$$

for every $a \in \mathscr{A}_{\tau}^{m,N}(\mathbb{R}^n \times \mathbb{R}^n)$.

Corollary 1.12 *Let* $m, \tau \in \mathbb{R}$ *and* $N \in \mathbb{N}_0 \cup \{\infty\}$ *with* $N > n + \tau$. *Additionally let* $(a_j)_{j\in\mathbb{N}}$ *be a bounded sequence in* $\mathscr{A}_{\tau}^{m,N}(\mathbb{R}^n \times \mathbb{R}^n)$, *i.e., for all* $\alpha, \beta \in \mathbb{N}_0^n$ *with* $|\alpha| \leq N$:

$$\left|\partial_y^\beta \partial_\eta^\alpha a_j(y,\eta)\right| \leq C_{\alpha,\beta}\langle\eta\rangle^m\langle y\rangle^\tau \qquad \text{for all } y,\eta \in \mathbb{R}^n \text{ and } j \in \mathbb{N}.$$

Moreover, there is an $a \in \mathscr{A}_\tau^{m,N}(\mathbb{R}^n \times \mathbb{R}^n)$ *such that*

$$\lim_{j\to\infty} \partial_\eta^\alpha \partial_y^\beta a_j(y,\eta) = \partial_\eta^\alpha \partial_y^\beta a(y,\eta) \qquad \text{for all } y, \eta \in \mathbb{R}^n$$

for each $\alpha, \beta \in \mathbb{N}_0^n$ *with* $|\alpha| \le N$. *Then*

$$\lim_{j\to\infty} Os\text{-}\iint e^{-iy\cdot\eta} a_j(y,\eta)dy đ\eta = Os\text{-}\iint e^{-iy\cdot\eta} a(y,\eta)dy đ\eta.$$

Theorem 1.13 *Let* $m, \tau \in \mathbb{R}$ *and* $N \in \mathbb{N}_0 \cup \{\infty\}$ *with* $N > n + \tau$. *For* $a \in \mathscr{A}_\tau^{m,N}(\mathbb{R}^n \times \mathbb{R}^n)$ *we have:*

$$Os\text{-}\iint e^{-i(y+y_0)\cdot(\eta+\eta_0)} a(y+y_0, \eta+\eta_0)dy đ\eta = Os\text{-}\iint e^{-iy\cdot\eta} a(y,\eta)dy đ\eta.$$

2 Pseudodifferential Operators

2.1 Properties of Pseudodifferential Operators

Equipped with the family of seminorms

$$\begin{aligned}
&|a|^m_{k, C^{\tilde{m},\tau} S^m_{\rho,0}(\mathbb{R}^n\times\mathbb{R}^n;M)} \\
&= \max_{|\alpha|\le k} \sup_{\xi\in\mathbb{R}^n} \|\partial_\xi^\alpha a(.,\xi)\|_{C^{\tilde{m},\tau}(\mathbb{R}^n)} \langle\xi\rangle^{-m+\rho|\alpha|} \\
&\quad + \max_{|\alpha|\le M} \sup_{\xi\in\mathbb{R}^n} \sup_{\eta\in\mathbb{R}^n\setminus\{0\}} \left\| \frac{\partial_\xi^\alpha a(x,\xi+\eta) - \partial_\xi^\alpha a(x,\xi)}{|\eta|^{M-\lfloor M\rfloor}} \right\|_{C^{\tilde{m},\tau}(\mathbb{R}^n_x)} \langle\xi\rangle^{-m+\rho|\alpha|}
\end{aligned}$$

for all $k \in \mathbb{N}_0$ with $k \le M$, the spaces $C^{\tilde{m},\tau} S^m_{\rho,0}(\mathbb{R}^n \times \mathbb{R}^n; M)$ are Fréchet spaces, where $\tilde{m} \in \mathbb{N}$, $0 \le \rho \le 1$, $0 < \tau < 1$ and $M \in [0,\infty]$.

Derivatives of non-smooth symbols have the next useful property:

Lemma 2.1 *Let* $\tilde{m} \in \mathbb{N}_0$, $0 < \tau < 1$, $M \in [0,\infty]$ *and* $0 < s \le \min\{M - \lfloor M\rfloor, \tau\}$ *if* $M \notin \mathbb{N}_0$ *and* $0 < s \le 1$ *else. Additionally let* $\mathscr{B} \subseteq C^{\tilde{m},\tau} S^0_{0,0}(\mathbb{R}^n \times \mathbb{R}^n; M)$ *be a bounded subset. Considering* $\alpha, \gamma \in \mathbb{N}_0^n$ *with* $|\alpha| \le \tilde{m}$, $|\gamma| < M$, *the set* $\{\partial_x^\alpha \partial_\xi^\gamma a : a \in \mathscr{B}\} \subseteq C^{0,s}(\mathbb{R}^n \times \mathbb{R}^n)$ *is bounded.*

Proof First we choose arbitrary $\alpha, \gamma \in \mathbb{N}_0^n$ with $|\alpha| \leq \tilde{m}$, $|\gamma| \leq M$ if $M \notin \mathbb{N}_0$ and $|\gamma| \leq M - 1$ else. Since $\mathscr{B} \subseteq C^{\tilde{m},s} S^0_{0,0}(\mathbb{R}^n \times \mathbb{R}^n; M)$ is bounded, we obtain

$$\{\partial_x^\alpha \partial_\xi^\gamma a : a \in \mathscr{B}\} \subseteq C_b^0(\mathbb{R}^n \times \mathbb{R}^n) \text{ is bounded and,} \tag{8}$$

$$\sup_{(x,\xi)\neq(y,\eta)} \frac{|\partial_x^\alpha \partial_\xi^\gamma a(x,\eta) - \partial_x^\alpha \partial_\xi^\gamma a(y,\eta)|}{|(x,\xi)-(y,\eta)|^s} \leq \sup_\eta \|\partial_\xi^\gamma a(.,\eta)\|_{C^{\tilde{m},\tau}(\mathbb{R}^n)} < C_\gamma \ \forall a \in \mathscr{B}. \tag{9}$$

If $|\gamma| < \lfloor M \rfloor$, we get by means of the fundamental theorem of calculus in the case $|\xi - \eta| < 1$ with $\xi \neq \eta$ and because of (8) for $|\xi - \eta| \geq 1$:

$$\sup_{(x,\xi)\neq(y,\eta)} \frac{|\partial_x^\alpha \partial_\xi^\gamma a(x,\xi) - \partial_x^\alpha \partial_\xi^\gamma a(x,\eta)|}{|(x,\xi)-(y,\eta)|^s} \leq C_\gamma \qquad \text{for all } a \in \mathscr{B}. \tag{10}$$

In case of $|\gamma| = \lfloor M \rfloor$, property $iv)$ of the definition of the non-smooth symbol-class provides for $|\xi - \eta| < 1$

$$\begin{aligned}\sup_{(x,\xi)\neq(y,\eta)} \frac{|\partial_x^\alpha \partial_\xi^\gamma a(x,\xi) - \partial_x^\alpha \partial_\xi^\gamma a(x,\eta)|}{|(x,\xi)-(y,\eta)|^s} &\leq \sup_{\xi\neq\eta} \frac{\|\partial_\xi^\gamma a(.,\xi) - \partial_\xi^\gamma a(.,\eta)\|_{C^{\tilde{m},\tau}(\mathbb{R}^n)}}{|\xi-\eta|^s} \\ &\leq \sup_{\xi\neq\eta} \frac{\|\partial_\xi^\gamma a(.,\xi) - \partial_\xi^\gamma a(.,\eta)\|_{C^{\tilde{m},\tau}(\mathbb{R}^n)}}{|\xi-\eta|^{M-\lfloor M \rfloor}} \\ &\leq C_\gamma \quad \forall a \in \mathscr{B}. \end{aligned} \tag{11}$$

Collecting the estimates (8),(9),(10), and (11) we finally obtain the claim. □

In the literature there are several boundedness results for smooth as well as for non-smooth pseudodifferential operators. Here we just mention those needed during this chapter.

Theorem 2.2 *Let* $p \in S^m_{1,0}(\mathbb{R}^n \times \mathbb{R}^n)$ *with* $m \in \mathbb{R}$. *Then*

$$p(x, D_x) : \mathcal{S}(\mathbb{R}^n) \to \mathcal{S}(\mathbb{R}^n)$$

is a bounded mapping. More precisely, for every $k \in \mathbb{N}_0$ *we can show*

$$|p(x, D_x) f|_{k,\mathcal{S}} \leq C_k |p|_k^{(m)} |f|_{\tilde{m},\mathcal{S}} \qquad \text{for all } f \in \mathcal{S}(\mathbb{R}^n),$$

where $\tilde{m} := \max\{0, m + 2(n+1) + k\}$ *if* $m \in \mathbb{Z}$ *and* $\tilde{m} := \max\{0, \lfloor m \rfloor + 2n + 3 + k\}$ *else.*

We refer to, e.g., [1, Theorem 3.6] for the proof. Now let $X \in \{C^{\tilde{m},\tau}, C_*^{\tilde{m}+\tau}\}$ with $\tilde{m} \in \mathbb{N}_0$ and $0 < \tau \leq 1$. For non-smooth pseudodifferential operators with symbols $a \in XS^m_{\rho,0}(\mathbb{R}^n \times \mathbb{R}^n; M)$ a similar result holds since the next estimate

$$\|e_\xi \cdot a(.,\xi)\|_X \leq C_{\tilde{m},\tau} \langle \xi \rangle^N \|a(.,\xi)\|_X \qquad \text{for all } \xi \in \mathbb{R}^n, \tag{12}$$

can be verified for some $N \in \mathbb{N}$ and $C_{\tilde{m},\tau} > 0$. This is done in the next remark which generalizes some cases of [3, Remark 3.3].

Remark 2.3 Let $X \in \{C^{\tilde{m},\tau}, C_*^{\tilde{m}+\tau}\}$ with $\tilde{m} \in \mathbb{N}_0$ and $0 < \tau \leq 1$. For $0 \leq \rho \leq 1$ and $M \in [0, \infty]$ we choose an arbitrary $a \in XS^m_{\rho,0}(\mathbb{R}^n \times \mathbb{R}^n; M)$. Then inequality (12) holds for $N = \tilde{m} + 2$ in the case $X = C_*^{\tilde{m}+\tau}$ and for $N = \tilde{m} + 1$ in the case $X = C^{\tilde{m},\tau}$. With the multiplication property

$$\|fg\|_{C^s_*} \leq C\|f\|_{C^s_*}\|g\|_{C^s_*} \qquad \text{for all } f, g \in C^s_*(\mathbb{R}^n),$$

and the embedding $C_b^{\tilde{m}+\lfloor\tau\rfloor+1}(\mathbb{R}^n) \hookrightarrow C_*^{\tilde{m}+\tau}(\mathbb{R}^n)$ at hand, we are in the position to prove the remark for $X = C_*^{\tilde{m}+\tau}$:

$$\|e_\xi \cdot a(.,\xi)\|_{C_*^{\tilde{m}+\tau}} \leq C_{\tilde{m},\tau}\|e_\xi\|_{C_b^{\tilde{m}+\lfloor\tau\rfloor+1}}\|a(.,\xi)\|_{C_*^{\tilde{m}+\tau}} \leq C_{\tilde{m},\tau}\langle\xi\rangle^{\tilde{m}+2}\|a(.,\xi)\|_{C_*^{\tilde{m}+\tau}}$$

for all $\xi \in \mathbb{R}^n$. It remains to prove the case $X = C^{\tilde{m},\tau}$. Using the mean value theorem in the case $|x_1 - x_2| \leq 1$, $x_1 \neq x_2$ we obtain

$$\max_{x_1 \neq x_2} \frac{|e^{ix_1\cdot\xi} - e^{ix_2\cdot\xi}|}{|x_1 - x_2|^\tau} \leq 2\langle\xi\rangle \qquad \text{for all } \xi \in \mathbb{R}^n.$$

On account of the previous inequality we are able to verify the next estimate:

$$\max_{|\alpha|\leq\tilde{m}} \sup_{x_1\neq x_2} \frac{|e^{ix_1\cdot\xi}\partial_x^\alpha a(x_1,\xi) - e^{ix_2\cdot\xi}\partial_x^\alpha a(x_2,\xi)|}{|x_1 - x_2|^\tau} \leq C_{\tilde{m},\tau}\langle\xi\rangle\|a(.,\xi)\|_{C^{\tilde{m},\tau}} \tag{13}$$

for all $\xi \in \mathbb{R}^n$. Moreover we are able to show

$$\|e_\xi \cdot a(.,\xi)\|_{C_b^{\tilde{m}}} \leq C_{\tilde{m}}\langle\xi\rangle^{\tilde{m}}\|a(.,\xi)\|_{C^{\tilde{m},\tau}} \qquad \text{for all } \xi \in \mathbb{R}^n. \tag{14}$$

A combination of the inequalities (13) and (14) yields

$$\|e_\xi \cdot a(.,\xi)\|_{C^{\tilde{m},\tau}} \leq C_{\tilde{m},\tau}\langle\xi\rangle^{\tilde{m}+1}\|a(.,\xi)\|_{C^{\tilde{m},\tau}} \qquad \text{for all } \xi \in \mathbb{R}^n.$$

The previous remark enables us to prove the next boundedness result which generalizes Lemma 3.4 of [3]:

Lemma 2.4 *Let $X \in \{C^{\tilde{m},\tau}, C_*^{\tilde{m}+\tau}\}$ with $\tilde{m} \in \mathbb{N}_0$ and $0 < \tau \leq 1$. Moreover let $m \in \mathbb{R}$, $M \in [0, \infty]$. Additionally let $0 \leq \rho \leq 1$. If p is an element of the Fréchet space $XS^m_{\rho,0}(\mathbb{R}^n \times \mathbb{R}^n; M)$, we obtain the continuity of $p(x, D_x) : \mathcal{S}(\mathbb{R}^n) \to X$.*

Proof Let $u \in \mathcal{S}(\mathbb{R}^n)$ be arbitrary. An application of $p \in XS^m_{\rho,0}(\mathbb{R}^n \times \mathbb{R}^n; M)$, $u \in \mathcal{S}(\mathbb{R}^n)$ and Remark 2.3 yields

$$\begin{aligned}\|p(x, D_x)u(x)\|_X &\leq \int \|e_\xi p(., \xi)\|_X |\hat{u}(\xi)| đ\xi \\ &\leq C \int \langle\xi\rangle^{-(n+1)} đ\xi \, |\hat{u}|_{\hat{m}+(n+1),\mathcal{S}} \\ &\leq C|u|_{\hat{m}+2(n+1),\mathcal{S}}\end{aligned}$$

for all $x \in \mathbb{R}^n$ and some $\hat{m} \in \mathbb{N}$.

□

In the case $X = C^{\tilde{m},\tau}$ this statement was proved in [13, Theorem 3.6]. We even can generalize the previous lemma for arbitrary Banach spaces X fulfilling $C_c^\infty(\mathbb{R}^n) \subseteq X \subseteq C^0(\mathbb{R}^n)$ and inequality (12), see, e.g., [3, Lemma 3.4] for the case $M \in \mathbb{N}_0 \cup \{\infty\}$. In view of the proof of Lemma 2.4 we easily see that we also get the boundedness of

$$\{p(x, D_x) : p \in \mathscr{B}\} \subseteq \mathscr{L}(\mathcal{S}(\mathbb{R}^n); X). \tag{15}$$

Such results on uniform boundedness of the operators are mostly not stated in the literature. Mostly only boundedness of a single operator for different function spaces is shown. It is often tedious to get similar results as (15) by means of verifying these proofs. Let us remark that in [3, Lemma 3.5] an argument is given that helps us to prove such results easily.

It is well known that pseudodifferential operators are bounded as maps between two Bessel potential spaces. These statements are summarized in the next four theorems. For the proof of the result in the smooth case, we refer to, e.g., [1, Theorem 5.20].

Theorem 2.5 *Let $m \in \mathbb{R}$, $p \in S^m_{1,0}(\mathbb{R}^n \times \mathbb{R}^n)$ and $1 < q < \infty$. Then $p(x, D_x)$ extends to a bounded linear operator*

$$p(x, D_x) : H_q^{s+m}(\mathbb{R}^n) \to H_q^s(\mathbb{R}^n) \qquad \textit{for all } s \in \mathbb{R}.$$

Theorem 2.6 *Let $m \in \mathbb{R}$, $0 < \rho \leq 1$ with $\rho > 0$ and $1 < p < \infty$. Additionally let $\tau > (1-\rho) \cdot \frac{n}{2}$ if $\rho < 1$ and $\tau > 0$ if $\rho = 1$, respectively. Moreover, let $M \in [0, \infty]$ with $M > n/2$ for $2 \leq p < \infty$ and $M > n/p$ else. Denoting $k_p := (1-\rho)n\,|1/2 - 1/p|$, let $\mathscr{B} \subseteq C_*^\tau S^{m-k_p}_{\rho,0}(\mathbb{R}^n \times \mathbb{R}^n; M)$ be a bounded subset.*

Then for each real number s with the property

$$(1-\rho)\frac{n}{p} - 1\tau < s < \tau$$

there is a constant $C_s > 0$*, independent of* $a \in \mathscr{B}$*, such that*

$$\|a(x, D_x)f\|_{H^s_p} \leq C_s \|f\|_{H^{s+m}_p} \qquad \text{for all } f \in H^{s+m}_p(\mathbb{R}^n) \text{ and } a \in \mathscr{B}.$$

Theorem 2.7 *Let* $m \in \mathbb{R}$ *and* $\tau > \frac{n}{2}$*. Moreover, let* $M \in [0, \infty]$ *with* $M > n/2$*. Additionally let* $a \in C^\tau_* S^m_{0,0}(\mathbb{R}^n \times \mathbb{R}^n; M)$*. Then for each real number* $s \in \left(\frac{n}{2} - \tau, \tau\right)$ *there is a constant* $C_s > 0$ *such that*

$$\|a(x, D_x)f\|_{H^s_2} \leq C_s \|f\|_{H^{s+m}_2} \qquad \text{for all } f \in H^{s+m}_2(\mathbb{R}^n).$$

Theorem 2.8 *Let* $m \in \mathbb{R}$, $M \in [0, \infty]$ *with* $M > n/2$ *and* $\tau > 0$*. Moreover let* P *be an element of* $OPC^\tau_* S^{m-n/2}_{0,0}(\mathbb{R}^n \times \mathbb{R}^n; M)$*. Then the operator*

$$P : H^{s+m}_2(\mathbb{R}^n) \to H^s_2(\mathbb{R}^n) \qquad \text{is continuous for all } -\tau < s < \tau.$$

The statements for non-smooth pseudodifferential operators instead are based on the results of Marschall, see [17, Theorem 2.7], [17, Theorem 4.2], [17, Theorem 2.1] and [17, Lemma 2.9]. Just the proof for the fact that the constant C_s is independent of the bounded set $\mathscr{B}$ was given in [3, Sect. 3.1]. If $\sharp\mathscr{B} = 1$, Theorem 2.6 also holds for $p = 1$ or $p = \infty$, cf. [17, Theorem 2.7 and Theorem 4.2].

We also will need the following estimate for pseudodifferential operators. It is a generalization of [3, Lemma 3.10].

Lemma 2.9 *Let* $s \in \mathbb{R}^+$ *with* $s \notin \mathbb{N}$, $m \in \mathbb{R}$ *and* $0 \leq \rho \leq 1$*. Additionally let* $M \in [0, \infty]$*. Moreover,* $\mathscr{B} \subseteq C^s S^m_{\rho,0}(\mathbb{R}^n \times \mathbb{R}^n; M)$ *should be a bounded subset and* $u \in \mathcal{S}(\mathbb{R}^n)$*. For every* $N \in \mathbb{N}_0$ *with* $2N \leq M$ *we have*

$$|a(x, D_x)u(x)| \leq C_{N,n}\langle x\rangle^{-2N} \qquad \text{for all } x \in \mathbb{R}^n \text{ and } a \in \mathscr{B}.$$

Note that $C_{N,n}$ *depends on* $u \in \mathcal{S}(\mathbb{R}^n)$*.*

Proof Let $N \in \mathbb{N}_0$ with $2N \leq M$. Choosing $M_{m,n} \in \mathbb{N}$ with $-M_{m,n} < -n - |m|$, we get for all $a \in \mathscr{B}$ and all $x \in \mathbb{R}^n$ by means of $u \in \mathcal{S}(\mathbb{R}^n)$ and the boundedness of $\mathscr{B} \subseteq C^s S^m_{\rho,0}(\mathbb{R}^n \times \mathbb{R}^n; M)$:

$$\left|\langle D_\xi\rangle^{2N}\left[a(x,\xi)\hat{u}(\xi)\right]\right| \leq C_{N,n}\langle\xi\rangle^{m-M_{m,n}} \in L^1(\mathbb{R}^n_\xi). \tag{16}$$

Here $C_{N,n}$ is independent of $x, \xi \in \mathbb{R}^n$ and $a \in \mathscr{B}$. On account of (16) and integration by parts with respect to ξ we conclude the claim:

$$\left|\langle x\rangle^{2N} a(x, D_x)u(x)\right| = \left|\int e^{ix\cdot\xi} \langle D_\xi\rangle^{2N} \left[a(x,\xi)\hat{u}(\xi)\right] đ\xi\right| \leq C_{N,n}$$

for all $a \in \mathscr{B}$ and $x \in \mathbb{R}^n$. □

2.2 Kernel Representation

Here we present the kernel representation of a non-smooth pseudodifferential operator $p(x, D_x)$, whose symbol is in the class $C^{\tilde{m},\tau} S^m_{1,0}(\mathbb{R}^n \times \mathbb{R}^n)$ with $\tilde{m} \in \mathbb{N}_0$, $0 < \tau \leq 1$. The results are special cases of the statements verified in [2, Sect. 3.2].

Theorem 2.10 *Let $p \in C^{\tilde{m},\tau} S^m_{1,0}(\mathbb{R}^n \times \mathbb{R}^n)$, where $\tilde{m} \in \mathbb{N}_0$, $0 < \tau \leq 1$ and $m \in \mathbb{R}$. Then there is a function $k : \mathbb{R}^n \times (\mathbb{R}^n \backslash \{0\}) \to \mathbb{C}$ such that $k(x, .) \in C^\infty(\mathbb{R}^n \backslash \{0\})$ for all $x \in \mathbb{R}^n$ and*

$$p(x, D_x)u(x) = \int k(x, x-y)u(y)dy \qquad \textit{for all } x \notin \textit{supp } u$$

for all $u \in \mathcal{S}(\mathbb{R}^n)$. Moreover, for every $\alpha \in \mathbb{N}_0^n$ and each $N \in \mathbb{N}_0$ the kernel k satisfies

$$\|\partial_z^\alpha k(., z)\|_X \leq \begin{cases} C_{\alpha,N}|z|^{-n-m-|\alpha|}\langle z\rangle^{-N} & \textit{if } n+m+|\alpha| > 0, \\ C_{\alpha,N}(1+|\log|z||)\langle z\rangle^{-N} & \textit{if } n+m+|\alpha| = 0, \\ C_{\alpha,N}\langle z\rangle^{-N} & \textit{if } n+m+|\alpha| < 0 \end{cases}$$

uniformly in $z \in \mathbb{R}^n \backslash \{0\}$.

Proof We are able to prove the statements in a similar way as in [1, Theorem 5.12] or [20, Chapter VI, § 4]. The main idea of the proof is to decompose

$$p(x, D_x)f = \sum_{j=0}^{\infty} p(x, D_x)\varphi_j(D_x)f \qquad \text{for all } f \in \mathcal{S}(\mathbb{R}^n),$$

where $(\varphi_j)_{j\in\mathbb{N}_0}$ is a dyadic partition of unity. The series converges in $C^{\tilde{m},\tau}(\mathbb{R}^n)$ due to Lemma 2.4. First of all we construct a kernel k_j of $p_j(x, D_x) := p(x, D_x)\varphi_j(D_x)$ for each $j \in \mathbb{N}_0$. This can be made in the same way as in the smooth case. We just have to use $\|\partial_\xi^\alpha p(., \xi)\|_{C^{\tilde{m},\tau}(\mathbb{R}^n)} \leq C_\alpha\langle\xi\rangle^{m-|\alpha|}$ instead of $|\partial_\xi^\alpha p(., \xi)| \leq C_\alpha\langle\xi\rangle^{m-|\alpha|}$ for all $\alpha \in \mathbb{N}_0^n$ and all $\xi \in \mathbb{R}^n$. Afterwards we use this kernel decompositions to construct the kernel of $p(x, D_x)$ as in the smooth case.

By means of the embedding $C^{\tilde{m},\tau}(\mathbb{R}^n) \subseteq C^0(\mathbb{R}^n)$ we get the absolute and uniform convergence of $k(x,z) = \sum_{j=0}^{\infty} k_j(x,z)$. □

Remark 2.11 If we even have $p \in S^m_{1,0}(\mathbb{R}^n \times \mathbb{R}^n)$ in the previous theorem, we can show that $k(.,z)$ is smooth for all $z \in \mathbb{R}^n$ while applying Theorem 2.10 for all $\tilde{m} \in \mathbb{N}$ and some $0 < \tau < 1$. This result already was shown in, e.g. ,[1, Theorem 5.12].

2.3 Double Symbols

The present subsection is devoted to the introduction of pseudodifferential operators with double symbols. They had been introduced in order to verify that the composition of two smooth pseudodifferential operators is a smooth pseudodifferential operator again. Pseudodifferential operators with double symbols are also a very important tool to show the characterization of non-smooth pseudodifferential operators.

Definition 2.12 Let $0 < s \leq 1$, $\tilde{m} \in \mathbb{N}_0$ and $m, m' \in \mathbb{R}$. Moreover, let $N \in \mathbb{N}_0 \cup \{\infty\}$ and $0 \leq \rho \leq 1$. Then the space of *non-smooth double (pseudodifferential) symbols* $C^{\tilde{m},s}S^{m,m'}_{\rho,0}(\mathbb{R}^n \times \mathbb{R}^n \times \mathbb{R}^n \times \mathbb{R}^n; N)$ is the set of all functions $p : \mathbb{R}^n_x \times \mathbb{R}^n_\xi \times \mathbb{R}^n_{x'} \times \mathbb{R}^n_{\xi'} \to \mathbb{C}$ such that

i) $\partial_\xi^\alpha \partial_{x'}^{\beta'} \partial_{\xi'}^{\alpha'} p \in C^{\tilde{m},s}(\mathbb{R}^n_x)$ and $\partial_x^\beta \partial_\xi^\alpha \partial_{x'}^{\beta'} \partial_{\xi'}^{\alpha'} p \in C^0(\mathbb{R}^n_x \times \mathbb{R}^n_\xi \times \mathbb{R}^n_{x'} \times \mathbb{R}^n_{\xi'})$,

ii) $\|\partial_\xi^\alpha \partial_{x'}^{\beta'} \partial_{\xi'}^{\alpha'} p(.,\xi,x',\xi')\|_{C^{\tilde{m},s}(\mathbb{R}^n)} \leq C_{\alpha,\beta',\alpha'} \langle\xi\rangle^{m-\rho|\alpha|} \langle\xi'\rangle^{m'-\rho|\alpha'|}$

for all $\xi, x', \xi' \in \mathbb{R}^n$ and arbitrary $\beta, \alpha, \beta', \alpha' \in \mathbb{N}_0^n$ with $|\beta| \leq \tilde{m}$ and $|\alpha| \leq N$. Here the constant $C_{\alpha,\beta',\alpha'}$ is independent of $\xi, x', \xi' \in \mathbb{R}^n$. In the case $N = \infty$ we write $C^{\tilde{m},s}S^{m,m'}_{\rho,0}(\mathbb{R}^n \times \mathbb{R}^n \times \mathbb{R}^n \times \mathbb{R}^n)$ instead of $C^{\tilde{m},s}S^{m,m'}_{\rho,0}(\mathbb{R}^n \times \mathbb{R}^n \times \mathbb{R}^n \times \mathbb{R}^n; \infty)$. Moreover, we define the set of semi-norms $\{|.|_k^{m,m'} : k \in \mathbb{N}_0\}$ by

$$|p|_k^{m,m'} = \max_{\substack{|\alpha|+|\beta'|+|\alpha'| \leq k \\ |\alpha| \leq N}} \sup_{\xi,x',\xi' \in \mathbb{R}^n} \|\partial_\xi^\alpha \partial_{x'}^{\beta'} \partial_{\xi'}^{\alpha'} p(.,\xi,x',\xi')\|_{C^{\tilde{m},s}(\mathbb{R}^n)} \langle\xi\rangle^{-(m-\rho|\alpha|)} \langle\xi'\rangle^{-(m'-\rho|\alpha'|)}.$$

Because of the previous definition, $p \in C^{\tilde{m},s}S^m_{\rho,0}(\mathbb{R}^n \times \mathbb{R}^N)$ is often called a *non-smooth single symbol.*

For every non-smooth double symbol we define the associated operator as follows:

Definition 2.13 Let $0 < s \leq 1$, $\tilde{m} \in \mathbb{N}_0$, $0 \leq \rho \leq 1$ and $m, m' \in \mathbb{R}$. Additionally let $N \in \mathbb{N}_0 \cup \{\infty\}$. Assuming a symbol $p \in C^{\tilde{m},s}S^{m,m'}_{\rho,0}(\mathbb{R}^n \times \mathbb{R}^n \times \mathbb{R}^n \times \mathbb{R}^n; N)$, we define the pseudodifferential operator $P = p(x, D_x, x', D_{x'})$ such that for all

$u \in \mathcal{S}(\mathbb{R}^n)$ and $x \in \mathbb{R}^n$

$$Pu(x) := \text{Os -}\iiiint e^{-i(y\cdot\xi + y'\cdot\xi')} p(x, \xi, x + y, \xi')u(x + y + y')dydy'đ\xi đ\xi'.$$

Note that we can verify the existence of the previous oscillatory integral by using the properties of such integrals. For more details, see [18, Lemma 4.64].

Now let $OPC^{\tilde{m},s}S^{m,m'}_{\rho,0}(\mathbb{R}^n \times \mathbb{R}^n \times \mathbb{R}^n \times \mathbb{R}^n; N)$ be the set of all non-smooth pseudodifferential operators whose double symbols are elements of the symbol-class $C^{\tilde{m},s}S^{m,m'}_{\rho,0}(\mathbb{R}^n \times \mathbb{R}^n \times \mathbb{R}^n \times \mathbb{R}^n; N)$.

Later on we will need a special subset of $C^{\tilde{m},s}S^{m,0}_{\rho,0}(\mathbb{R}^n \times \mathbb{R}^n \times \mathbb{R}^n \times \mathbb{R}^n; N)$: For $0 < s \leq 1$, $\tilde{m} \in \mathbb{N}_0$, $N \in \mathbb{N}_0 \cup \{\infty\}$ and $m \in \mathbb{R}$ the set of all $p \in C^{\tilde{m},s}S^{m,0}_{\rho,0}(\mathbb{R}^n \times \mathbb{R}^n \times \mathbb{R}^n \times \mathbb{R}^n; N)$ which are independent of ξ' is denoted by $C^{\tilde{m},s}S^{m}_{\rho,0}(\mathbb{R}^n \times \mathbb{R}^n \times \mathbb{R}^n; N)$. The associated pseudodifferential operators $p(x, D_x, x')$ to elements of this subclass of double symbols are defined via

$$p(x, D_x, x') := p(x, D_x, x', D_{x'}).$$

We call the set of all non-smooth pseudodifferential operators whose double symbols are in the set $C^{\tilde{m},s}S^{m}_{\rho,0}(\mathbb{R}^n \times \mathbb{R}^n \times \mathbb{R}^n; N)$ by $OPC^{\tilde{m},s}S^{m}_{\rho,0}(\mathbb{R}^n \times \mathbb{R}^n \times \mathbb{R}^n; N)$.

We can prove the following representation for pseudodifferential operators of the symbol-class $C^{\tilde{m},s}S^{m}_{\rho,0}(\mathbb{R}^n \times \mathbb{R}^n \times \mathbb{R}^n; N)$ applied on a Schwartz function, see, e.g., [3, Lemma 3.13]:

Lemma 2.14 *Let $0 < s < 1$, $\tilde{m} \in \mathbb{N}_0$, $0 \leq \rho \leq 1$, $m \in \mathbb{R}$ and $N \in \mathbb{N}_0 \cup \{\infty\}$. Considering $a \in C^{\tilde{m},s}S^{m}_{\rho,0}(\mathbb{R}^n \times \mathbb{R}^n \times \mathbb{R}^n; N)$, we obtain for all $u \in \mathcal{S}(\mathbb{R}^n)$:*

$$a(x, D_x, x')u(x) = Os\text{ -}\iint e^{i(x-y)\cdot\xi} a(x, \xi, y)u(y)dyđ\xi \qquad \textit{for all } x \in \mathbb{R}^n.$$

Proof Let $u \in \mathcal{S}(\mathbb{R}^n)$ and $x \in \mathbb{R}^n$ be arbitrary. Then

$$\langle y' \rangle^{-2l} \langle D_{\xi'} \rangle^{2l} a(x, \xi, z)u(z + y') \in \mathscr{A}^{m_+, N}_{-2n-2}(\mathbb{R}^{2n}_{(z,y')} \times \mathbb{R}^{2n}_{(\xi,\xi')}).$$

With Theorem 1.9, Corollary 1.12, Theorem 1.13, and Theorem 1.10 at hand, we get

$$a(x, D_x, x')u(x)$$
$$= \text{Os -}\iiiint e^{-i(y\cdot\xi + y'\cdot\xi')} \langle y' \rangle^{-2l} \langle D_{\xi'} \rangle^{2l} [a(x, \xi, x + y)u(x + y + y')]dydy'đ\xi đ\xi'$$

$$= \text{Os-}\iiiint e^{-i(z-x)\cdot\xi} e^{-iy'\cdot\xi'} a(x,\xi,z)\langle y'\rangle^{-2l}\langle D_{\xi'}\rangle^{2l} u(z+y')dzdy'đ\xi đ\xi'$$
$$= \text{Os-}\iint e^{i(x-z)\cdot\xi} a(x,\xi,z)\left[\text{Os-}\iint e^{-iy'\cdot\xi'}\langle y'\rangle^{-2l}\langle D_{\xi'}\rangle^{2l} u(z+y')dy'đ\xi'\right]dzđ\xi.$$

By means of

$$\langle y'\rangle^{-2l}\langle D_{\xi'}\rangle^{2l} u(z+y') \in \mathcal{S}(\mathbb{R}^n_{y'}) \subseteq \mathscr{A}^0_{-k}(\mathbb{R}^n_{y'} \times \mathbb{R}^n_{\xi'})$$

we are able to apply Theorem 1.11 and Theorem 1.13 and get

$$\text{Os-}\iint e^{-iy'\cdot\xi'}\langle y'\rangle^{-2l}\langle D_{\xi'}\rangle^{2l} u(z+y')dy'đ\xi' = \text{Os-}\iint e^{-i(\tilde{z}-z)\cdot\xi'} u(\tilde{z})d\tilde{z}đ\xi' = u(z).$$

For the proof of the last equality, we refer to [1, Example 3.11]. Combining all these results we conclude the proof. □

As a direct consequence of the definition of double symbols we obtain the next remark.

Remark 2.15 Let $0 < s < 1$, $m \in \mathbb{R}$, $\tilde{m} \in \mathbb{N}_0$ and $N \in \mathbb{N}_0 \cup \{\infty\}$. The boundedness of the subset $\mathscr{B} \subseteq C^{\tilde{m},s} S^m_{\rho,0}(\mathbb{R}^n \times \mathbb{R}^n \times \mathbb{R}^n; N)$, $0 \leq \rho \leq 1$, implies the boundedness of

$$\left\{\partial_x^\delta \partial_\xi^\gamma a : a \in \mathscr{B}\right\} \subseteq C^{\tilde{m}-|\delta|,s} S^{m-\rho|\gamma|}_{\rho,0}(\mathbb{R}^n \times \mathbb{R}^n \times \mathbb{R}^n; N-|\gamma|)$$

for each $\gamma, \delta \in \mathbb{N}_0^n$ with $|\delta| \leq \tilde{m}$ and $|\gamma| \leq N$.

This remark was already verified in [3, Remark 3.14].

3 Characterization of Non-Smooth Pseudodifferential Operators

Throughout the whole section $(\varphi_j)_{j\in\mathbb{N}_0}$ is an arbitrary but fixed dyadic partition of unity on $\mathbb{R}^n$, that is a partition of unity with

$$\operatorname{supp}\varphi_0 \subseteq \overline{B_2(0)} \quad \text{and} \quad \operatorname{supp}\varphi_j \subseteq \{\xi \in \mathbb{R}^n : 2^{j-1} \leq |\xi| \leq 2^{j+1}\}$$

for all $j \in \mathbb{N}$.

3.1 Pointwise Convergence in $C^{m,s}S^0_{0,0}$

As an ingredient to show the characterization of non-smooth pseudodifferential operators, we need the following statement: Each bounded set $(p_\varepsilon)_{\varepsilon>0}$ of the symbol-class $C^{m,s}S^0_{0,0}(\mathbb{R}^n \times \mathbb{R}^n; M)$ contains a sequence which converges pointwise in $C^{m,s}S^0_{0,0}(\mathbb{R}^n \times \mathbb{R}^n; M-1)$. To this end we use

Lemma 3.1 *Let $m \in \mathbb{N}_0$, $0 < s \leq 1$ and $(p_\varepsilon)_{\varepsilon>0} \subseteq C^{m,s}(\mathbb{R}^n)$ be a bounded set. Then there is a sequence $(p_{\varepsilon_k})_{k\in\mathbb{N}} \subseteq (p_\varepsilon)_{\varepsilon>0}$ with $\varepsilon_k \to 0$ for $k \to \infty$ and a $p \in C^{m,s}(\mathbb{R}^n)$ such that for all $\beta \in \mathbb{N}_0^n$ with $|\beta| \leq m$*

$$\partial_x^\beta p_{\varepsilon_k} \xrightarrow{k\to\infty} \partial_x^\beta p$$

converges uniformly on each compact set $K \subseteq \mathbb{R}^n$.

Proof It is sufficient to prove the claim for each $\overline{B_j(0)}$, $j \in \mathbb{N}$. Due to the boundedness of $(p_\varepsilon|_{\overline{B_j(0)}})_{\varepsilon>0} \subseteq C^{m,s}(\overline{B_j(0)})$ and the compactness of the embedding $C^{m,s}(\overline{B_j(0)}) \subseteq C^m(\overline{B_j(0)})$ we get by a diagonal sequence argument the existence of a sequence $(p_{\varepsilon_k})_{k\in\mathbb{N}} \subseteq (p_\varepsilon)_{\varepsilon>0}$ with $\varepsilon_k \to 0$ for $k \to \infty$ and of unique functions $p_{B_j(0)} \in C^m(\overline{B_j(0)})$ such that

$$p_{\varepsilon_k} \xrightarrow{k\to\infty} p_{B_j(0)} \qquad \text{in } C^m(\overline{B_j(0)}) \text{ for all } j \in \mathbb{N}.$$

We define $p : \mathbb{R}^n \to \mathbb{C}$ via $p(x) := p_{B_j(0)}(x)$ for all $x \in \overline{B_j(0)}$ and each $j \in \mathbb{N}$. This implies the uniform convergence of

$$\partial_x^\beta p_{\varepsilon_k} \xrightarrow{k\to\infty} \partial_x^\beta p \qquad \text{on } \overline{B_j(0)}$$

for all $j \in \mathbb{N}$ and $\beta \in \mathbb{N}_0^n$ with $|\beta| \leq m$. The definition of p provides $p \in C^m(\overline{B_j(0)})$. The boundedness of $(p_\varepsilon)_{\varepsilon>0} \subseteq C^{m,s}(\mathbb{R}^n)$ and the pointwise convergence of $\partial_x^\beta p_\varepsilon \to \partial_x^\beta p$ if $\varepsilon \to 0$ for all $\beta \in \mathbb{N}_0^n$ with $|\beta| \leq m$ yields $p \in C^{m,s}(\mathbb{R}^n)$. □

With the previous result at hand, which was already shown in [3, Lemma 4.1] we can verify the next claim, see [3, Lemma 4.2]:

Lemma 3.2 *Let $m \in \mathbb{N}_0$ and $0 < s \leq 1$. Furthermore, let $(\partial_x^\beta p_\varepsilon)_{\varepsilon>0} \subseteq C^{0,s}(\mathbb{R}^n_x \times \mathbb{R}^n_\xi)$ be a bounded set for all $\beta \in \mathbb{N}_0^n$ with $|\beta| \leq m$. Then there is a sequence $(p_{\varepsilon_k})_{k\in\mathbb{N}} \subseteq (p_\varepsilon)_{\varepsilon>0}$ with $\varepsilon_k \to 0$ for $k \to \infty$ and a $p \in C^{0,s}(\mathbb{R}^n \times \mathbb{R}^n)$ such that for all $\beta \in \mathbb{N}_0^n$ with $|\beta| \leq m$ we have*

i) $\partial_x^\beta p \in C^{0,s}(\mathbb{R}^n_x \times \mathbb{R}^n_\xi)$,

ii) $\partial_x^\beta p_{\varepsilon_k} \xrightarrow{k\to\infty} \partial_x^\beta p$ *converges uniformly on each compact set $K \subseteq \mathbb{R}^n \times \mathbb{R}^n$.*

Proof It is sufficient to show the claim for all sets $\overline{B_j(0) \times B_i(0)}$, $i, j \in \mathbb{N}$. Since the set $(\partial_x^\beta p_\varepsilon)_{\varepsilon>0}$ is bounded in $C^{0,s}(\mathbb{R}^n_x \times \mathbb{R}^n_\xi)$, we iteratively conclude from Lemma 3.1 the existence of a sequence $(p_{\varepsilon_k})_{k\in\mathbb{N}}$ of $(p_\varepsilon)_{\varepsilon>0}$ and of functions $q_\beta \in C^{0,s}(\mathbb{R}^n_x \times \mathbb{R}^n_\xi)$ such that

$$\partial_x^\beta p_{\varepsilon_k} \xrightarrow{k\to\infty} q_\beta \qquad \text{uniformly in } \overline{B_j(0) \times B_i(0)} \tag{17}$$

for all $i, j \in \mathbb{N}$ and $\beta \in \mathbb{N}_0^n$ with $|\beta| \leq m$. Choosing an arbitrary but fixed $\xi \in \mathbb{R}^n$, (17) implies the uniformly convergence of

$$\partial_x^\beta p_{\varepsilon_k}(.,\xi) \xrightarrow{k\to\infty} q_\beta(.,\xi) \tag{18}$$

in $\overline{B_j(0)}$ for all $\beta \in \mathbb{N}_0^n$ with $|\beta| \leq m$ and all $j \in \mathbb{N}$. Hence $(p_{\varepsilon_k}(.,\xi))_{k\in\mathbb{N}}$ is a Cauchy sequence in $C^m(\overline{B_j(0)})$. Due to the completeness of $C^m(\overline{B_j(0)})$ we have the convergence of $(p_{\varepsilon_k}(.,\xi))_{k\in\mathbb{N}}$ to $\tilde{p}$ in $C^m(\overline{B_j(0)})$. Consequently we obtain for all $\beta \in \mathbb{N}_0^n$ with $|\beta| \leq m$ and each $j \in \mathbb{N}$:

$$\partial_x^\beta p_{\varepsilon_k}(.,\xi) \xrightarrow{k\to\infty} \partial_x^\beta \tilde{p} \qquad \text{in } C^0(\overline{B_j(0)}). \tag{19}$$

Because of the uniqueness of the strong limit we get together with (18) that $\partial_x^\beta \tilde{p} = q_\beta(.,\xi)$ for each $\beta \in \mathbb{N}_0^n$ with $|\beta| \leq m$. Thus with $p(x,\xi) := q_0(x,\xi)$ for all $x, \xi \in \mathbb{R}^n$ the claim holds. □

All verified results enable us to show the main theorem of this subsection, which generalizes [3, Theorem 4.3]:

Theorem 3.3 *Let $\tilde{m} \in \mathbb{N}_0$, $M \in [0,\infty]$ and $0 < s \leq 1$. Additionally, let $(p_\varepsilon)_{\varepsilon>0}$ be a bounded set in $C^{\tilde{m},s}S^0_{0,0}(\mathbb{R}^n \times \mathbb{R}^n; M)$. Then there is a sequence $(p_{\varepsilon_l})_{l\in\mathbb{N}} \subseteq (p_\varepsilon)_{\varepsilon>0}$ with $\varepsilon_l \to 0$ for $l \to \infty$ and a function $p : \mathbb{R}^n_x \times \mathbb{R}^n_\xi \to \mathbb{C}$ such that for all $\alpha, \beta \in \mathbb{N}_0^n$ with $|\beta| \leq \tilde{m}$ and $|\alpha| < M$ we get*

i) *$\partial_x^\beta \partial_\xi^\alpha p$ exists and $\partial_x^\beta \partial_\xi^\alpha p \in C^{0,\tau}(\mathbb{R}^n \times \mathbb{R}^n)$ for some $0 < \tau < \min\{s, M - \lfloor M \rfloor\}$, if $M \notin \mathbb{N}$ and $\tau \in (0,1)$ else*

ii) *$\partial_x^\beta \partial_\xi^\alpha p_{\varepsilon_l} \xrightarrow{l\to\infty} \partial_x^\beta \partial_\xi^\alpha p$ is uniformly convergent on each compact set $K \subseteq \mathbb{R}^n \times \mathbb{R}^n$.*

In particular $p \in C^{\tilde{m},s}S^0_{0,0}(\mathbb{R}^n \times \mathbb{R}^n; \lceil M \rceil - 1)$.

Proof It is sufficient to prove the claim for $\overline{B_j(0) \times B_j(0)}$, $j \in \mathbb{N}$. Applying Lemma 2.1 we get for all $\beta, \gamma \in \mathbb{N}_0^n$ with $|\beta| \leq \tilde{m}$ and $|\gamma| < M$ the boundedness of the set $(\partial_x^\beta \partial_\xi^\gamma p_\varepsilon)_{\varepsilon>0} \subseteq C^{0,\tau}(\mathbb{R}^n \times \mathbb{R}^n)$ for some $0 < \tau < \min\{s, M - \lfloor M \rfloor\}$ if $M \notin \mathbb{N}_0 \cup \{\infty\}$ and $\tau \in (0,1)$ else. Thus by Lemma 3.2 we inductively obtain the existence of a sequence $(p_{\varepsilon_l})_{l\in\mathbb{N}} \subseteq (p_\varepsilon)_{\varepsilon>0}$ with $\varepsilon_l \to 0$ for $l \to \infty$ and functions $q_\alpha \in C^{0,\tau}(\mathbb{R}^n \times \mathbb{R}^n)$ with the following properties: For all $j \in \mathbb{N}$ and $\alpha, \beta \in \mathbb{N}_0^n$

with $|\beta| \leq \tilde{m}$ and $|\alpha| < M$ we have $\partial_x^\beta q_\alpha \in C^{0,\tau}(\mathbb{R}^n_x \times \mathbb{R}^n_\xi)$ and

$$\partial_x^\beta \partial_\xi^\alpha p_{\varepsilon_l} \xrightarrow{l\to\infty} \partial_x^\beta q_\alpha \tag{20}$$

converges uniformly on $\overline{B_j(0) \times B_j(0)}$. Now we choose an arbitrary but fixed $k \in \mathbb{N}_0$ with $k < M$ and $x \in \mathbb{R}^n$. The boundedness of $(\partial_\xi^\gamma p_{\varepsilon_l})_{l\in\mathbb{N}} \subseteq C^{0,\tau}(\mathbb{R}^n \times \mathbb{R}^n)$ for all $\gamma \in \mathbb{N}_0^n$ with $|\gamma| \leq k$ leads to

$$\|p_{\varepsilon_l}(x,.)\|_{C^{k,\tau}(\mathbb{R}^n)} \leq \max_{\substack{\gamma\in\mathbb{N}_0^n \\ |\gamma|\leq k}} \|\partial_\xi^\gamma p_{\varepsilon_l}\|_{C^{0,\tau}(\mathbb{R}^n\times\mathbb{R}^n)} \leq C_k$$

for all $x \in \mathbb{R}^n$ and $l \in \mathbb{N}$. By means of Lemma 3.1 we obtain via a diagonal sequence argument the existence of a subsequence of $(p_{\varepsilon_l})_{l\in\mathbb{N}}$ again denoted by $(p_{\varepsilon_l})_{l\in\mathbb{N}}$ and of a function $\tilde{p} \in C^{\lceil M\rceil-1}(\mathbb{R}^n)$ with the property

$$\partial_\xi^\gamma p_{\varepsilon_l}(x,\xi) \xrightarrow{l\to\infty} \partial_\xi^\gamma \tilde{p}(\xi) \qquad \text{pointwise for all } x \in \mathbb{R}^n \tag{21}$$

and every $\gamma \in \mathbb{N}_0^n$ with $|\gamma| \leq \lceil M\rceil - 1$. On account of (20) and (21) the uniqueness of the limit gives us $q_\alpha(x,.) = \partial_\xi^\alpha \tilde{p}$. This implies $p(x,.) := q_0(x,.) \in C^{\lceil M\rceil-1}(\mathbb{R}^n)$ for all $x \in \mathbb{R}^n$, (i) and (ii). Additionally the boundedness of $(p_{\varepsilon_l})_{\varepsilon>0} \subseteq C^{\tilde{m},s}S^0_{0,0}(\mathbb{R}^n \times \mathbb{R}^n; M)$ provides for all $\alpha \in \mathbb{N}_0^n$ with $|\alpha| < M$ and each fixed $\xi \in \mathbb{R}^n$

$$\|\partial_\xi^\alpha p_{\varepsilon_l}(.,\xi)\|_{C^{\tilde{m},s}(\mathbb{R}^n)} \leq C_\alpha.$$

Hence we get due to Lemma 3.1 via a diagonal sequence argument the existence of a subsequence of $(p_{\varepsilon_l})_{l\in\mathbb{N}}$ again denoted by $(p_{\varepsilon_l})_{l\in\mathbb{N}}$ and of a function $\hat{p}_\alpha \in C^{\tilde{m},s}(\mathbb{R}^n)$ with the property

$$\partial_\xi^\alpha p_{\varepsilon_l}(x,\xi) \xrightarrow{l\to\infty} \hat{p}_\alpha(x) \qquad \text{pointwise for all } \xi \in \mathbb{R}^n \tag{22}$$

and every $\alpha \in \mathbb{N}_0^n$ with $|\alpha| < M$. The uniqueness of the limit, (20) and (22) provides the equality $\partial_\xi^\alpha p(.,\xi) = q_\alpha(.,\xi) = \hat{p}_\alpha$ for each $\alpha \in \mathbb{N}_0^n$ with $|\alpha| < M$. Consequently p is a non-smooth symbol of the symbol-class $C^{\tilde{m},s}S^0_{0,0}(\mathbb{R}^n \times \mathbb{R}^n; \lceil M\rceil - 1)$. □

3.2 *Properties of the Operator T_ε*

Apart from the results of Sect. 3.1 an approximation $(T_\varepsilon)_{\varepsilon\in(0,1]}$ of $T \in \mathcal{A}^{0,M}_{0,0}(\tilde{m}, q)$ is needed in order to prove the characterization of non-smooth pseudodifferential operators. Thereby the approximation operators $(T_\varepsilon)_{\varepsilon\in(0,1]}$ have to satisfy the following conditions:

- $T_\varepsilon : \mathcal{S}'(\mathbb{R}^n) \to \mathcal{S}(\mathbb{R}^n)$ is continuous,
- The iterated commutators of T_ε are uniformly bounded with respect to ε as maps from $L^q(\mathbb{R}^n)$ to $L^q(\mathbb{R}^n)$,
- T_ε converges pointwise to T if $\varepsilon \to 0$.

All results of the present subsection are taken from [3, Sect. 4.3].

Throughout the whole subsection we assume: Let $1 < q < \infty$ and $M \in \mathbb{N}_0 \cup \{\infty\}$ be arbitrary. Additionally, let $T \in \mathcal{A}^{0,M}_{0,0}(\tilde{m}, q)$ with $\tilde{m} \in \mathbb{N}_0$ and $\varphi \in C_0^\infty(\mathbb{R}^n)$ with $\varphi(x) = 1$ for all $|x| \leq \frac{1}{2}$ and $\varphi(x) = 0$ for all $|x| \geq 1$. For all $0 < \varepsilon \leq 1$ we define the pseudodifferential operators

$$P_\varepsilon := \tilde{p}_\varepsilon(x, D_x) \qquad \text{and} \qquad Q_\varepsilon := q_\varepsilon(x, D_x),$$

with the symbols $\tilde{p}_\varepsilon(x, \xi) := \varphi(\varepsilon x)$ and $q_\varepsilon(x, \xi) := \varphi(\varepsilon \xi)$. Then the sets $\{\tilde{p}_\varepsilon | 0 < \varepsilon \leq 1\}$ and $\{q_\varepsilon | 0 < \varepsilon \leq 1\}$ are bounded subsets of $S^0_{1,0}(\mathbb{R}^n \times \mathbb{R}^n)$. We remark that for all $u \in \mathcal{S}(\mathbb{R}^n)$ we have $P_\varepsilon u = \tilde{p}_\varepsilon u$. Moreover we get the continuity of $P_\varepsilon : C^\infty(\mathbb{R}^n) \to C_0^\infty(\mathbb{R}^n)$ by means of the continuity of multiplication operators with C_0^∞-functions. Furthermore, we define

$$T_\varepsilon := P_\varepsilon Q_\varepsilon T P_\varepsilon Q_\varepsilon.$$

Since $P_\varepsilon Q_\varepsilon : \mathcal{S}(\mathbb{R}^n)' \to \mathcal{S}(\mathbb{R}^n)$ is continuous, $T_\varepsilon : \mathcal{S}(\mathbb{R}^n)' \to \mathcal{S}(\mathbb{R}^n)$ is continuous, too. Later on we will need the next statements:

Lemma 3.4 *For all $u \in L^q(\mathbb{R}^n)$ we have the following convergence:*

$$L^q - \lim_{\varepsilon \to 0} T_\varepsilon u = Tu.$$

Proof With the theorem of Banach-Steinhaus at hand, we can easily show

$$Q_\varepsilon u \xrightarrow{\varepsilon\to 0} u \qquad \text{and} \qquad P_\varepsilon u \xrightarrow{\varepsilon\to 0} u \qquad \text{in } L^q(\mathbb{R}^n). \tag{23}$$

For more details, see [18, Proof of Lemma 5.27]. By means of (23) and Theorem 2.5 we get for all $u \in L^q(\mathbb{R}^n)$:

$$\|P_\varepsilon Q_\varepsilon u - u\|_{L^q(\mathbb{R}^n)} \leq C\|Q_\varepsilon u - u\|_{L^q(\mathbb{R}^n)} + \|P_\varepsilon u - u\|_{L^q(\mathbb{R}^n)} \xrightarrow{\varepsilon\to 0} 0.$$

An application of Theorem 2.5 gives us for all $u \in L^q(\mathbb{R}^n)$:

$$\begin{aligned} \|T_\varepsilon u - Tu\|_{L^q(\mathbb{R}^n)} &\leq \|P_\varepsilon Q_\varepsilon T P_\varepsilon Q_\varepsilon u - P_\varepsilon Q_\varepsilon T u\|_{L^q(\mathbb{R}^n)} + \|P_\varepsilon Q_\varepsilon T u - Tu\|_{L^q(\mathbb{R}^n)} \\ &\leq C\|P_\varepsilon Q_\varepsilon u - u\|_{L^q(\mathbb{R}^n)} + \|P_\varepsilon Q_\varepsilon T u - Tu\|_{L^q(\mathbb{R}^n)} \xrightarrow{\varepsilon\to 0} 0. \end{aligned}$$

□

Lemma 3.5 *Let $\alpha, \beta \in \mathbb{N}_0^n$ with $|\beta| \le \tilde{m}$ and $|\alpha| \le M$. Then*

$$\| \operatorname{ad}(-ix)^\alpha \operatorname{ad}(D_x)^\beta T_\varepsilon \|_{\mathscr{L}(L^q(\mathbb{R}^n))} \le C_{\alpha,\beta} \qquad \text{for all } 0 < \varepsilon \le 1.$$

Proof Let $\alpha, \beta \in \mathbb{N}_0^n$ with $|\beta| \le \tilde{m}$ and $|\alpha| \le M$. We define

$$R^{\beta_1,\beta_2,\beta_3}_{\alpha_1,\alpha_2,\alpha_3} := \left[\operatorname{ad}(D_x)^{\beta_1} P_\varepsilon\right]\left[\operatorname{ad}(-ix)^{\alpha_1} Q_\varepsilon\right] T^{\alpha_2,\beta_2}\left[\operatorname{ad}(D_x)^{\beta_3} P_\varepsilon\right]\left[\operatorname{ad}(-ix)^{\alpha_3} Q_\varepsilon\right],$$

where $T^{\alpha_2,\beta_2} := \operatorname{ad}(-ix)^{\alpha_2} \operatorname{ad}(D_x)^{\beta_2} T$. Then we obtain for all $u \in \mathcal{S}(\mathbb{R}^n)$

$$\operatorname{ad}(-ix)^\alpha \operatorname{ad}(D_x)^\beta T_\varepsilon u = \sum_{\substack{\alpha_1+\alpha_2+\alpha_3=\alpha \\ \beta_1+\beta_2+\beta_3=\beta}} C_{\alpha_1,\alpha_2,\beta_1,\beta_2} R^{\beta_1,\beta_2,\beta_3}_{\alpha_1,\alpha_2,\alpha_3} u.$$

Due to Remark 1.5 we get $\operatorname{ad}(D_x)^\gamma P_\varepsilon \in OPS^0_{1,0}$ and $\operatorname{ad}(-ix)^\delta Q_\varepsilon \in OPS^{-|\delta|}_{1,0} \subseteq OPS^0_{1,0}$ for each $\gamma, \delta \in \mathbb{N}_0^n$. On account of Theorem 2.5, the boundedness of $\{\tilde{p}_\varepsilon | 0 < \varepsilon \le 1\}$ and $\{q_\varepsilon | 0 < \varepsilon \le 1\}$ in $S^0_{1,0}(\mathbb{R}^n \times \mathbb{R}^n)$ and of $T \in \mathcal{A}^{0,M}_{0,0}(\tilde{m}, q)$ we obtain

$$\| \operatorname{ad}(-ix)^\alpha \operatorname{ad}(D_x)^\beta T_\varepsilon u \|_{L^q} \le C_{\alpha,\beta,q} \|u\|_{L^q} \qquad \text{for all } u \in \mathcal{S}(\mathbb{R}^n). \qquad \square$$

Proposition 3.6 *Let $g \in \mathcal{S}(\mathbb{R}^n)$ and $0 < \varepsilon \le 1$. For each $y \in \mathbb{R}^n$ we define $g_y := \tau_y(g)$. Moreover, we define*

$$p_{\varepsilon,0}(x, \xi, y) := e^{-ix\cdot\xi} T_\varepsilon(e_\xi g_y)(x) \qquad \text{for all } (x, \xi, y) \in \mathbb{R}^n \times \mathbb{R}^n \times \mathbb{R}^n.$$

Then $p_{\varepsilon,0} \in C^\infty(\mathbb{R}^n \times \mathbb{R}^n \times \mathbb{R}^n)$.

For the proof of Proposition 3.6, we still need:

Definition 3.7 For $k \in \mathbb{N}_0$ we define the normed space $L^q_k(\mathbb{R}^n)$ as

$$L^q_k(\mathbb{R}^n) := \left\{ f \in L^q(\mathbb{R}^n) : \|f\|_{L^q_k} := \|\langle x\rangle^{k+1} f(x)\|_{L^q(\mathbb{R}^n_x)} < \infty \right\}.$$

Sketch of the proof of Proposition 3.6 Let $k \in \mathbb{N}_0$ be arbitrary but fixed. For every $x, \xi \in \mathbb{R}^n$, $f \in C^{k+1}_b(\mathbb{R}^n)$ and each $h \in L^q_k(\mathbb{R}^n)$ we define $\delta_x(f) := f(x)$ and $M_\xi(h) := e_\xi h$. Additionally we define the functions $\tilde{\delta} : \mathbb{R}^n \times \mathbb{R}^n \times \mathbb{R}^n \to \mathscr{L}(C^{k+1}_b(\mathbb{R}^n), \mathbb{C})$, $\tilde{G} : \mathbb{R}^n \times \mathbb{R}^n \times \mathbb{R}^n \to L^q_k(\mathbb{R}^n)$ and $\tilde{M} : \mathbb{R}^n \times \mathbb{R}^n \times \mathbb{R}^n \to \mathscr{L}(L^q_k(\mathbb{R}^n), L^q(\mathbb{R}^n))$ by

$$\tilde{\delta}(x, y, \xi) := \delta_x, \quad \tilde{G}(x, y, \xi) := g_y, \quad \tilde{M}(x, y, \xi) := M_\xi \qquad \text{for all } x, y, \xi \in \mathbb{R}^n.$$

One can show that $\tilde{G}$ is a smooth function and that $\tilde{\delta}$, $\tilde{M}$ are k-times continuously differentiable, cf. [18, Proposition 5.33 and Proposition 5.34]. On account of the product rule we get

$$\tilde{M}(x, y, \xi) \circ \tilde{G}(x, y, \xi) \in C^k(\mathbb{R}^n_x \times \mathbb{R}^n_y \times \mathbb{R}^n_\xi, L^q(\mathbb{R}^n)). \tag{24}$$

Since T_ε is continuous as map from $\mathcal{S}'(\mathbb{R}^n)$ to $\mathcal{S}(\mathbb{R}^n)$, we have

$$T_\varepsilon \in \mathscr{L}(L^q(\mathbb{R}^n), C_b^{k+1}(\mathbb{R}^n))$$

and hence we obtain

$$T_\varepsilon(\tilde{M}(x, y, \xi) \circ \tilde{G}(x, y, \xi)) \in C^k(\mathbb{R}^n_x \times \mathbb{R}^n_y \times \mathbb{R}^n_\xi, C_b^{k+1}(\mathbb{R}^n))$$

due to (24). Applying the product rule again yields

$$p_{\varepsilon,0}(x, y, \xi) = e^{-ix\cdot\xi}\tilde{\delta}(x, y, \xi) \circ T_\varepsilon(\tilde{M}(x, y, \xi) \circ \tilde{G}(x, y, \xi)) \in C^k(\mathbb{R}^n_x \times \mathbb{R}^n_y \times \mathbb{R}^n_\xi).$$

□

We refer to [18, Proposition 5.31] for more details.

3.3 Characterization of Pseudodifferential Operators with Symbols in $C^s S^m_{0,0}$

Besides the pointwise convergence result of Sect. 3.1 and the results of Sect. 3.2 also a formula for representing an operator with a non-smooth double symbol as an operator with a non-smooth single symbol is needed to derive the characterization of non-smooth pseudodifferential operators. We achieve this result by a careful adaption of the known symbol-reduction result for smooth pseudodifferential operators, cf. [15, Theorem 2.5] by using the extended properties of the oscillatory integrals presented in Sect. 1.3. This is the reason why we just give a short sketch of the proof of the next result here.

Theorem 3.8 *Let $N \in \mathbb{N}_0 \cup \{\infty\}$ with $N > n$. We define $\tilde{N} := N - (n + 1)$. For each element a of the bounded set $\mathscr{B} \subseteq C^{\tilde{m},s} S^m_{0,0}(\mathbb{R}^n \times \mathbb{R}^n \times \mathbb{R}^n; N)$ with $\tilde{m} \in \mathbb{N}_0$, $m \in \mathbb{R}$ and $0 < s < 1$, we define the function $a_L : \mathbb{R}^n \times \mathbb{R}^n \to \mathbb{C}$ by*

$$a_L(x, \xi) := Os\text{-}\iint e^{-iy\cdot\eta} a(x, \eta + \xi, x + y) dy đ\eta$$

for all $x, \xi \in \mathbb{R}^n$. Then $\{a_L : a \in \mathscr{B}\} \subseteq C^{\tilde{m},s} S^m_{0,0}(\mathbb{R}^n \times \mathbb{R}^n; \tilde{N})$ is bounded and we have for every $a \in \mathscr{B}$ and $u \in \mathcal{S}(\mathbb{R}^n)$

$$a(x, D_x, x')u = a_L(x, D_x)u. \tag{25}$$

Proof The first task is to verify by means of the properties of the oscillatory integral, Theorem of Fubini and the Theorem of dominated convergence, that the set $\{a_L(x,\xi) : a \in \mathscr{B}\}$ is a bounded set of non-smooth single symbols of the symbol-class $C^{\tilde{m},s}S^m_{0,0}(\mathbb{R}^n_x \times \mathbb{R}^n_\xi; \tilde{N})$. Afterwards one can show by means of an integral representation of the operator $a(x, D_x, x')$ applied on a Schwartz function u and the properties of the oscillatory integral that the equality (25) is true. □

For more details, we refer to [3, Sect. 4.2]. Moreover, a calculation of an integral representation for a non-smooth pseudodifferential operator with double symbol, an application of Remark 1.7 and Remark 1.8 and of integration by parts with respect to ξ yields the next lemma, cf. [3, Lemma 4.4]:

Lemma 3.9 *Let $s > 0$, $s \notin \mathbb{N}_0$ and $m, m' \in \mathbb{R}$. Additionally let $N \in \mathbb{N}_0 \cup \{\infty\}$ and $l' \in \mathbb{N}_0$ with $l' \leq N$. Furthermore, let $\mathscr{B} \subset C^s S^{m,m'}_{0,0}(\mathbb{R}^n \times \mathbb{R}^n \times \mathbb{R}^n \times \mathbb{R}^n; N)$ be bounded and $u \in \mathcal{S}(\mathbb{R}^n)$. Assuming $p \in C^s S^{m,m'}_{0,0}(\mathbb{R}^n \times \mathbb{R}^n \times \mathbb{R}^n \times \mathbb{R}^n; N)$, we denote $P := p(x, D_x, x', D_{x'})$. Then we obtain the existence of a constant C, independent of $x \in \mathbb{R}^n$ and $p \in \mathscr{B}$, such that*

$$|Pu(x)| \leq C\langle x\rangle^{-l'} \qquad \textit{for all } x \in \mathbb{R}^n.$$

With the previous results at hand we now are in the position to show the characterization of pseudodifferential operators with symbols of the symbol-class $C^s S^0_{0,0}(\mathbb{R}^n \times \mathbb{R}^n; M)$. Symbol reducing smooth pseudodifferential operators enable us to extend this result to non-smooth pseudodifferential operators of the class $C^s S^m_{0,0}(\mathbb{R}^n \times \mathbb{R}^n; M)$, $m \in \mathbb{N}_0$. In all cases the following problem arises: Non-smooth pseudodifferential operators with coefficients in a Hölder space are in general not continuous as a map from $H^m_q(\mathbb{R}^n)$ to $L^q(\mathbb{R}^n)$. However each element of $\mathcal{A}^{m,M}_{0,0}(\tilde{m}, q)$ is linear and bounded as a map from $H^m_q(\mathbb{R}^n)$ to $L^q(\mathbb{R}^n)$. Hence we only can get a characterization of those non-smooth pseudodifferential operators which are linear and bounded as maps from $H^m_q(\mathbb{R}^n)$ to $L^q(\mathbb{R}^n)$ by means of this ansatz. Note that the proofs of this subsection are based on the main idea of the proof in the smooth case by Ueberberg [22] and are taken from [3, Sect. 4.4].

Theorem 3.10 *Let $1 < q < \infty$ and $m \in \mathbb{N}_0$ with $m > n/q$. Additionally let $M \in \mathbb{N} \cup \{\infty\}$ with $M > n+1$. We define $\tilde{M} := M - (n+1)$. Considering $T \in \mathcal{A}^{0,M}_{0,0}(m, q)$, we get for all $0 < \tau \leq m - n/q$ with $\tau \notin \mathbb{N}_0$*

$$T \in OPC^\tau S^0_{0,0}(\mathbb{R}^n \times \mathbb{R}^n; \tilde{M} - 1) \cap \mathscr{L}(L^q(\mathbb{R}^n)).$$

Proof Let $\tau \in (0, m - n/q]$ with $\tau \notin \mathbb{N}$ be arbitrary but fixed. Let T_ε, $\varepsilon \in (0, 1]$, be as in Sect. 3.2. Then $T_\varepsilon : \mathcal{S}'(\mathbb{R}^n) \to \mathcal{S}(\mathbb{R}^n)$ is continuous. The proof of this theorem is divided into three different parts. First we write T_ε as a pseudodifferential operator with a double symbol. In step two we reduce the double symbol to an ordinary symbol p_ε of T_ε. Finally, we conclude the proof in part three. Here we use the pointwise convergence of a subsequence of $(p_\varepsilon)_{\varepsilon>0}$ to get a symbol p with the property $p(x, D_x)u = Tu$ for all $u \in \mathcal{S}(\mathbb{R}^n)$.

We begin with step one: Since $T_\varepsilon : \mathcal{S}'(\mathbb{R}^n) \to \mathcal{S}(\mathbb{R}^n)$ is continuous, Theorem 1.2 gives us the existence of a Schwartz-kernel $t_\varepsilon \in \mathcal{S}(\mathbb{R}^n \times \mathbb{R}^n)$ of T_ε. Thus

$$T_\varepsilon u(x) = \int t_\varepsilon(x, y) u(y) dy \qquad \text{for all } u \in \mathcal{S}(\mathbb{R}^n) \text{ and all } x \in \mathbb{R}^n. \tag{26}$$

Now we choose $u, g \in \mathcal{S}(\mathbb{R}^n)$ with $g(0) = 1$ and $g(-x) = g(x)$ for all $x \in \mathbb{R}$. We define $g_y : \mathbb{R}^n \to \mathbb{C}$ for $y \in \mathbb{R}^n$ by $g_y := \tau_y(g)$. Next let $x \in \mathbb{R}^n$ be arbitrary, but fixed. Then we define

$$h(z) := u(z) g_z(x) \qquad \text{for all } z \in \mathbb{R}^n.$$

Using the inversion formula, cf., e.g., [1, Example 3.11], we obtain

$$u(x) = h(x) = \mathrm{Os}\text{-}\iint e^{i(x-y)\cdot\xi} h(y) dy đ\xi = \mathrm{Os}\text{-}\iint e^{i(x-y)\cdot\xi} u(y) g_y(x) dy đ\xi.$$

If we first insert the previous equality in (26) and use the definition of the oscillatory integrals, integration by parts with respect to y and Lebesgue's theorem afterwards, we get

$$\begin{aligned} T_\varepsilon u(x) &= \int t_\varepsilon(x, z) \left[\mathrm{Os}\text{-}\iint e^{i(z-y)\cdot\xi} u(y) g_y(z) dy đ\xi \right] dz \\ &= \lim_{\alpha \to 0} \int t_\varepsilon(x, z) \cdot \iint e^{-iy\cdot\xi} e^{iz\cdot\xi} u(y) g_y(z) \chi(\alpha y, \alpha\xi) dy đ\xi \, dz \\ &= \lim_{\alpha \to 0} \iint e^{-iy\cdot\xi} \chi(\alpha y, \alpha\xi) \left[T_\varepsilon(e_\xi g_y)(x) \right] u(y) dy đ\xi, \end{aligned}$$

where $\chi \in \mathcal{S}(\mathbb{R}^n \times \mathbb{R}^n)$ with $\chi(0, 0) = 1$. Defining $p_{\varepsilon,0} : \mathbb{R}^n \times \mathbb{R}^n \times \mathbb{R}^n \to \mathbb{C}$ by

$$p_{\varepsilon,0}(x, \xi, y) := e^{-ix\cdot\xi} T_\varepsilon(e_\xi g_y)(x) \qquad \text{for all } x, \xi, y \in \mathbb{R}^n,$$

we conclude

$$T_\varepsilon u(x) = \mathrm{Os}\text{-}\iint e^{i(x-y)\cdot\xi} p_{\varepsilon,0}(x, \xi, y) u(y) dy đ\xi.$$

Here $p_{\varepsilon,0}$ is the double symbol of T_ε, cf. Lemma 2.14, as we will see in step two.

Secondly we want to construct for all $0 < \varepsilon \leq 1$ symbols

$$p_\varepsilon \in C^\tau S^0_{0,0}(\mathbb{R}^n \times \mathbb{R}^n; \tilde{M}),$$

with

i) $T_\varepsilon u = p_\varepsilon(x, D_x) u$ for all $u \in \mathcal{S}(\mathbb{R}^n)$,
ii) $(p_\varepsilon)_{0 < \varepsilon \leq 1}$ is a bounded set of $C^\tau S^0_{0,0}(\mathbb{R}^n \times \mathbb{R}^n; \tilde{M})$.

Since $T_\varepsilon : \mathcal{S}'(\mathbb{R}^n) \to \mathcal{S}(\mathbb{R}^n)$ is linear and continuous and because of Proposition 3.6, we can apply Lemma 1.6 and Lemma 3.5 and get for $\alpha, \gamma \in \mathbb{N}_0^n$ with $|\alpha| \leq M$:

$$
\begin{aligned}
&\left\| \partial_\xi^\alpha D_y^\gamma p_{\varepsilon,0}(.,\xi,y) \right\|_{C^\tau}^q \\
&\leq \left\| \partial_\xi^\alpha D_y^\gamma p_{\varepsilon,0}(.,\xi,y) \right\|_{H_q^m}^q \\
&\leq \sum_{|\beta|\leq m} \left\| \partial_\xi^\alpha D_x^\beta D_y^\gamma p_{\varepsilon,0}(x,\xi,y) \right\|_{L^q(\mathbb{R}_x^n)}^q \\
&\leq \sum_{|\beta|\leq m} \sum_{\beta_1+\beta_2=\beta} \left\| C_{\beta_1,\beta_2} \left[\mathrm{ad}(-ix)^\alpha \, \mathrm{ad}(D_x)^{\beta_1} T_\varepsilon \right] \left(e^{ix\cdot\xi} D_x^{\beta_2+\gamma} g_y \right)(x) \right\|_{L^q(\mathbb{R}_x^n)}^q \\
&\leq C_{\alpha,m,\gamma}
\end{aligned}
$$

for all $\xi, y \in \mathbb{R}^n$ and $0 < \varepsilon \leq 1$. Hence $\{p_{\varepsilon,0} : 0 < \varepsilon \leq 1\} \subseteq C^\tau S^0_{0,0}(\mathbb{R}^n \times \mathbb{R}^n \times \mathbb{R}^n; M)$ is bounded. Now we define

$$
p_\varepsilon(x,\xi) := \mathrm{Os}\text{-}\iint e^{-iy\cdot\eta} p_{\varepsilon,0}(x,\xi+\eta,x+y)\,dy\,đ\eta \qquad \text{for all } x,\xi \in \mathbb{R}^n.
$$

An application of Theorem 3.8 and Theorem 3.8 yields the properties i) and ii). So we can turn to step three now.

On account of ii) it is possible to apply Lemma 3.3 which yields the existence of a sequence $(p_{\varepsilon_k})_{k\in\mathbb{N}}$ of $(p_\varepsilon)_{0<\varepsilon\leq 1}$ with $\varepsilon_k \to 0$ if $k \to \infty$ such that

$$
p_{\varepsilon_k} \xrightarrow{k\to\infty} p \qquad \text{pointwise,} \tag{27}
$$

where $p \in C^\tau S^0_{0,0}(\mathbb{R}^n \times \mathbb{R}^n; \tilde{M} - 1)$. Let $u \in \mathcal{S}(\mathbb{R}^n)$ be arbitrary. Because of (27) and the boundedness of $(p_{\varepsilon_k})_{k\in\mathbb{N}} \subseteq C^\tau S^0_{0,0}(\mathbb{R}^n \times \mathbb{R}^n; \tilde{M})$, we get

$$
p_{\varepsilon_k}(x, D_x)u \xrightarrow{k\to\infty} p(x, D_x)u \tag{28}
$$

pointwise due to Lebesgue's theorem. Choosing $N \in \mathbb{N}$ with $n < 2N \leq M$ we get by Lemma 3.9:

$$
\begin{aligned}
|p_{\varepsilon_k}(x,D_x)u(x) - p(x,D_x)u(x)|^q &\leq \left(|p_{\varepsilon_k}(x,D_x)u(x)| + \lim_{k\to\infty} |p_{\varepsilon_k}(x,D_x)u(x)| \right)^q \\
&\leq C_{N,n}\langle x\rangle^{-2Nq} \in L^1(\mathbb{R}_x^n)
\end{aligned}
$$

for all $k \in \mathbb{N}$. Together with (28) we can apply Lebesgue's theorem and obtain

$$\|p_{\varepsilon_k}(x, D_x)u - p(x, D_x)u\|_{L^q(\mathbb{R}^n)}^q = \int_{\mathbb{R}^n} |p_{\varepsilon_k}(x, D_x)u(x) - p(x, D_x)u(x)|^q dx \xrightarrow{k\to\infty} 0.$$

Together with i) and Lemma 3.4 we conclude

$$p(x, D_x)u = L^q - \lim_{k\to\infty} p_{\varepsilon_k}(x, D_x)u = L^q - \lim_{k\to\infty} T_{\varepsilon_k}u = Tu.$$

□

As already mentioned, order reducing operators now enable us to extend the previous characterization to the class $C^s S^m_{0,0}$ for general m:

Lemma 3.11 *Let $m \in \mathbb{R}$, $1 < q < \infty$, $\tilde{m} \in \mathbb{N}_0$ with $\tilde{m} > n/q$. Additionally let $M \in \mathbb{N}_0 \cup \{\infty\}$ with $M > n + 1$. We define $\tilde{M} := M - (n + 1)$. Considering $T \in \mathcal{A}^{m,M}_{0,0}(\tilde{m}, q)$ we have for $s \in (0, \tilde{m} - n/q]$ with $s \notin \mathbb{N}_0$:*

$$T \in OPC^s S^m_{0,0}(\mathbb{R}^n \times \mathbb{R}^n; \tilde{M} - 1) \cap \mathscr{L}(H^m_q(\mathbb{R}^n), L^q(\mathbb{R}^n)).$$

Proof Let $s \in (0, \tilde{m} - n/q]$ with $s \notin \mathbb{N}_0$ and $\delta \in \mathbb{N}_0^n$. Moreover we define for every $\alpha \in \mathbb{R}$ the order reducing pseudodifferential operator $\Lambda^\alpha := \lambda^\alpha(D_x)$, where $\lambda^\alpha(\xi) := \langle\xi\rangle^\alpha \in S^{|\alpha|}_{1,0}(\mathbb{R}^n \times \mathbb{R}^n)$. Due to Remark 1.5 and Theorem 2.5 we get that

$$\operatorname{ad}(-ix)^\delta \Lambda^{-m} : L^q(\mathbb{R}^n) \to H^{m+|\delta|}_q(\mathbb{R}^n) \subseteq H^m_q(\mathbb{R}^n) \text{ is continuous.} \qquad (29)$$

Now let $l \in \mathbb{N}_0$, $\alpha_1, \ldots, \alpha_l \in \mathbb{N}_0^n$ and $\beta_1, \ldots, \beta_l \in \mathbb{N}_0^n$ such that $|\beta| \le \tilde{m}$ and $|\alpha| \le M$, where $\beta := \beta_1 + \ldots + \beta_l$ and $\alpha := \alpha_1 + \ldots + \alpha_l$. Since $\operatorname{ad}(-ix)^{\tau_2} \operatorname{ad}(D_x)^\delta \Lambda^{-m} \equiv 0$ for every $\tau_2, \delta \in \mathbb{N}_0^n$ with $|\delta| \neq 0$ due to Remark 1.5, we can iteratively show

$$\begin{aligned}
&\operatorname{ad}(-ix)^{\alpha_1} \operatorname{ad}(D_x)^{\beta_1} \ldots \operatorname{ad}(-ix)^{\alpha_l} \operatorname{ad}(D_x)^{\beta_l}(T\Lambda^{-m}) \\
&= \sum_{\substack{\gamma_1+\delta_1=\alpha_1 \\ \vdots \\ \gamma_l+\delta_l=\alpha_l}} C_{\gamma_1,\ldots,\gamma_l} [\operatorname{ad}(-ix)^{\gamma_1} \operatorname{ad}(D_x)^{\beta_1} \ldots \operatorname{ad}(-ix)^{\gamma_l} \operatorname{ad}(D_x)^{\beta_l} T][\operatorname{ad}(-ix)^\delta \Lambda^{-m}],
\end{aligned}$$

where δ is defined by $\delta := \delta_1 + \ldots + \delta_l$. Combining (29) and $T \in \mathcal{A}^{m,M}_{0,0}(\tilde{m}, q)$ we obtain the continuity of

$$\operatorname{ad}(-ix)^{\alpha_1} \operatorname{ad}(D_x)^{\beta_1} \ldots \operatorname{ad}(-ix)^{\alpha_l} \operatorname{ad}(D_x)^{\beta_l}(T\Lambda^{-m}) : L^q(\mathbb{R}^n) \to L^q(\mathbb{R}^n).$$

Therefore $T\Lambda^{-m} \in \mathcal{A}^{0,M}_{0,0}(\tilde{m}, q)$. If we use Theorem 3.10, we get

$$T\Lambda^{-m} \in OPC^s S^0_{0,0}(\mathbb{R}^n \times \mathbb{R}^n; \tilde{M} - 1) \cap \mathscr{L}(L^q(\mathbb{R}^n)).$$

On account of $\Lambda^m \in OPS^m_{1,0}(\mathbb{R}^n \times \mathbb{R}^n)$ and Theorem 2.5 we have

$$T \in OPC^s S^m_{0,0}(\mathbb{R}^n \times \mathbb{R}^n; \tilde{M} - 1) \cap \mathscr{L}(H^m_q(\mathbb{R}^n), L^q(\mathbb{R}^n)). \qquad \square$$

3.4 Characterization of Pseudodifferential Operators with Symbols in $C^s S^m_{1,0}$

Pseudodifferential operators of the class $C^s S^m_{1,0}(\mathbb{R}^n \times \mathbb{R}^n)$ often appear in the field of nonlinear partial differential equations. Due to Example 0.2, these operators are elements of the set $\mathcal{A}^m_{1,0}(\lfloor s \rfloor, q)$ with $1 < q < \infty$. This subsection is dedicated for proving the following result: Elements of the set $\mathcal{A}^{m,M}_{1,0}(\tilde{m}, q)$ are non-smooth pseudodifferential operators of the order m whose coefficients are in a Hölder space if $\tilde{m}$ is sufficiently large. Since $\mathcal{A}^{m,M}_{1,0}(\tilde{m}, q) \subseteq \mathcal{A}^{m,M}_{0,0}(\tilde{m}, q)$, we can use the characterization of pseudodifferential operators belonging to the class $C^s S^m_{0,0}(\mathbb{R}^n \times \mathbb{R}^n, M)$ in order to get the main result of this section. Just let us mention that all statements are taken from [3, Sect. 4.5].

Theorem 3.12 *Let $m \in \mathbb{R}$, $1 < q < \infty$ and $\tilde{m} \in \mathbb{N}_0$ with $\tilde{m} > n/q$. Additionally let $M \in \mathbb{N}_0$ with $M > n+1$. We define $\tilde{M} := M-(n+1)$. Assuming $P \in \mathcal{A}^{m,M}_{1,0}(\tilde{m}, q)$, we obtain for all $\tau \in (0, \tilde{m} - n/q]$ with $\tau \notin \mathbb{N}_0$:*

$$P \in OPC^\tau S^m_{1,0}(\mathbb{R}^n \times \mathbb{R}^n; \tilde{M} - 1) \cap \mathscr{L}(H^m_q(\mathbb{R}^n), L^q(\mathbb{R}^n)).$$

Proof Let $\tilde{m} - n/q \geq \tau > 0$ with $\tau \notin \mathbb{N}_0$ and $P \in \mathcal{A}^{m,M}_{1,0}(\tilde{m}, q)$. Due to Lemma 0.3 we have $P \in \mathcal{A}^{m,M}_{1,0}(\tilde{m}, q) \subseteq \mathcal{A}^{m,M}_{0,0}(\tilde{m}, q)$. Hence we get by means of Lemma 3.11:

$$P \in OPC^\tau S^m_{0,0}(\mathbb{R}^n \times \mathbb{R}^n; \tilde{M} - 1) \cap \mathscr{L}(H^m_q(\mathbb{R}^n), L^q(\mathbb{R}^n)).$$

Let $\alpha \in \mathbb{N}^n_0$ with $|\alpha| \leq \tilde{M} - 1$. Then $\text{ad}(-ix)^\alpha P \in \mathcal{A}^{m-|\alpha|, M-|\alpha|}_{1,0}(\tilde{m}, q)$. Because of Lemma 0.3 and Lemma 3.11, we obtain

$$\text{ad}(-ix)^\alpha P \in OPC^\tau S^{m-|\alpha|}_{0,0}(\mathbb{R}^n \times \mathbb{R}^n; \tilde{M} - |\alpha| - 1).$$

Due to Remark 1.5 the symbol of $\text{ad}(-ix)^\alpha P$ is $\partial^\alpha_\xi p(x, \xi)$ if p is the symbol of P. Hence

$$\|\partial^\alpha_\xi p(., \xi)\|_{C^\tau(\mathbb{R}^n)} \leq C_\alpha \langle \xi \rangle^{m-|\alpha|} \qquad \text{for all } \xi \in \mathbb{R}^n.$$

Consequently p is an element of $C^\tau S^m_{1,0}(\mathbb{R}^n \times \mathbb{R}^n; \tilde{M} - 1)$. $\square$

For $\tilde{M} - 1 > \max\{n/2, n/q\}$, $1 < q < \infty$ Theorem 2.6 provides that every operator of $OPC^\tau S^m_{1,0}(\mathbb{R}^n \times \mathbb{R}^n; \tilde{M} - 1)$ with $\tau > 0$ and $m \in \mathbb{R}$ is an element of $\mathscr{L}(H^m_q(\mathbb{R}^n), L^q(\mathbb{R}^n))$. This implies

$$\begin{aligned} &OPC^\tau S^m_{1,0}(\mathbb{R}^n \times \mathbb{R}^n; \tilde{M} - 1) \cap \mathscr{L}(H^m_q(\mathbb{R}^n), L^q(\mathbb{R}^n)) \\ = \ &OPC^\tau S^m_{1,0}(\mathbb{R}^n \times \mathbb{R}^n; \tilde{M} - 1). \end{aligned}$$

Having a look at the characterization of non-smooth pseudodifferential operators again we see that unfortunately we lose some regularity with respect to $\tilde{m}$. In [2, Sect. 4] we were able to improve the results of Theorem 3.10 and Theorem 3.12, so that no regularity with respect to the first variable is lost.

4 Spectral Invariance

4.1 The Inverse of a Pseudodifferential Operator in the Symbol-Class $C^\tau S^0_{0,0}$

The goal of the present subsection is to verify the following theorem:

Theorem 4.1 *Let $\tilde{m} \in \mathbb{N}_0$, $0 < \tau < 1$ and $M \in \mathbb{N} \cup \{\infty\}$ with*

$$\tilde{M} := M - (n + 1) > 0.$$

Additionally let $N \in \mathbb{N}_0 \cup \{\infty\}$ such that $N - M > n/2$. We assume

$$\hat{m} := \max\{k \in \mathbb{N}_0 : \tilde{m} + \tau - k > n/2\} > n/2.$$

For each $p \in C^{\tilde{m},\tau} S^0_{0,0}(\mathbb{R}^n \times \mathbb{R}^n; N)$ with $p(x, D_x)^{-1} \in \mathscr{L}(L^2(\mathbb{R}^n))$ we get

$$p(x, D_x)^{-1} \in OPC^s S^0_{0,0}(\mathbb{R}^n \times \mathbb{R}^n; \tilde{M} - 1)$$

for all $s \in (0, \hat{m} - n/2]$ with $s \notin \mathbb{N}$.

In the smooth case Ueberberg already has shown a similar result, cf. [22, Theorem 4.3]:

Theorem 4.2 *Let $1 < q < \infty$ and $0 \leq \delta \leq \rho \leq 1$ with $\delta < 1$.*

i) *For $p \in S^0_{\rho,\delta}(\mathbb{R}^n \times \mathbb{R}^n)$ with $p(x, D_x)^{-1} \in \mathscr{L}(L^2(\mathbb{R}^n))$ we obtain $p(x, D_x)^{-1} \in OPS^0_{\rho,\delta}(\mathbb{R}^n \times \mathbb{R}^n)$.*
ii) *For $p \in S^0_{1,\delta}(\mathbb{R}^n \times \mathbb{R}^n)$ with $p(x, D_x)^{-1} \in \mathscr{L}(L^q(\mathbb{R}^n))$ we get $p(x, D_x)^{-1} \in OPS^0_{1,\delta}(\mathbb{R}^n \times \mathbb{R}^n)$.*

Luckily the main idea of the proof in the smooth case can also be taken to show Theorem 4.1: We apply the characterization of pseudodifferential operators. Hence we need to verify the boundedness of certain iterated commutators of $p(x, D_x)^{-1}$. Since the iterated commutators of $p(x, D_x)$ fulfill these mapping properties, we try to write the iterated commutators of $p(x, D_x)^{-1}$ as a sum and compositions of $p(x, D_x)^{-1}$ and the iterated commutators of $p(x, D_x)$. Thereby we have to take care of the following fact: Non-smooth pseudodifferential operators are in general not bounded as operators from $\mathcal{S}(\mathbb{R}^n)$ to $\mathcal{S}(\mathbb{R}^n)$ like the smooth ones. Consequently we have to prove the formal identities for the iterated commutators rigorously.

Remark 4.3 (Formal Identities for the Iterated Commutators) Let $m, s \in \mathbb{R}$, $1 < q < \infty$ and $M, \tilde{m} \in \mathbb{N}_0$ with $\tilde{m} + M \geq 1$. We assume that $P \in \mathscr{L}(H_q^{s+m}, H_q^s)$ with $P^{-1} \in \mathscr{L}(H_q^s, H_q^{s+m})$ and

$$\mathrm{ad}(-ix)^{\alpha_1}\,\mathrm{ad}(D_x)^{\beta_1} \ldots \mathrm{ad}(-ix)^{\alpha_l}\,\mathrm{ad}(D_x)^{\beta_l} P \in \mathscr{L}(H_q^{s+m}, H_q^s)$$

for all $l \in \mathbb{N}$, $\alpha_1, \ldots, \alpha_l \in \mathbb{N}_0^n$ and $\beta_1, \ldots, \beta_l \in \mathbb{N}_0^n$ with$|\alpha_1| + \ldots + |\alpha_l| \leq M$, $|\beta_1| + \ldots + |\beta_l| \leq \tilde{m}$ and $|\alpha_j + \beta_j| = 1$ for all $j \in \{1, \ldots, l\}$. For arbitrary $\alpha, \beta \in \mathbb{N}_0^n$ with $|\alpha + \beta| = 1$ we have $\mathrm{ad}(-ix)^\alpha\,\mathrm{ad}(D_x)^\beta P^{-1} : \mathcal{S}(\mathbb{R}^n) \to \mathcal{S}'(\mathbb{R}^n)$. We consider $|\beta| = 0$ and $\alpha = e_j$ for $j \in \{1, \ldots, n\}$ first. On account of $\mathrm{ad}(-ix_j)P \in \mathscr{L}(H_q^{s+m}, H_q^s)$, we know that

$$\mathrm{ad}(-ix_j)Pu = -ix_jPu + P(ix_ju) \in H_q^s(\mathbb{R}^n) \qquad \text{for all } u \in \mathcal{S}(\mathbb{R}^n). \tag{30}$$

If $u \in \mathcal{S}(\mathbb{R}^n) \subseteq H_q^{m+s}(\mathbb{R}^n)$, we obtain $P(ix_ju) \in H_q^s(\mathbb{R}^n)$. Together with (30) this implies

$$-ix_jPu \in H_q^s(\mathbb{R}^n) \qquad \text{for all } u \in \mathcal{S}(\mathbb{R}^n). \tag{31}$$

Now we define $\mathscr{D} := \{Pu : u \in \mathcal{S}(\mathbb{R}^n)\} \subseteq H_q^s(\mathbb{R}^n)$. To show the density of $\mathscr{D}$ in $H_q^s(\mathbb{R}^n)$ we choose an arbitrary $v \in H_q^s(\mathbb{R}^n)$. On account of $P^{-1} \in \mathscr{L}(H_q^s, H_q^{s+m})$ we have $u := P^{-1}v \in H_q^{s+m}(\mathbb{R}^n)$ and therefore $v = Pu$. Considering a sequence $(u_j)_{j\in\mathbb{N}_0} \subseteq \mathcal{S}(\mathbb{R}^n)$, which converges to u in $H_q^{s+m}(\mathbb{R}^n)$, we define $v_j := Pu_j$ for each $j \in \mathbb{N}_0$. Due to $P \in \mathscr{L}(H_q^{s+m}, H_q^s)$ the sequence $(v_j)_{j\in\mathbb{N}}$ converges to v. This implies the density of $\mathscr{D}$ in $H_q^s(\mathbb{R}^n)$. Next we define the operator $Q : \mathscr{D} \to H_q^{s+m}(\mathbb{R}^n)$ by $Qu := -ix_jP^{-1}u + P^{-1}(ix_ju)$ for all $u \in \mathscr{D}$. Due to (31) Q is well-defined and we obtain for all $u \in \mathcal{S}(\mathbb{R}^n)$:

$$Q(Pu) = -ix_ju + P^{-1}(ix_jPu) = -P^{-1}[\mathrm{ad}(-ix_j)P]u. \tag{32}$$

With $\mathrm{ad}(-ix_j)P \in \mathscr{L}(H_q^{s+m}, H_q^s)$ and $P^{-1} \in \mathscr{L}(H_q^s, H_q^{s+m})$ we get

$$\|Q(Pu)\|_{H_q^{s+m}} \leq C\|Pu\|_{H_q^s}$$

for all $u \in \mathcal{S}(\mathbb{R}^n)$. Due to the density of $\mathscr{D}$ in $H^s_q(\mathbb{R}^n)$ this implies

$$Q \in \mathscr{L}(H^s_q, H^{s+m}_q).$$

As a direct consequence we obtain

$$\mathrm{ad}(-ix_j)P^{-1} \in \mathscr{L}(H^s_q, H^{s+m}_q)$$

since $Qu = \mathrm{ad}(-ix_j)P^{-1}u$ for all $u \in \mathcal{S}(\mathbb{R}^n)$. Together with $\mathscr{D} \subseteq H^s_q(\mathbb{R}^n)$ and (32) we get

$$[\mathrm{ad}(-ix_j)P^{-1}]Pu = -P^{-1}[\mathrm{ad}(-ix_j)P]u \qquad \text{for all } u \in \mathcal{S}(\mathbb{R}^n).$$

On account of $[\mathrm{ad}(-ix_j)P^{-1}]P \in \mathscr{L}(H^{s+m}_q)$ and $P^{-1}[\mathrm{ad}(-ix_j)P] \in \mathscr{L}(H^{s+m}_q)$ the previous equality holds for all $u \in H^{s+m}_q(\mathbb{R}^n)$. The surjectivity of

$$P \in \mathscr{L}(H^{s+m}_q; H^s_q)$$

yields for all $v \in H^s_q(\mathbb{R}^n)$:

$$\begin{aligned} \mathrm{ad}(-ix)^\alpha \, \mathrm{ad}(D_x)^\beta P^{-1}v = [\mathrm{ad}(-ix_j)P^{-1}]v &= -P^{-1}[\mathrm{ad}(-ix_j)P]P^{-1}v \\ &= -P^{-1}[\mathrm{ad}(-ix)^\alpha \, \mathrm{ad}(D_x)^\beta P]P^{-1}v. \end{aligned} \tag{33}$$

In case $\beta = e_j$, $j \in \{1, \dots, n\}$ and $|\alpha| = 0$ we get (33) for all $u \in \mathcal{S}(\mathbb{R}^n)$ in the same way as before. Moreover, let $l \in \mathbb{N}$, $\alpha_1, \dots, \alpha_l \in \mathbb{N}^n_0$ and $\beta_1, \dots, \beta_l \in \mathbb{N}^n_0$ with $|\alpha_j + \beta_j| = 1$ for all $j \in \{1, \dots, l\}$, $|\alpha_1| + \ldots + |\alpha_l| \leq M$ and $|\beta_1| + \ldots + |\beta_l| \leq \tilde{m}$. Denoting $\alpha := \alpha_1 + \ldots + \alpha_l$ and $\beta := \beta_1 + \ldots + \beta_l$ we get by mathematical induction with respect to l:

$$\mathrm{ad}(-ix)^{\alpha_1} \mathrm{ad}(D_x)^{\beta_1} \ldots \mathrm{ad}(-ix)^{\alpha_l} \mathrm{ad}(D_x)^{\beta_l} P^{-1} = \sum_{\substack{(\alpha^1_1+\ldots+\alpha^1_l)+\ldots+(\alpha^l_1+\ldots+\alpha^l_l)=\alpha \\ (\beta^1_1+\ldots+\beta^1_l)+\ldots+(\beta^l_1+\ldots+\beta^l_l)=\beta}} R_{\alpha^1_1,\ldots,\alpha^l_l,\beta^1_1,\ldots,\beta^l_l}$$

where

$$\begin{aligned} &R_{\alpha^1_1,\ldots,\alpha^l_l,\beta^1_1,\ldots,\beta^l_l} := \\ &C_{\alpha^1_1,\ldots,\alpha^l_l,\beta^1_1,\ldots,\beta^l_l} P^{-1} \left[\mathrm{ad}(-ix)^{\alpha^1_l} \mathrm{ad}(D_x)^{\beta^1_l} \ldots \mathrm{ad}(-ix)^{\alpha^1_1} \mathrm{ad}(D_x)^{\beta^1_1} P \right] P^{-1} \\ &\quad \circ \ldots \circ \left[\mathrm{ad}(-ix)^{\alpha^l_l} \mathrm{ad}(D_x)^{\beta^l_l} \ldots \mathrm{ad}(-ix)^{\alpha^l_1} \mathrm{ad}(D_x)^{\beta^l_1} P \right] P^{-1}. \end{aligned}$$

Proof of Theorem 4.1 Let $l \in \mathbb{N}_0$, $\alpha_1, \dots, \alpha_l \in \mathbb{N}^n_0$ and $\beta_1, \dots, \beta_l \in \mathbb{N}^n_0$ with $|\alpha_j + \beta_j| = 1$ for all $j \in \{1, \dots, n\}$ and $|\beta_1 + \ldots + \beta_l| \leq \hat{m}$ be arbitrary. On account of Remark 1.5 and Theorem 2.7 we get $p(x, D_x) \in \mathcal{A}^0_{0,0}(\hat{m}, 2)$. By means of $p(x, D_x) \in \mathcal{A}^0_{0,0}(\hat{m}, 2)$ and $P^{-1} \in \mathscr{L}(L^2(\mathbb{R}^n))$ we can apply Remark 4.3 and for $P := p(x, D_x)$ we get:

$$\mathrm{ad}(-ix)^{\alpha_1}\, \mathrm{ad}(D_x)^{\beta_1} \ldots \mathrm{ad}(-ix)^{\alpha_l}\, \mathrm{ad}(D_x)^{\beta_l} P^{-1} = \sum_{\substack{(\alpha_1^1+\ldots+\alpha_l^1)+\ldots+(\alpha_1^l+\ldots+\alpha_l^l)=\alpha \\ (\beta_1^1+\ldots+\beta_l^1)+\ldots+(\beta_1^l+\ldots+\beta_l^l)=\beta}} R_{\alpha_1^1,\ldots,\alpha_l^l,\beta_1^1,\ldots,\beta_l^l}$$

where $R_{\alpha_1^1,\ldots,\alpha_l^l,\beta_1^1,\ldots,\beta_l^l}$ are defined as in Remark 4.3. $P \in \mathcal{A}^0_{0,0}(\hat{m}, 2)$ and $P^{-1} \in \mathscr{L}(L^2(\mathbb{R}^n))$ yield together with the previous equality $P^{-1} \in \mathcal{A}^0_{0,0}(\hat{m}, 2)$. For $0 < s \leq \hat{m} - n/2$, $s \notin \mathbb{N}$, Theorem 3.10 provides the claim. □

We can show the next result in the same way as Theorem 4.1 if we use Theorem 2.8 instead of Theorem 2.7:

Lemma 4.4 *Let $\tilde{m} \in \mathbb{N}_0$ with $\tilde{m} > n/2$ and $0 < \tau < 1$. Additionally let $M \in \mathbb{N} \cup \{\infty\}$ such that $\tilde{M} := M - (n + 1) > 0$. We choose $N \in \mathbb{N}_0 \cup \{\infty\}$ such that $N - M > n/2$. For every non-smooth symbol $p \in C^{\tilde{m},\tau} S^{-n/2}_{0,0}(\mathbb{R}^n \times \mathbb{R}^n; N)$ with $p(x, D_x)^{-1} \in \mathscr{L}(L^2(\mathbb{R}^n))$ we get*

$$p(x, D_x)^{-1} \in OPC^s S^{-n/2}_{0,0}(\mathbb{R}^n \times \mathbb{R}^n; \tilde{M} - 1)$$

for all $s \in (0, \tilde{m} - n/2]$ with $s \notin \mathbb{N}$.

4.2 Properties of Difference Quotients

The investigation of a spectral invariance result for pseudodifferential operators $P \in C^\tau S^0_{1,0}(\mathbb{R}^n \times \mathbb{R}^n)$, $\tau > 0$ is one of the main goals of this chapter. Apart from the characterization the main tool for verifying this result again are the formal identities for the iterated commutators of P^{-1}, cf. Remark 4.3. But now $\mathrm{ad}(-ix)^\alpha \mathrm{ad}(D_x)^\beta P$, $|\alpha| \neq 0$ are pseudodifferential operators of negative order $-|\alpha|$. Consequently an application of $\mathrm{ad}(-ix)^\alpha \mathrm{ad}(D_x)^\beta P$, $|\alpha| \neq 0$ increases the order of the Bessel potential space. Hence besides $P^{-1} \in \mathscr{L}(L^q(\mathbb{R}^n))$ we also need $P^{-1} \in \mathscr{L}(H^{-s}_q(\mathbb{R}^n))$ for certain $s \in \mathbb{N}_0$. Thus we need to prove $P^{-1} \in \mathscr{L}(H^{-s}_q(\mathbb{R}^n))$ with the help of the assumptions. In the smooth case, Ueberberg used the following fact for the proof: The commutator of two smooth pseudodifferential operators is again a smooth pseudodifferential operator. Unfortunately in general this is not true in the non-smooth case. However the tool of difference quotients enables us to get a similar result in the non-smooth case. The tool of difference quotients is presented in this subsection. All results are taken from [2, Sect. 5.2].

Definition 4.5 Let $h \in \mathbb{R}\backslash\{0\}$ and $j \in \{1, \ldots, n\}$. For $u \in H^s_p(\mathbb{R}^n)$ with $s \in \mathbb{R}$ and $1 < p < \infty$ we define the *difference quotient* of u by

$$\partial^h_{x_j} u := h^{-1}\{u(. + he_j) - u\}.$$

Next we summarize some useful properties of difference quotients:

Lemma 4.6 *Let $m \in \mathbb{R}$, $\tilde{m} \in \mathbb{N}$, $0 < \tau < 1$ and $M \in \mathbb{N}_0 \cup \{\infty\}$. Considering a non-smooth symbol $p \in C^{\tilde{m},\tau} S^m_{1,0}(\mathbb{R}^n \times \mathbb{R}^n; M)$, we get for all $j \in \{1, \ldots, n\}$:*

i) $\left\{\partial^h_{x_j} p(x, \xi) : h \in \mathbb{R}\backslash\{0\}\right\} \subseteq C^{\tilde{m}-1,\tau} S^m_{1,0}(\mathbb{R}^n \times \mathbb{R}^n; M)$ *is bounded,*
ii) $[\partial^h_{x_j}, p(x, D_x)]u(x) = \left[\left(\partial^{-h}_{x_j} p\right)(x, D_x)u\right](x + he_j)$ *for all* $u \in \mathcal{S}(\mathbb{R}^n)$, $x \in \mathbb{R}^n$ *and* $h \in \mathbb{R}\backslash\{0\}$.
iii) *Additionally let $M \in \mathbb{N}_0 \cup \{\infty\}$ with $M > n/2$ for $q \geq 2$ and $M > n/q$ else and $s \in \mathbb{R}$ with $|s| < \tilde{m} - 1 + \tau$. Then we have for some $C > 0$:*

$$\|[\partial^h_{x_j}, p(x, D_x)]u\|_{H^s_q} \leq C\|u\|_{H^{s+m}_q} \qquad \text{for all } u \in H^{s+m}_q(\mathbb{R}^n), h \in \mathbb{R}\backslash\{0\}.$$

Proof Due to the fundamental theorem of calculus we get claim i). In order to verify claim ii) let $u \in \mathcal{S}(\mathbb{R}^n)$ be arbitrary. An application of $\frac{e^{ihe_j\cdot\xi}-1}{h}\hat{u}(\xi) = \widehat{\partial^h_{x_j} u}(\xi)$ yields for all $x \in \mathbb{R}^n$:

$$\partial^h_{x_j}[p(x, D_x)u(x)] = \left[\left(\partial^{-h}_{x_j} p\right)(x, D_x)u\right](x + he_j) + \left[p(x, D_x)\left(\partial^h_{x_j} u\right)\right](x).$$

Hence ii) holds. Finally we can verify iii) by means of Lemma 4.6 i), Theorem 2.6 and the density of $\mathcal{S}(\mathbb{R}^n)$ in $H^{s+m}_q(\mathbb{R}^n)$. □

Theorem 4.7 (Difference Quotients and Weak Derivatives)

i) *We suppose $1 < p < \infty$ and $u \in H^{s+1}_p(\mathbb{R}^n)$ with $s \in \mathbb{R}$. Then there is a constant C, independent of $h \in \mathbb{R}\backslash\{0\}$ and $u \in H^{s+1}_p(\mathbb{R}^n)$, such that*

$$\|\partial^h_{x_j} u\|_{H^s_p} \leq C\|\partial_{x_j} u\|_{H^s_p}$$

for all $j \in \{1, \ldots, n\}$ and all $h \in \mathbb{R}\backslash\{0\}$.
ii) *Let $1 < p < \infty$ and $u \in H^s_p(\mathbb{R}^n)$ with $s \in \mathbb{R}$. Additionally we assume*

$$\|\partial^h_{x_j} u\|_{H^s_p} \leq C \qquad \text{for all } j \in \{1, \ldots, n\} \text{ and } h \in \mathbb{R}\backslash\{0\}.$$

Then $u \in H^{s+1}_p(\mathbb{R}^n)$ and $\|\partial_{x_j} u\|_{H^s_p} \leq C$.

Note that assertion ii) is false for $p = 1$ while i) also holds for $p = 1$.

Proof The proof of case $s = 0$ is essentially the same as that one of Theorem 5.8.3 in [11]. Assuming an arbitrary $s \in \mathbb{R}\backslash\{0\}$ and $u \in H^{s+1}_p(\mathbb{R}^n)$ we know that

$\langle D_x\rangle^s u \in H^1_p(\mathbb{R}^n)$. Hence an application of Lemma 4.6 and case $s = 0$ provides for each $j \in \{1, \ldots, n\}$:

$$\|\partial^h_{x_j} u\|_{H^s_p} = \|\partial^h_{x_j}\langle D_x\rangle^s u\|_{L^p} \le C\|\partial_{x_j}\langle D_x\rangle^s u\|_{L^p} = C\|\partial_{x_j} u\|_{H^s_p}$$

for all $h \in \mathbb{R}\backslash\{0\}$ and $u \in H^{s+1}_p(\mathbb{R}^n)$. Similar to $i)$ we obtain case $s \in \mathbb{R}$ of $ii)$ as a consequence of case $s = 0$ and Lemma 4.6. □

With the previous theorem at hand we can show the following proposition:

Proposition 4.8 *Let $k \in \mathbb{N}_0$, $r \in \mathbb{R}$ and $1 < q < \infty$. Moreover, let P be an operator, which fulfills for all $s \in \{r, r+1, \ldots, r+k\}$ the properties*

i) $P \in \mathscr{L}(H^s_q, H^s_q)$,
ii) $P \in \mathscr{L}(H^{r+k+1}_q, H^{r+k+1}_q)$,
iii) $\{[P, \partial^h_{x_j}] : h \in \mathbb{R}\backslash\{0\}\} \subseteq \mathscr{L}(H^s_q, H^s_q)$ *is bounded for all* $j \in \{1, \ldots, n\}$,
iv) $P^{-1} \in \mathscr{L}(H^r_q, H^r_q)$.

Then $P^{-1} \in \mathscr{L}(H^s_q, H^s_q)$ for each $s \in \{r, r+1, \ldots, r+k+1\}$.

Proof We prove the claim by mathematical induction with respect to s. In case $s = r$ there is nothing to show. For $s \in \{r, r+1, \ldots, r+k\}$ we choose an arbitrary $j \in \{1, \ldots, n\}$ and a function $f \in H^{s+1}_q(\mathbb{R}^n) \subseteq H^s_q(\mathbb{R}^n)$. The induction hypothesis provides the existence of a $u \in H^s_q(\mathbb{R}^n)$ with $u = P^{-1}f$. Due to $P \in \mathscr{L}(H^s_q, H^s_q)$, we get $Pu \in H^s_q(\mathbb{R}^n)$ and consequently $\partial^h_{x_j}(Pu) \in H^s_q(\mathbb{R}^n)$. Similarly we get $P(\partial^h_{x_j}u) \in H^s_q(\mathbb{R}^n)$. An application of P^{-1} to $P(\partial^h_{x_j}u) = [P, \partial^h_{x_j}]u + \partial^h_{x_j}(Pu)$, the induction hypothesis, the assumptions, and Theorem 4.7 $i)$ yield

$$\begin{aligned}\|\partial^h_{x_j} u\|_{H^s_q} &= \|P^{-1}\{[P, \partial^h_{x_j}]u + \partial^h_{x_j}(Pu)\}\|_{H^s_q} \le C\|[P, \partial^h_{x_j}]u\|_{H^s_q} + C\|\partial^h_{x_j} f\|_{H^s_q}\\ &\le C\|u\|_{H^s_q} + C\|\partial_{x_j} f\|_{H^s_q} \le C \qquad \text{for all } h \in \mathbb{R}\backslash\{0\}, u \in H^s_q(\mathbb{R}^n).\end{aligned}$$

Therefore Theorem 4.7 $ii)$ provides $u \in H^{s+1}_q(\mathbb{R}^n)$ which proves the surjectivity of the linear, bounded, and injective operator $P : H^{s+1}_q(\mathbb{R}^n) \to H^{s+1}_q(\mathbb{R}^n)$. Then $P^{-1} \in \mathscr{L}(H^{s+1}_q, H^{s+1}_q)$ by means of the bounded inverse theorem. □

The previous proposition enables us to verify the central result of this subsection:

Theorem 4.9 *Let $1 < q < \infty$, $0 < \tau < 1$, $\tilde{m} \in \mathbb{N}$ and $N \in \mathbb{N}_0 \cup \{\infty\}$ with $N > n/2$ for $q \ge 2$ and $N > n/q$ else. We define $k := \max\{l \in \mathbb{N}_0 : r + l < \tilde{m} + \tau\}$ for one $r \in \mathbb{R}$ with $|r| < \tilde{m} + \tau$. Considering $p \in C^{\tilde{m},\tau} S^0_{1,0}(\mathbb{R}^n \times \mathbb{R}^n; N)$, where $p(x, D_x)^{-1} \in \mathscr{L}(H^r_q, H^r_q)$, we obtain*

$$p(x, D_x)^{-1} \in \mathscr{L}(H^s_q, H^s_q) \qquad \textit{for all } s \in [-r-k, r+k]. \tag{34}$$

Proof On account of Theorem 2.6 and Lemma 4.6 we can apply Proposition 4.8 and get the claim for all $s \in \{r, \ldots, r+k\}$. With $(\partial_{x_j}^h)^* = -\partial_{x_j}^{-h}$ at hand we have $[P^*, \partial_{x_j}^h] = [P, \partial_{x_j}^{-h}]^*$. An application of Proposition 4.8 to P^* provides the claim for all $s \in \{-r-k, \ldots, r-1\}$. Then the claim follows for all $s \in [-r-k, r+k]$ by means of interpolation. □

4.3 Spectral Invariance of Pseudodifferential Operators in the Symbol-Class $C^{\tilde{m},\tau} S^0_{1,0}$

We are now in the position to show the next spectral invariance result. All statements have been shown in [2, Sect. 5.3].

Theorem 4.10 *Let* $1 < q_0 < \infty$, $0 < \tau < 1$ *and* $\tilde{m}, \hat{m} \in \mathbb{N}_0$ *with* $\tilde{m} \geq \hat{m} > n/q_0$. *Additionally let* $M \in \mathbb{N}_0$ *with* $n < M \leq \tilde{m} - \hat{m}$. *We define* $\tilde{M} := M - (n+1)$. *Furthermore, let* $N \in \mathbb{N} \cup \{\infty\}$ *with* $N - M > n/2$ *if* $q_0 \geq 2$ *and* $N - M > n/q_0$ *else. Considering* $p \in C^{\tilde{m},\tau} S^0_{1,0}(\mathbb{R}^n \times \mathbb{R}^n; N)$, *where* $p(x, D_x)^{-1} \in \mathscr{L}(H^r_{q_0}, H^r_{q_0})$ *for one* $|r| < \tilde{m} + \tau$, *we get for all* $0 < s \leq \hat{m} - n/q$ *with* $s \notin \mathbb{N}$

$$p(x, D_x)^{-1} \in OPC^s S^0_{1,0}(\mathbb{R}^n \times \mathbb{R}^n; \tilde{M} - 1).$$

In case $\tilde{M} - 1 > n/\tilde{q}$ *for some* $1 < \tilde{q} \leq 2$, *we even have*

$$p(x, D_x)^{-1} \in \mathscr{L}(L^q, L^q) \qquad \text{for all } q \in [\tilde{q}; \infty) \cup \{q_0\}.$$

Proof An application of Theorem 4.9 provides the boundedness of

$$p(x, D_x)^{-1} \in \mathscr{L}(H^{-\iota}_{q_0}, H^{-\iota}_{q_0}) \qquad \text{for all } \iota \in \{0, \ldots, M\}. \tag{35}$$

Let $l \in \mathbb{N}_0$, $\alpha_1, \ldots, \alpha_l \in \mathbb{N}^n_0$ and $\beta_1, \ldots, \beta_l \in \mathbb{N}^n_0$ with $|\alpha_j| + |\beta_j| = 1$ for all $j \in \{1, \ldots, l\}$, $|\alpha| \leq M$ and $|\beta| \leq \hat{m}$ where $\alpha := \alpha_1 + \ldots + \alpha_l$ and $\beta := \beta_1 + \ldots + \beta_l$. Then Remark 1.5 and Theorem 2.6 yield for all $\iota \in \{0, \ldots, M - |\alpha|\}$:

$$\mathrm{ad}(-ix)^{\alpha_l}\, \mathrm{ad}(D_x)^{\beta_l} \ldots \mathrm{ad}(-ix)^{\alpha_1}\, \mathrm{ad}(D_x)^{\beta_1} p(x, D_x) \in \mathscr{L}(H^{-\iota-|\alpha|}_{q_0}, H^{-\iota}_{q_0}). \tag{36}$$

Setting $P := p(x, D_x)$ we get due to Remark 4.3

$$\mathrm{ad}(-ix)^{\alpha_1}\, \mathrm{ad}(D_x)^{\beta_1} \ldots \mathrm{ad}(-ix)^{\alpha_l}\, \mathrm{ad}(D_x)^{\beta_l} P^{-1} = \sum_{\substack{(\alpha_1^1+\ldots+\alpha_l^1)+\ldots+(\alpha_1^l+\ldots+\alpha_l^l)=\alpha \\ (\beta_1^1+\ldots+\beta_l^1)+\ldots+(\beta_1^l+\ldots+\beta_l^l)=\beta}} R_{\alpha_1^1,\ldots,\alpha_l^l,\beta_1^1,\ldots,\beta_l^l}$$

where $R_{\alpha_1^1,\dots,\alpha_l^l,\beta_1^1,\dots,\beta_l^l}$ is defined as in Remark 1.5. Together with (35) and (36) we get that P^{-1} is an element of $\mathcal{A}_{1,0}^{0,M}(\hat{m}, q_0)$. By means of Theorem 3.12 we obtain for each $0 < s \leq \hat{m} - n/q_0$ with $s \notin \mathbb{N}$:

$$p(x, D_x)^{-1} \in C^s S^0_{1,0}(\mathbb{R}^n \times \mathbb{R}^n; \tilde{M} - 1).$$

Finally, considering $\tilde{M} - 1 > n/\tilde{q}$ for some $1 < \tilde{q} \leq 2$ we obtain for every $q \in [\tilde{q}, \infty)$ due to Theorem 2.6 the boundedness of $P^{-1} : L^q(\mathbb{R}^n) \to L^q(\mathbb{R}^n)$. □

One may wonder why the previous result is called spectral invariance result. The reason of this is that we can easily show the next corollary by means of Theorem 4.10. For more details, we refer to [18, Corollary 6.12].

Corollary 4.11 *Let the assumptions of Theorem* 4.10 *hold. Additionally we choose an arbitrary but fixed* $\tilde{q} \in (1, 2]$ *fulfilling the conditions of Theorem* 4.10 *and denote*

$$P_{L^q} := p(x, D_x) : L^q(\mathbb{R}^n) \to L^q(\mathbb{R}^n) \qquad \text{for all } \tilde{q} \leq q < \infty.$$

Then $\sigma(P_{L^q}) = \sigma(P_{L^r})$ *for all* $\tilde{q} \leq q, r < \infty$.

We can also ask ourselves whether it is possible to verify that $p(x, D_x)^{-1}$ is even an element of $OPC^s S^0_{1,0}(\mathbb{R}^n \times \mathbb{R}^n)$ with $0 < s \leq \hat{m} - n/q_0$, $s \notin \mathbb{N}$ in the case that all assumptions of the Theorem 4.10 hold and additionally $p(x, D_x) \in OPC^{\tilde{m},\tau} S^0_{1,0}(\mathbb{R}^n \times \mathbb{R}^n)$. Unfortunately in general this is not the case as the next example shows:

Example 4.12 Let $s > 0$, $1 < q_0 < \infty$ and $\tau > s + \lfloor n/q_0 \rfloor + n + 4$. Additionally let $p(\xi) \in S^0_{1,0}(\mathbb{R}^n_x \times \mathbb{R}^n_\xi)$ be a symbol which is not constantly equal to zero and where $p(D_x)^{-1} \in \mathscr{L}(L^{q_0}(\mathbb{R}^n), L^{q_0}(\mathbb{R}^n))$. Moreover let $a \in C^\tau(\mathbb{R}^n)$ such that there is no open set $U \subseteq \mathbb{R}^n$, $U \neq \varnothing$ with $a|_U \in C^\infty(U)$ and there are two constants $c, C > 0$ with $C > a(x) > c$ for all $x \in \mathbb{R}^n$. Then the operator $T := a(x)p(D_x) \in C^\tau S^0_{1,0}(\mathbb{R}^n \times \mathbb{R}^n)$ fulfills all assumptions of Theorem 4.10 for $M = n + 3$ and $\hat{m} := \lfloor \tau \rfloor - (n + 3)$. Consequently $T^{-1} \in OPC^s S^0_{1,0}(\mathbb{R}^n \times \mathbb{R}^n; \tilde{M} - 1)$, where $\tilde{M} := M - (n + 1)$, but $T^{-1} \notin OPC^s S^0_{1,0}(\mathbb{R}^n \times \mathbb{R}^n)$ with $s \in (0, \hat{m} - n/q_0]$.

Proof First we define $b(x) := (a(x))^{-1}$ for all $x \in \mathbb{R}^n$. Then we have $b \in C^\tau(\mathbb{R}^n)$ and $T \in C^\tau S^0_{1,0}(\mathbb{R}^n \times \mathbb{R}^n)$. Due to Theorem 4.2 the operator $p(D_x)^{-1}$ is an element of $OPS^0_{1,0}(\mathbb{R}^n \times \mathbb{R}^n)$. Hence $T^{-1} = p(D_x)^{-1} b(x)$. In particular the boundedness of b and $p(D_x)^{-1} \in \mathscr{L}(L^{q_0}(\mathbb{R}^n), L^{q_0}(\mathbb{R}^n))$ imply $T^{-1} \in \mathscr{L}(L^{q_0}(\mathbb{R}^n), L^{q_0}(\mathbb{R}^n))$. Therefore all assumptions of Theorem 4.10 are fulfilled for $M = n + 3$ and $\hat{m} := \lfloor \tau \rfloor - (n + 3)$. Let $\tilde{\tau} \in (0, \hat{m} - n/q_0]$. Assuming $T^{-1} \in C^{\tilde{\tau}} S^0_{1,0}(\mathbb{R}^n \times \mathbb{R}^n)$ there is a kernel $\tilde{k} : \mathbb{R}^n \times \mathbb{R}^n \setminus \{0\} \to \mathbb{C}$ such that $\tilde{k}(x, .) \in C^\infty(\mathbb{R}^n \setminus \{0\})$ for each $x \in \mathbb{R}^n$ and

$$T^{-1} f(x) = \int \tilde{k}(x, x - y) f(y) dy \tag{37}$$

for all $f \in \mathcal{S}(\mathbb{R}^n)$ and $x \notin \operatorname{supp} f$ due to Theorem 2.10. An application of Remark 2.11 provides the existence of a kernel $k \in C^\infty(\mathbb{R}^n \backslash \{0\})$ such that

$$p(D_x)^{-1}u(x) = \int k(x-y)u(y)dy \qquad \text{for all } x \notin \operatorname{supp} u \tag{38}$$

for all $u \in \mathcal{S}(\mathbb{R}^n)$. Now let $(\delta_\varepsilon)_{\varepsilon>0} \subseteq C_0^\infty(\mathbb{R}^n)$ be a Dirac family, i.e., for all $\varepsilon > 0$ we have $\delta_\varepsilon \geq 0$, $\int \delta_\varepsilon(x)dx = 1$ and $\lim_{\varepsilon \to 0} \int_{|x|\geq d} \delta_\varepsilon(x)dx = 0$ for every $d > 0$. Then $\delta_\varepsilon * b \in C^\infty(\mathbb{R}^n)$ for each $\varepsilon > 0$. The boundedness of b, Theorem 10.7 in [12] and $\partial_x^\alpha \delta_\varepsilon \in C_0^\infty(\mathbb{R}^n) \subseteq L^1(\mathbb{R}^n)$ provides for every $\alpha \in \mathbb{N}_0^n$

$$\left|\partial_x^\alpha(\delta_\varepsilon * b)(x)\right| \leq \int \left|(\partial_x^\alpha \delta_\varepsilon)(y)\right| |b(x-y)|\, dy \leq \|b\|_{L^\infty} \|\partial_x^\alpha \delta_\varepsilon\|_{L^1} \leq C_{\alpha,\varepsilon}$$

for all $x \in \mathbb{R}^n$. In case $|\alpha| = 0$ the constant $C_{\alpha,\varepsilon}$ is even independent of $\varepsilon > 0$. In particular $\delta_\varepsilon * b \in C_b^\infty(\mathbb{R}^n)$ for every $\varepsilon > 0$ and therefore $(\delta_\varepsilon * b)f \in \mathcal{S}(\mathbb{R}^n)$ for all $f \in \mathcal{S}(\mathbb{R}^n)$. Additionally we obtain for all $f \in \mathcal{S}(\mathbb{R}^n)$ with $x \notin \operatorname{supp} f$ the existence of a constant C, independent of $\varepsilon > 0$, such that

$$|k(x-y)(\delta_\varepsilon * b)(y)f(y)| \leq C|k(x-y)f(y)| \in L^1(\mathbb{R}_y^n)$$

and

$$(\delta_\varepsilon * b)(y)f(y) \xrightarrow{\varepsilon \to 0} b(y)f(y) \qquad \text{for all } y \in \mathbb{R}^n. \tag{39}$$

Using (37), $p(D_x)^{-1} \in \mathscr{L}(L^2(\mathbb{R}^n))$ and (39) first we obtain together with (38) and an application of Lebesgue's theorem for all $f \in \mathcal{S}(\mathbb{R}^n)$:

$$\begin{aligned} \int \tilde{k}(x, x-y)f(y)dy &= T^{-1}f(x) = p(D_x)^{-1}\left[\lim_{\varepsilon \to 0} (\delta_\varepsilon * b)(x)f(x)\right] \\ &= \lim_{\varepsilon \to 0} \int k(x-y)(\delta_\varepsilon * b)(y)f(y)dy = \int k(x-y)b(y)f(y)dy \end{aligned} \tag{40}$$

for all $x \notin \operatorname{supp} f$. Now we fix $x \in \mathbb{R}^n$ such that $\tilde{k}(x, .)$ is not constantly equal to zero. An application of the fundamental lemma of calculus of variations yields

$$k(x-y)b(y) = \tilde{k}(x, x-y) \qquad \text{for all } y \in \mathbb{R}^n \backslash \{x\}$$

since $k(x-y)$, $\tilde{k}(x, x-y)$ and $b(y)$ are continuous with respect to $y \in \mathbb{R}^n \backslash \{x\}$. By means of a change of variables we obtain

$$k(z) = a(x-z)\tilde{k}(x, z) \in C^\infty(\mathbb{R}_z^n \backslash \{0\}). \tag{41}$$

Now we choose $z \in \mathbb{R}^n \backslash \{0\}$ with $\tilde{k}(x, z) \neq 0$. Due to $\tilde{k}(x, .) \in C^\infty(\mathbb{R}^n \backslash \{0\})$, there is some $\delta > 0$ such that $\tilde{k}(x, \tilde{z}) \neq 0$ for all $\tilde{z} \in B_\delta(z)$ and $0 \notin B_\delta(z)$. Together with $\tilde{k}(x, .) \in C^\infty(\mathbb{R}^n \backslash \{0\})$ and (41) we obtain $a \in C^\infty(B_\delta(x - z))$. This is a contradiction to the choice of a. Therefore $T^{-1} \notin C^s S^0_{1,0}(\mathbb{R}^n \times \mathbb{R}^n)$. □

Finally let us remark that in [2, Sect. 5.4] Theorem 4.10 was improved for non-smooth pseudodifferential operators of the order zero with coefficients in the so-called uniformly local Sobolev space $W^{\tilde{m},q}_{uloc}(\mathbb{R}^n)$.

References

1. H. Abels, *Pseudodifferential and Singular Integral Operators: An Introduction with Applications* (De Gruyter, Berlin/Boston, 2012)
2. H. Abels, C. Pfeuffer, Spectral invariance of non-smooth pseudo-differential operators. Integr. Equ. Oper. Theory **86**(1), 41–70 (2016)
3. H. Abels, C. Pfeuffer, Characterization of non-smooth pseudodifferential operators. J. Fourier Anal. Appl. **24**, 371 (2018). https://doi.org/10.1007/s00041-017-9529-7
4. H. Amann, Vector-valued distributions and Fourier multipliers. Unpublished manuscript (2003). https://www.math.uzh.ch/amann/books.html
5. R. Beals, Characterization of pseudodifferential operators and applications. Duke Math. J. **44**(1), 45–57 (1977)
6. R. Beals, Weighted distribution spaces and pseudodifferential operators. J. Anal. Math. **39**, 131–187 (1981)
7. J. Bergh, J. Lofstrom, *Interpolation spaces*, An introduction (Springer-Verlag, Berlin, 1976)
8. H.O. Cordes, On pseudodifferential operators and smoothness of special Lie-group representations. Manuscripta Math. **28**, 1–3, 51–69 (1979)
9. H.O. Cordes, On some C^* -algebras and Freche*t** -algebras of pseudodifferential operators. Proc. Symp. Pure Math. AMS **43**, 79–104 (1985)
10. R.R. Coifman, Y. Meyer, *Au dela des operateurs pseudo-differentiels* (Société mathématique de France, Paris, 1978)
11. L.C. Evans, *Partial Differential Equations*, 2nd edn. (Providence, American Mathematical Society, 2010)
12. W. Kaballo, *Einführung in die Analysis III* (Spektrum Akademischer Verlag, Heidelberg, 1999)
13. D. Kopp, *Pseudodifferentialoperatoren mit nichtglatten Koeffizienten auf Mannigfaltigkeiten*, Diploma thesis (University Regensburg, Regensburg, 2011)
14. V.D. Kryakvin, Characterization of pseudodifferential operators in Hölder-Zygmund spaces. Diff. Equat. **49**(3), 306–312 (2013)
15. H. Kumano-Go, *Pseudo-Differential Operators* (MIT Press, Cambridge, 1974)
16. H.G. Leopold, E. Schrohe, Spectral invariance for algebras of pseudodifferential operators on Besov-Triebel-Lizorkin spaces. Manuscripta Math. **78**, 99–110 (1993)
17. J. Marschall, Pseudodifferential operators with nonregular symbols of the class $S^m_{\rho\delta}$. Commun. Part. Diff. Eq. **12**(8), 921–965 (1987)
18. C. Pfeuffer, Characterization of Non-Smooth Pseudodifferential Operators, Ph.D., Universität Regensburg, 2015
19. E. Schrohe, Boundedness and spectral invariance for standard pseudodifferential operators on anisotropically weighted L^p-Sobolev spaces. Integr. Equ. Oper. Theory **13**(2), 271–284 (1990)
20. E.M. Stein, *Harmonic Analysis* (Princeton University Press, Princeton 1993)
21. F. Treves, *Topological Vector Spaces and Distributions and Kernels* (Academic Press New York, San Francisco, 1967)
22. J. Ueberberg, Zur Spektralinvarianz von Algebren von Pseudodifferentialoperatoren in der L^p-Theorie. Manuscripta Math. **61**(4), 459–475 (1988)

Fredholmness and Ellipticity of ΨDOs on $B^s_{pq}(\mathbb{R}^n)$ and $F^s_{pq}(\mathbb{R}^n)$

Pedro T. P. Lopes

Abstract We give a condition under which a pseudodifferential operator with symbol in $S^m(\mathbb{R}^n \times \mathbb{R}^n)$ cannot be a Fredholm operator when acting on suitable Besov and Triebel-Lizorkin spaces. As a corollary, we show that, if a classical pseudodifferential operator on $\mathbb{R}^n$ is Fredholm in one of these spaces, then this operator must be elliptic.

Keywords Pseudodifferential operators · Spectral invariance · Besov and Triebel-Lizorkin spaces

Mathematics Subject Classification (2000) 35S05, 46E35, 35S15

1 Introduction

Motivated by applications in non-linear PDEs, the spectral invariance of the algebra of pseudodifferential boundary value problems with conical singularities in L_p-spaces was proved recently in [9]. In this work, first the equivalence of the Fredholm property and ellipticity was proved, then a simple argument using parametrices led to the conclusion that the inverses of pseudodifferential operators acting on suitable function spaces are again contained in the pseudodifferential algebra.

It is interesting to note that, although the strategy is quite standard, some new challenges have appeared. In fact, whenever we are working with boundary value problems on L_p-spaces, the Besov spaces appear naturally as the spaces of traces of functions belonging to the Bessel potential spaces. Therefore, the proof that Fredholm property implies ellipticity required the extension, among other things, of the usual symbol reproducing argument, which is used by many authors [3, 5, 12, 16], to the Besov spaces.

P. T. P. Lopes (✉)
Instituto de Matemática e Estatística, Universidade de São Paulo, São Paulo, SP, Brazil
e-mail: pplopes@ime.usp.br

S. Molahajloo, M. W. Wong (eds.), *Analysis of Pseudo-Differential Operators*, Trends in Mathematics, https://doi.org/10.1007/978-3-030-05168-6_3

In this contribution, we show how to use some of the arguments contained in [8, 9, 13] to provide a condition under which a pseudodifferential operator in the Hömander class $S^m(\mathbb{R}^n \times \mathbb{R}^n)$, also called a ΨDO, is not Fredholm when acting on suitable Besov and Triebel-Lizorkin spaces. In particular, for classical pseudodifferential operators, our result implies that the Fredholm property in these spaces implies ellipticity. The converse in $\mathbb{R}^n$ is clearly not true, as the Laplacian operator shows us. This extends some of the results of Dasgupta [3], see also Wong [16] and [4], for classical symbols on these more general classes of spaces.

We provide complete proofs of our statements. Many of the arguments are very similar to the ones that can be found in [8, 9, 13]. We hope that the new result presented here can also be seen as an illustration of these methods. We do not include some important spaces such as $B^s_{\infty\infty}(\mathbb{R}^n)$. However our methods allow all Besov spaces $B^s_{pq}(\mathbb{R}^n)$ and Triebel-Lizorkin spaces $F^s_{pq}(\mathbb{R}^n)$ for $s \in \mathbb{R}$, $1 < p, q < \infty$.

Recently some interesting results related to ellipticity and the Fredholm property were proved. We mention here a few of them. Kryakvin [6], for instance, has proved spectral invariance in Hölder-Zygmund spaces and in spaces of variable smoothness, as can be seen in Kryakvin and Omarova [7]. Results for non-smooth pseudodifferential operators were obtained by Abels and Pfeuffer [1]. We can also mention some results on the circle by Molahajloo and Wong [11] and, in the analytic setting, by Cabral and Melo [2].

2 Pseudodifferential Operators on $\mathbb{R}^n$ and Function Spaces

In this paper, we use the multi-index notation: if $\mathbb{N}_0 = \{0, 1, 2, 3, \ldots\}$ and $\alpha = (\alpha_1, \ldots, \alpha_n) \in \mathbb{N}_0^n$, then $x^\alpha = x_1^{\alpha_1} \ldots x_n^{\alpha_n}$ and $\partial_x^\alpha = \partial_{x_1}^{\alpha_1} \ldots \partial_{x_n}^{\alpha_n}$. For any Banach spaces E and F, we denote by $\mathcal{B}(E, F)$ the set of all bounded operators from E to F and $\mathcal{B}(E) := \mathcal{B}(E, E)$. The Schwartz space of smooth functions in $\mathbb{R}^n$ such that $\sup_{x\in\mathbb{R}^n} \left| x^\alpha \partial_x^\beta u(x) \right| < \infty$ for all $\alpha, \beta \in \mathbb{N}_0^n$ is denoted by $\mathcal{S}(\mathbb{R}^n)$. We denote by $\mathcal{S}'(\mathbb{R}^n)$ the space of all tempered distributions. The Fourier transform $\mathcal{F} : \mathcal{S}(\mathbb{R}^n) \to \mathcal{S}(\mathbb{R}^n)$ in our convention is $\mathcal{F}u(\xi) = \hat{u}(\xi) = \int_{\mathbb{R}^n} e^{-ix\xi} u(x)\, dx$, where $x\xi := x_1\xi_1 + \ldots + x_n\xi_n$. Its inverse is given by $\mathcal{F}^{-1}u(x) = \frac{1}{(2\pi)^n} \int_{\mathbb{R}^n} e^{ix\xi} u(\xi)\, d\xi$. In many estimates, we use the function $\langle . \rangle : \mathbb{R}^n \to \mathbb{R}$ defined by $\langle \xi \rangle = \sqrt{1 + |\xi|^2}$, where $\xi \in \mathbb{R}^n \mapsto |\xi| \in \mathbb{R}$ is the Euclidean norm.

Definition 1 The space of symbols $S^m(\mathbb{R}^n \times \mathbb{R}^n)$, $m \in \mathbb{R}$, is the set of all functions $a \in C^\infty(\mathbb{R}^n \times \mathbb{R}^n)$ with the following property: For all $\alpha, \beta \in \mathbb{N}_0^n$, there is a constant $C_{\alpha\beta} > 0$ such that

$$\left| \partial_x^\beta \partial_\xi^\alpha a(x, \xi) \right| \le C_{\alpha\beta} \langle \xi \rangle^{m-|\alpha|}, \ (x, \xi) \in \mathbb{R}^n \times \mathbb{R}^n.$$

We say that $a \in S^m(\mathbb{R}^n \times \mathbb{R}^n)$ is an elliptic symbol if there are constants $C > 0$ and $R > 0$ such that, for all $x \in \mathbb{R}^n$ and $\xi \in \mathbb{R}^n$, $|\xi| > R$, we have

$$|a(x,\xi)| \geq C \langle \xi \rangle^m .$$

An important subset of $S^m(\mathbb{R}^n \times \mathbb{R}^n)$ is the class of classical symbols.

Definition 2 The space of classical symbols $S^m_{cl}(\mathbb{R}^n \times \mathbb{R}^n)$, $m \in \mathbb{R}$, is the set of all $a \in S^m(\mathbb{R}^n \times \mathbb{R}^n)$ for which there is a sequence of functions $\{a_{(m-j)}\}_{j \in \mathbb{N}_0}$ in $C^\infty(\mathbb{R}^n \times (\mathbb{R}^n \setminus \{0\}))$, such that

1) $a_{(m-j)}(x, t\xi) = t^{m-j} a_{(m-j)}(x,\xi)$ for all $t > 0$ and $(x,\xi) \in \mathbb{R}^n \times (\mathbb{R}^n \setminus \{0\})$.
2) For any $\chi \in C^\infty(\mathbb{R}^n)$ such that $\chi(\xi) = 0$ in a neighborhood of the origin and $\chi(\xi) = 1$ outside a compact set, we have $(x,\xi) \mapsto \chi(\xi) a_{(m-j)}(x,\xi) \in S^{m-j}_{1,0}(\mathbb{R}^n \times \mathbb{R}^n)$, $\forall j \in \mathbb{N}_0$, and

$$(x,\xi) \mapsto a(x,\xi) - \sum_{j=0}^{N-1} \chi(\xi) a_{(m-j)}(x,\xi) \in S^{m-N}\left(\mathbb{R}^n \times \mathbb{R}^n\right), \ \forall N \in \mathbb{N}_0 \setminus \{0\}.$$

The sequence $\{a_{(m-j)}\}_{j \in \mathbb{N}_0}$ is called an asymptotic expansion of a. It is uniquely determined by a and we write $a \sim \sum_{j=0}^{\infty} a_{(m-j)}$ to indicate it.

Proposition 3 *A classical symbol $a \in S^m_{cl}(\mathbb{R}^n \times \mathbb{R}^n)$ with asymptotic expansion $a \sim \sum_{j=0}^{\infty} a_{(m-j)}$ is elliptic if and only if there is a constant $C > 0$ such that $\left|a_{(m)}(x,\xi)\right| \geq C|\xi|^m$, for all $(x,\xi) \in \mathbb{R}^n \times (\mathbb{R}^n \setminus \{0\})$. Due to the homogeneity of the symbol $a_{(m)}$, this is equivalent to $\left|a_{(m)}(x,\xi)\right| \geq C$, for all $x \in \mathbb{R}^n$ and $\xi \in \mathbb{R}^n$ such that $|\xi| = 1$.*

Proof The proof follows easily by noting that, for $|\xi| \geq 1$, we have

$$a(x,\xi) = a_{(m)}(x,\xi) + r(x,\xi),$$

where $r \in S^{m-1}(\mathbb{R}^n \times \mathbb{R}^n)$. □

For each symbol, we define a pseudodifferential operator on $\mathbb{R}^n$.

Definition 4 Let $a \in S^m(\mathbb{R}^n \times \mathbb{R}^n)$, $m \in \mathbb{R}$. The pseudodifferential operator $op(a) : \mathcal{S}(\mathbb{R}^n) \to \mathcal{S}(\mathbb{R}^n)$ is defined by

$$op(a)u(x) = \frac{1}{(2\pi)^n} \int_{\mathbb{R}^n} e^{ix\xi} a(x,\xi)\hat{u}(\xi)\, d\xi.$$

If $a \in S^m(\mathbb{R}^n \times \mathbb{R}^n)$ is an elliptic symbol, we say that $op(a)$ is an elliptic pseudodifferential operator. Sometimes the term ΨDOs is used as a short way to refer to the pseudodifferential operators.

The set of all pseudodifferential operators is an algebra with the composition, due to the following proposition [16].

Proposition 5 *Let $a \in S^m(\mathbb{R}^n \times \mathbb{R}^n)$ and $b \in S^{\tilde{m}}(\mathbb{R}^n \times \mathbb{R}^n)$. Then $op(a)\,op(b) = op(c)$, where $c \in S^{m+\tilde{m}}(\mathbb{R}^n \times \mathbb{R}^n)$ and $c = ab \mod S^{m+\tilde{m}-1}(\mathbb{R}^n \times \mathbb{R}^n)$. If a and b are classical, then c is also a classical symbol.*

These operators have continuous extensions to Besov and Triebel-Lizorkin spaces. We recall here the definition of these spaces.

Definition 6 Let us fix a function $\varphi_0 \in C_c^\infty(\mathbb{R}^n)$ such that its support is contained in $\{\xi \in \mathbb{R}^n;\ |\xi| < 2\}$ and $\varphi_0(\xi) = 1$ for all ξ contained in a neighborhood of the unit ball. We define functions $\varphi_j \in C_c^\infty(\mathbb{R}^n)$, $j \geq 1$, by $\varphi_j(\xi) = \varphi_0\left(2^{-j}\xi\right) - \varphi_0\left(2^{-j+1}\xi\right)$ and the sets $K_0 := \{\xi \in \mathbb{R}^n;\ |\xi| \leq 2\}$ and $K_j := \left\{\xi \in \mathbb{R}^n;\ 2^{j-1} \leq |\xi| \leq 2^{j+1}\right\}$, for all $j \geq 1$. The sequence of functions $\left\{\varphi_j\right\}_{j \in \mathbb{N}_0}$ is called a dyadic partition of unity.

Associated to a dyadic partition of unity, we can always define operators $\varphi_j(D) : \mathcal{S}(\mathbb{R}^n) \to \mathcal{S}(\mathbb{R}^n)$ by $\varphi_j(D)\,u = op\left(\varphi_j\right)u = \mathcal{F}^{-1}\left(\varphi_j \mathcal{F}(u)\right)$.

Definition 7 Let $s \in \mathbb{R}$ and $1 < p, q < \infty$.

1) The Besov space $B^s_{pq}(\mathbb{R}^n)$ is the space of all tempered distributions $u \in \mathcal{S}'(\mathbb{R}^n)$ such that

$$\|u\|_{B^s_{pq}(\mathbb{R}^n)} := \left(\sum_{j=0}^{\infty} 2^{jsq} \left\|\varphi_j(D)\,u\right\|^q_{L^p(\mathbb{R}^n)}\right)^{\frac{1}{q}} < \infty.$$

   It is a Banach space with norm $\|.\|_{B^s_{pq}(\mathbb{R}^n)}$.

2) The Triebel-Lizorkin space $F^s_{pq}(\mathbb{R}^n)$ is the space of all tempered distributions $u \in \mathcal{S}'(\mathbb{R}^n)$ such that

$$\|u\|_{F^s_{pq}(\mathbb{R}^n)} := \left\|\left(\sum_{j=0}^{\infty} 2^{jsq} \left|\varphi_j(D)\,u\right|^q\right)^{\frac{1}{q}}\right\|_{L^p(\mathbb{R}^n)} < \infty.$$

   It is a Banach space with norm $\|.\|_{F^s_{pq}(\mathbb{R}^n)}$.

Remark 8 We note that

i) If $q = 2$ and $1 < p < \infty$, then

$$F^s_{p2}\left(\mathbb{R}^n\right) = H^s_p\left(\mathbb{R}^n\right) := \left\{u \in \mathcal{S}'\left(\mathbb{R}^n\right);\ \mathcal{F}^{-1}\left(\langle\xi\rangle^s\,\hat{u}(\xi)\right) \in L^p\left(\mathbb{R}^n\right)\right\}.$$

ii) We can define Besov and Triebel-Lizorkin spaces for p or q in $]0, 1]$ and also for p or q equal to ∞ (the Triebel-Lizorkin is usually only defined for $p < \infty$). This includes interesting spaces such as $B^s_{\infty\infty}(\mathbb{R}^n)$, which, for $s > 0$, are the Hölder-Zygmund spaces. Unfortunately outside $]1, \infty[$ our arguments do not work. Therefore we will not treat these cases here.

iii) With a different choice of dyadic partition of unity, we obtain the same spaces with equivalent norms.

The above spaces have the following important properties, see [15, Sections 2.4.2 and 2.4.7]:

Proposition 9 *The following properties hold:*

1) *The set $\mathcal{S}(\mathbb{R}^n)$ is dense in $B^s_{pq}(\mathbb{R}^n)$ and in $F^s_{pq}(\mathbb{R}^n)$, for all $s \in \mathbb{R}$ and $1 < p, q < \infty$.*
2) *The dual of $B^s_{pq}(\mathbb{R}^n)$ can be identified with $B^{-s}_{p'q'}(\mathbb{R}^n)$ using the dual pair $(f, g) := \int_{\mathbb{R}^n} f(x) g(x) dx$, for $f, g \in \mathcal{S}(\mathbb{R}^n)$. The same can be said of the dual of $F^s_{pq}(\mathbb{R}^n)$ and $F^{-s}_{p'q'}(\mathbb{R}^n)$, where $\frac{1}{p} + \frac{1}{p'} = 1$ and $\frac{1}{q} + \frac{1}{q'} = 1$.*
For all $\theta \in]0, 1[$, $s_1, s_2 \in \mathbb{R}$, we have
3) *$\left(B^{s_1}_{pq_1}(\mathbb{R}^n), B^{s_2}_{pq_2}(\mathbb{R}^n)\right)_{\theta,q} = B^s_{pq}(\mathbb{R}^n)$, $1 < p, q, q_1, q_2 < \infty$ and $s = (1-\theta) s_1 + \theta s_2$.*
4) *$\left(F^{s_1}_{pq_1}(\mathbb{R}^n), F^{s_2}_{pq_2}(\mathbb{R}^n)\right)_{\theta,q} = B^s_{pq}(\mathbb{R}^n)$, where $1 < p, q, q_1, q_2 < \infty$ and $s = (1-\theta) s_1 + \theta s_2$.*
5) *$\left[F^{s_1}_{p_1q_1}(\mathbb{R}^n), F^{s_2}_{p_2q_2}(\mathbb{R}^n)\right]_{\theta} = F^s_{pq}(\mathbb{R}^n)$, where $1 < p_1, p_2, q_1, q_2 < \infty$, $s = (1-\theta) s_1 + \theta s_2$, $\frac{1}{p} = (1-\theta)\frac{1}{p_1} + \theta\frac{1}{p_2}$ and $\frac{1}{q} = (1-\theta)\frac{1}{q_1} + \theta\frac{1}{q_2}$.*

In particular, the following corollary is important for our results:

Corollary 10 *If $1 < p, q < \infty$, then, for each $\theta \in]0, 1[$, there is a constant $C_\theta > 0$ such that*

$$\max\left\{\|u\|_{B^0_{pq}(\mathbb{R}^n)}, \|u\|_{F^0_{pq}(\mathbb{R}^n)}\right\} \leq C_\theta \|u\|^{1-\theta}_{L_p(\mathbb{R}^n)} \|u\|^{\theta}_{H^1_p(\mathbb{R}^n)}, \ \forall u \in \mathcal{S}\left(\mathbb{R}^n\right).$$

Proof The fact that $\left(L_p(\mathbb{R}^n), H^1_p(\mathbb{R}^n)\right)_{\theta,q} = \left(F^0_{p2}(\mathbb{R}^n), F^1_{p2}(\mathbb{R}^n)\right)_{\theta,q} = B^\theta_{pq}(\mathbb{R}^n)$ implies that

$$\|u\|_{B^0_{pq}(\mathbb{R}^n)} \leq \|u\|_{B^\theta_{pq}(\mathbb{R}^n)} \leq C_\theta \|u\|^{1-\theta}_{L_p(\mathbb{R}^n)} \|u\|^{\theta}_{H^1_p(\mathbb{R}^n)}, \ \forall u \in \mathcal{S}\left(\mathbb{R}^n\right).$$

Moreover, as $\left[L_p(\mathbb{R}^n), H^1_p(\mathbb{R}^n)\right]_\theta = \left[F^0_{p2}(\mathbb{R}^n), F^1_{p2}(\mathbb{R}^n)\right]_\theta = F^\theta_{p2}(\mathbb{R}^n)$, we conclude that

$$\|u\|_{F^0_{pq}(\mathbb{R}^n)} \leq \|u\|_{F^\theta_{p2}(\mathbb{R}^n)} \leq C_\theta \|u\|^{1-\theta}_{L_p(\mathbb{R}^n)} \|u\|^{\theta}_{H^1_p(\mathbb{R}^n)}.$$

The estimates with the exponents θ and $1-\theta$ follow from usual results of Interpolation Theory, see, for instance, Lunardi [10, Corollary 1.1.7]. □

We can now state precisely the continuity of the pseudodifferential operators [14, Section 6.2.2].

Theorem 11 *Let $a \in S^m(\mathbb{R}^n \times \mathbb{R}^n)$, $m \in \mathbb{R}$. For each $1 < p, q < \infty$ and $s \in \mathbb{R}$, the operator $op(a)$ extends to a continuous operator on Besov spaces $op(a) : B^s_{pq}(\mathbb{R}^n) \to B^{s-m}_{pq}(\mathbb{R}^n)$ and on Triebel-Lizorkin spaces $op(a) : F^s_{pq}(\mathbb{R}^n) \to F^{s-m}_{pq}(\mathbb{R}^n)$.*

3 Main Result

The main result of this contribution is the following:

Theorem 12 *Let $a \in S^m(\mathbb{R}^n \times \mathbb{R}^n)$. If there is a sequence $\{(y_k, \eta_k)\}_{k\in\mathbb{N}_0}$ in $\mathbb{R}^n \times \mathbb{R}^n$ such that $\lim_{k\to\infty} |\eta_k| = \infty$ and $\lim_{k\to\infty} |\eta_k|^{-m+r} a(y_k, \eta_k) = 0$, for some $r > 0$, then, for all $s \in \mathbb{R}$ and $1 < p, q < \infty$, the operators $op(a) : B^s_{pq}(\mathbb{R}^n) \to B^{s-m}_{pq}(\mathbb{R}^n)$ and $op(a) : F^s_{pq}(\mathbb{R}^n) \to F^{s-m}_{pq}(\mathbb{R}^n)$ are not Fredholm.*

Corollary 13 *Let $a \in S^m_{cl}(\mathbb{R}^n \times \mathbb{R}^n)$ be a classical symbol. If for some $1 < p, q < \infty$ and $s \in \mathbb{R}$ the operator $op(a) : B^s_{pq}(\mathbb{R}^n) \to B^{s-m}_{pq}(\mathbb{R}^n)$ or $op(a) : F^s_{pq}(\mathbb{R}^n) \to F^{s-m}_{pq}(\mathbb{R}^n)$ is Fredholm, then a is elliptic.*

Proof In fact, if a is not an elliptic operator, then, by Proposition 3, there is a sequence $\left\{\left(y_k, \tilde{\xi}_k\right)\right\}_{k\in\mathbb{N}_0}$ such that $\left|\tilde{\xi}_k\right| = 1$ and $\left|a_{(m)}\left(y_k, \tilde{\xi}_k\right)\right| < \frac{1}{k+1}$. This means that

$$\left|(k+1)^{-m+r} a_{(m)}\left(y_k, (k+1)\tilde{\xi}_k\right)\right| < (k+1)^{r-1}.$$

Let us define $\eta_k = (k+1)\tilde{\xi}_k$. Then $\lim_{k\to\infty} |\eta_k| = \infty$ and, for all $r < 1$, we have $\lim_{k\to\infty} \left||\eta_k|^{-m+r} a_{(m)}(y_k, \eta_k)\right| = 0$. The sequence $\{(y_k, \eta_k)\}_{k\in\mathbb{N}_0}$ satisfies the conditions of Theorem 12. Therefore the operators $op(a) : B^s_{pq}(\mathbb{R}^n) \to B^{s-m}_{pq}(\mathbb{R}^n)$ and $op(a) : F^s_{pq}(\mathbb{R}^n) \to F^{s-m}_{pq}(\mathbb{R}^n)$ cannot be Fredholm. □

Remark 14 The argument of the proof of Corollary 13 cannot be extended to nonclassical symbols. In fact, let $\chi \in C^\infty(\mathbb{R}^n)$ be a smooth function that is equal to 0 in a neighborhood of 0 and equal to 1 outside a compact set. We define

$$a(\xi) = \frac{\chi(\xi)}{\ln(|\xi|)} \in S^0\left(\mathbb{R}^n\right).$$

This symbol is not elliptic, as $\lim_{|\xi|\to\infty} a(\xi) = 0$. However for any $r > 0$ and any sequence $\{\eta_k\}_{k\in\mathbb{N}_0}$ in $\mathbb{R}^n$ such that $\lim_{k\to\infty} |\eta_k| = \infty$, we have $\lim_{k\to\infty} |\eta_k|^r |a(\eta_k)| = \infty$. Hence there is no sequence that satisfies the conditions of Theorem 12.

We need some preparation for the proof of Theorem 12. In [9], very similar results to the next lemmas were proved, but for the spaces $B^s_{pp}(\mathbb{R}^n)$, $1 < p < \infty$, and for parameter-dependent symbols.

Lemma 15 *For each $1 < p, q < \infty$, there is a constant $C > 0$ such that*

$$\frac{1}{C}\max\left\{\|u\|_{B^0_{pq}(\mathbb{R}^n)}, \|u\|_{F^0_{pq}(\mathbb{R}^n)}\right\} \leq \|u\|_{L_p(\mathbb{R}^n)} \leq C\max\left\{\|u\|_{B^0_{pq}(\mathbb{R}^n)}, \|u\|_{F^0_{pq}(\mathbb{R}^n)}\right\},$$

whenever $u \in \mathcal{S}(\mathbb{R}^n)$ is a function with supp $(\mathcal{F}(u)) \subset \cup_{k=m}^{m+2} K_k$, for some constant $m \in \mathbb{N}_0$, where the sets $\{K_j\}_{j\in\mathbb{N}_0}$ are as in Definition 6. The constants C depend on p and q, but do not depend on m.

Proof First we note that if $u \in \mathcal{S}(\mathbb{R}^n)$ is such that $\operatorname{supp}(\mathcal{F}(u)) \subset \cup_{k=m}^{m+2} K_k$, for some constant $m \in \mathbb{N}_0$, then $\mathcal{F}(u)(\xi) = \sum_{j=m-1}^{m+3} \varphi_j(\xi)\mathcal{F}(u)(\xi)$, using the convention that $\varphi_{-1}(\xi) = 0$. Hence $u = \sum_{j=m-1}^{m+3} \varphi_j(D)u$ and

$$\|u\|_{B^0_{p,q}(\mathbb{R}^n)} = \left(\sum_{j=m-1}^{m+3} \|\varphi_j(D)u\|^q_{L^p(\mathbb{R}^n)}\right)^{\frac{1}{q}}.$$

Due to the equivalence of norms in finite dimensional vector spaces, we know that $\frac{1}{C_q}\sum_{j=1}^5 |a_j| \leq \left(\sum_{j=1}^5 |a_j|^q\right)^{\frac{1}{q}} \leq C_q \sum_{j=1}^5 |a_j|$, for all $(a_1, \ldots, a_5) \in \mathbb{C}^5$ and $1 < q < \infty$, where $C_q > 0$ is a constant that depends only on q. We also note that $\varphi_j(\xi) = \varphi_1\left(2^{-j+1}\xi\right)$, for $j \geq 1$. Therefore, as $\varphi_j(D)u = \mathcal{F}^{-1}(\varphi_j) * u$, Young's inequality implies that $\|\varphi_j(D)\|_{\mathcal{B}(L^p(\mathbb{R}^n))}$ is uniformly bounded for $j \in \mathbb{N}_0$.

Therefore, the result follows from

$$\|u\|_{L_p(\mathbb{R}^n)} = \left\|\sum_{j=m-1}^{m+3} \varphi_j(D)u\right\|_{L_p(\mathbb{R}^n)} \leq C_1\left(\sum_{j=m-1}^{m+3} \|\varphi_j(D)u\|^q_{L_p(\mathbb{R}^n)}\right)^{\frac{1}{q}}$$

$$\leq C_2 \sum_{j=m-1}^{m+3} \|\varphi_j(D)u\|_{L_p(\mathbb{R}^n)} \leq C_3\|u\|_{L_p(\mathbb{R}^n)}.$$

Similarly, for the Triebel-Lizorkin space we have

$$\|u\|_{F^0_{pq}(\mathbb{R}^n)} = \left\|\left(\sum_{j=m-1}^{m+3} |\varphi_j(D)u|^q\right)^{\frac{1}{q}}\right\|_{L^p(\mathbb{R}^n)}$$

and

$$\|u\|_{L_p(\mathbb{R}^n)} \leq \left\| \sum_{j=m-1}^{m+3} |\varphi_j(D)u| \right\|_{L_p(\mathbb{R}^n)} \leq C_1 \left\| \left(\sum_{j=m-1}^{m+3} |\varphi_j(D)u|^q \right)^{\frac{1}{q}} \right\|_{L_p(\mathbb{R}^n)}$$

$$\leq C_2 \sum_{j=m-1}^{m+3} \left\| \varphi_j(D)u \right\|_{L^p(\mathbb{R}^n)} \leq C_3 \|u\|_{L^p(\mathbb{R}^n)},$$

□

Definition 16 Let us fix $0 < \tau < \frac{1}{3}$. For all $s \in \mathbb{R}$ and $(y, \eta) \in \mathbb{R}^n \times \mathbb{R}^n$, we define the bijections $R_s = R_s(y, \eta) : \mathcal{S}(\mathbb{R}^n) \to \mathcal{S}(\mathbb{R}^n)$ by

$$R_s u(x) = s^{\frac{\tau n}{p}} e^{isx\eta} u\left(s^\tau (x - y)\right).$$

Lemma 17 *Let* $\mathcal{K} := \left\{u \in \mathcal{S}(\mathbb{R}^n);\ supp(\mathcal{F}(u)) \subset \left\{\xi \in \mathbb{R}^n;\ \frac{1}{2} < |\xi| < 1\right\}\right\}$. *Then, for all* $1 < p, q < \infty$*:*

1) There are constants $D_{pq} > 0$ *and* $s_0 > 0$ *such that*

$$\frac{1}{D_{pq}} \|u\|_{B^0_{pq}(\mathbb{R}^n)} \leq \|R_s u\|_{B^0_{pq}(\mathbb{R}^n)} \leq D_{pq} \|u\|_{B^0_{pq}(\mathbb{R}^n)}$$

and

$$\frac{1}{D_{pq}} \|u\|_{F^0_{pq}(\mathbb{R}^n)} \leq \|R_s u\|_{F^0_{pq}(\mathbb{R}^n)} \leq D_{pq} \|u\|_{F^0_{pq}(\mathbb{R}^n)},$$

for all $u \in \mathcal{K}$ *and* $s > s_0$. *The constants* D_{pq} *depend on* p *and* q *but not on* (y, η). *If* $|\eta| \geq 1$, *then we can choose* $s_0 = 2^{\frac{1}{1-\tau}}$.

2) If $u \in \mathcal{K}$, *then* $\lim_{s\to\infty} R_s u = 0$ *weakly in* $B^0_{pq}(\mathbb{R}^n)$ *and in* $F^0_{pq}(\mathbb{R}^n)$.

Proof

1) A simple computation shows us that the Fourier transform of $R_s u$ is given by

$$\mathcal{F}(R_s u)(\xi) = s^{\frac{\tau n}{p} - n\tau} e^{-iy(\xi - s\eta)} \hat{u}\left(s^{-\tau}(\xi - s\eta)\right). \tag{3.1}$$

Therefore if ξ is such that $\mathcal{F}(R_s u)(\xi) \neq 0$, then $\frac{1}{2} < \left|s^{-\tau}(\xi - s\eta)\right| < 1$. If $\eta = 0$, then $\operatorname{supp}(\mathcal{F}(R_s u)) \subset \left\{\xi \in \mathbb{R}^n;\ \frac{1}{2}s^\tau < |\xi| < s^\tau\right\}$. If $\eta \neq 0$, then $\operatorname{supp}(\mathcal{F}(R_s(u))) \subset \left\{\xi;\ \frac{1}{2}s|\eta| < |\xi| < 2s|\eta|\right\}$, for all $s > s_0$, where s_0 is chosen such that $s^\tau < s\frac{1}{2}|\eta|$, for all $s > s_0$. This implies that, for $s \geq s_0$, there is a constant $m \in \mathbb{N}_0$ such that $\operatorname{supp}(\mathcal{F}(R_s u)) \subset \cup_{k=m}^{m+2} K_k$, where the sets K_k are as in Definition 6.

Now, it is a simple computation to show that $\|R_s u\|_{L_p(\mathbb{R}^n)} = \|u\|_{L_p(\mathbb{R}^n)}$, for every $u \in \mathcal{S}(\mathbb{R}^n)$. Therefore, using Lemma 15, we conclude that there is a constant $C_{pq} > 0$ such that

$$\|u\|_{B^0_{pq}(\mathbb{R}^n)} \le C_{pq}\|u\|_{L_p(\mathbb{R}^n)} = C_{pq}\|R_s u\|_{L_p(\mathbb{R}^n)} \le C^2_{pq}\|R_s u\|_{B^0_{pq}(\mathbb{R}^n)}$$

and

$$\|R_s u\|_{B^0_{pq}(\mathbb{R}^n)} \le C_{pq}\|R_s u\|_{L_p(\mathbb{R}^n)} = C_{pq}\|u\|_{L_p(\mathbb{R}^n)} \le C^2_{pq}\|u\|_{B^0_{pq}(\mathbb{R}^n)}.$$

The same argument holds for $F^0_{pq}(\mathbb{R}^n)$.

2) For every u and $v \in \mathcal{S}(\mathbb{R}^n)$, we have

$$\begin{aligned}\left|\int_{\mathbb{R}^n}(R_s u)(x)\, v(x)\, dx\right| &= \left|\int_{\mathbb{R}^n} s^{\frac{\tau n}{p}} e^{isx\eta} u\left(s^\tau (x-y)\right) v(x)\, dx\right| \\ &= \left|s^{\frac{\tau n}{p}-n\tau}\int_{\mathbb{R}^n} e^{is^{1-\tau}w\eta} u\left(w - s^\tau y\right) v\left(s^{-\tau} w\right) dw\right| \\ &\le s^{\frac{\tau n}{p}-n\tau}\|v\|_{L^\infty(\mathbb{R}^n)}\|u\|_{L^1(\mathbb{R}^n)}.\end{aligned}$$

Hence $\lim_{s\to\infty}\int_{\mathbb{R}^n}(R_s u)(x)\, v(x)\, dx = 0$. If $u \in \mathcal{K}$, then $R_s u$ is uniformly bounded in $B^0_{pq}(\mathbb{R}^n)$. As $\mathcal{S}(\mathbb{R}^n)$ is dense in $B^0_{pq}(\mathbb{R}^n)$ and the dual of $B^0_{pq}(\mathbb{R}^n)$ can be identified with $B^0_{p'q'}(\mathbb{R}^n)$ for $\frac{1}{p} + \frac{1}{p'} = 1$ and $\frac{1}{q} + \frac{1}{q'} = 1$ due to Proposition 9, we obtain the result. □

Lemma 18 *Let $(y, \eta) \in \mathbb{R}^n \times \mathbb{R}^n$, $s \in \mathbb{R}$ and $a \in S^m(\mathbb{R}^n \times \mathbb{R}^n)$. Then the operator $R_s = R_s(y, \eta)$ is such that*

$$R_s^{-1} op(a) R_s u(x) = op(a_s) u(x), \tag{3.2}$$

where $a_s(x, \xi) = a\left(y + s^{-\tau}x, s\eta + s^\tau \xi\right)$.

Proof First, we note that the operator R_s is invertible with inverse given by

$$R_s^{-1} u(x) = s^{-\frac{\tau n}{p}} e^{-is(y+s^{-\tau}x)\eta} u\left(y + s^{-\tau}x\right). \tag{3.3}$$

A direct computation using Eqs. (3.1), (3.3), and Definition 4 shows (3.2). □

Lemma 19 *Let $a \in S^0(\mathbb{R}^n \times \mathbb{R}^n)$ and $\{(y_k, \eta_k) \in \mathbb{R}^n \times \mathbb{R}^n\}_{k\in\mathbb{N}_0}$ be a sequence such that $\lim_{k\to\infty}|\eta_k| = \infty$. If $u \in \mathcal{S}(\mathbb{R}^n)$, then setting $s_k = |\eta_k|$ and $R_k = R_{s_k}\left(y_k, \frac{\eta_k}{|\eta_k|}\right)$, we have*

1) For every $r > 0$

$$\lim_{k\to\infty}|\eta_k|^{-r}\left\|R_k^{-1} op(a) R_k u\right\|_{L_p(\mathbb{R}^n)} = 0. \tag{3.4}$$

2) *If* $\lim_{k\to\infty} |\eta_k|^r\, a(y_k,\eta_k) = 0$, *for some* $r > 0$, *then, for* $1 < p < \infty$ *and* $\tilde{r} < \min\{r,\tau\}$, *we have*

$$\lim_{k\to\infty} |\eta_k|^{\tilde{r}} \left\| R_k^{-1} op(a) R_k u \right\|_{H_p^1(\mathbb{R}^n)} = 0. \tag{3.5}$$

3) *If* $\lim_{k\to\infty} |\eta_k|^r\, a(y_k,\eta_k) = 0$, *for some* $r > 0$, *then, for* $1 < p, q < \infty$,

$$\lim_{k\to\infty} \|op(a) R_k u\|_{B^0_{pq}(\mathbb{R}^n)} = 0. \tag{3.6}$$

The item 3) is also true for $F^0_{pq}(\mathbb{R}^n)$ *in place of* $B^0_{pq}(\mathbb{R}^n)$.

Remark 20 Without loss of generality, we will always suppose that $s_k \geq 2^{\frac{1}{1-\tau}}$.

Proof For the proof, we always denote $a_{s_k}(x,\xi) := a\left(y_k + s_k^{-\tau}x, \eta_k + s_k^{\tau}\xi\right)$.

1) We note that

$$\left| |\eta_k|^{-r}\, a\left(y_k + s_k^{-\tau}x, s_k\frac{\eta_k}{|\eta_k|} + s_k^{\tau}\xi\right)\right| \leq \|a\|_{L^\infty(\mathbb{R}^n\times\mathbb{R}^n)} |\eta_k|^{-r}. \tag{3.7}$$

In particular, the pointwise limit is equal to zero, when $k \to \infty$. Moreover, for all $x \in \mathbb{R}^n$, we have

$$\begin{aligned}&\lim_{k\to\infty} |\eta_k|^{-r}\, op\left(a_{s_k}\right) u(x)\\ &= \lim_{k\to\infty} \frac{1}{(2\pi)^n} \int_{\mathbb{R}^n} e^{ix\xi} |\eta_k|^{-r}\, a\left(y_k + s_k^{-\tau}x, \eta_k + s_k^{\tau}\xi\right) \left(\mathcal{F}_{x\to\xi}u\right)(\xi)\, d\xi = 0.\end{aligned}$$

In fact, the integrand goes to zero due to (3.7). As $\left(\mathcal{F}_{x\to\xi}u\right)(\xi) \in L^1\left(\mathbb{R}^n_\xi\right)$, the result follows from the dominated convergence theorem.

In order to prove the limit of Eq. (3.4), it is enough to prove that the functions $\left\{x \in \mathbb{R}^n \mapsto \left||\eta_k|^{-r}\, op\left(a_{s_k}\right) u(x)\right|^p\right\}_{k\in\mathbb{N}_0}$ are dominated by an integrable function. This can be seen using integration by parts:

$$\begin{aligned}&|\eta_k|^{-r} x^\gamma \left(op\left(a_{s_k}\right) u(x)\right)\\ &= (-1)^{|\gamma|} \sum_{\sigma\leq\gamma} \binom{\gamma}{\sigma} |\eta_k|^{-r} \frac{1}{(2\pi)^n} \int_{\mathbb{R}^n} e^{ix\xi} D_\xi^\sigma \left(a\left(y_k + s_k^{-\tau}x, \eta_k + s_k^{\tau}\xi\right)\right)\\ &D_\xi^{\gamma-\sigma} \left(\mathcal{F}_{x\to\xi}u\right)(\xi)\, d\xi.\end{aligned}$$

If $\sigma \neq 0$, we have

$$\begin{aligned}
& s_k^{-r} \left| D_\xi^\sigma \left(a \left(y + s_k^{-\tau} x, \eta_k + s_k^\tau \xi \right) \right) \right| \\
& \leq s_k^{-r+\tau|\sigma|} \left| \left(D_\xi^\sigma a \right) \left(y + s_k^{-\tau} x, \eta_k + s_k^\tau \xi \right) \right| \leq C s_k^{-r+\tau|\sigma|} \left\langle \eta_k + s_k^\tau \xi \right\rangle^{-|\sigma|} \\
& \leq C s_k^{-r+2\tau|\sigma|-|\sigma|} \langle \xi \rangle^{|\sigma|}, \qquad (3.8)
\end{aligned}$$

where we have used Petree's inequality in the last inequality.

The function $\xi \in \mathbb{R}^n \mapsto \langle \xi \rangle^M D_\xi^\gamma \mathcal{F}_{x \to \xi} u(\xi)$ is integrable for all $M \in \mathbb{Z}$ and $\gamma \in \mathbb{N}_0^n$. We conclude that there is a constant C_γ, that depends on γ but not on x, such that

$$\left| s_k^{-r} x^\gamma op\left(a_{s_k}\right) u(x) \right| \leq C_\gamma.$$

Therefore, for all $N \in \mathbb{N}_0$, there is a constant $C_N \geq 0$ such that

$$s_k^{-r} \left| op\left(a_{s_k}\right) u(x) \right| \leq C_N \langle x \rangle^{-N}.$$

This concludes the proof of the limit of Eq. (3.4) by the dominated convergence theorem.

2) The argument is similar to item 1). We first note that

$$\begin{aligned}
& \left| |\eta_k|^{\tilde{r}} a \left(y_k + s_k^{-\tau} x, s_k \frac{\eta_k}{|\eta_k|} + s_k^\tau \xi \right) \right| \\
& \leq |\eta_k|^{\tilde{r}} \left| a\left(y_k + s_k^{-\tau} x, \eta_k + s_k^\tau \xi \right) - a(y_k, \eta_k) \right| + |\eta_k|^{\tilde{r}} \left| a(y_k, \eta_k) \right| \\
& \leq \sum_{j=1}^n \left(\int_0^1 s_k^{\tilde{r}-\tau} |x_j| \left| \partial_{x_j} a\left(y_k + t s_k^{-\tau} x, \eta_k + t s_k^\tau \xi \right) \right| dt \right. \\
& \left. + s_k^{\tilde{r}+\tau} \int_0^1 \left| \xi_j \partial_{\xi_j} a\left(y_k + t s_k^{-\tau} x, \eta_k + t s_k^\tau \xi \right) \right| dt \right) + |\eta_k|^{\tilde{r}-r} \left(|\eta_k|^r \left| a(y_k, \eta_k) \right| \right) \\
& \leq \sum_{j=1}^n \left(C_1 s_k^{\tilde{r}-\tau} |x_j| + C_2 \frac{s_k^{\tilde{r}+2\tau}}{\langle \eta_k \rangle} |\xi_j| \langle \xi \rangle \right) + \sup_{k \in \mathbb{N}_0} \left(|\eta_k|^r \left| a(y_k, \eta_k) \right| \right) |\eta_k|^{\tilde{r}-r} \\
& \leq C \left(\langle x \rangle s_k^{\tilde{r}-\tau} + \langle \xi \rangle^2 s_k^{\tilde{r}+2\tau-1} + s_k^{\tilde{r}-r} \right). \qquad (3.9)
\end{aligned}$$

We conclude that

$$|\eta_k|^{\tilde{r}} \left| a \left(y_k + s_k^{-\tau} x, s_k \frac{\eta_k}{|\eta_k|} + s_k^\tau \xi \right) \right| \leq C \langle x \rangle \langle \xi \rangle^2 s_k^{-\sigma}, \qquad (3.10)$$

where $\sigma := \min\{r - \tilde{r}, 1 - \tilde{r} - 2\tau, \tau - \tilde{r}\} > 0$, as $\tilde{r} < \tau < \frac{1}{3}$.

We now consider, for $x \in \mathbb{R}^n$,

$$s_k^{\tilde{r}} op\left(a_{s_k}\right) u\left(x\right) = \frac{1}{(2\pi)^n} \int_{\mathbb{R}^n} e^{ix\xi} s_k^{\tilde{r}} a\left(y_k + s_k^{-\tau} x, \eta_k + s_k^{\tau}\xi\right) \left(\mathcal{F}_{x\to\xi} u\right)(\xi)\, d\xi.$$

Equation (3.10) implies that the above integrand goes pointwise to zero. Moreover, as $\langle x\rangle \langle\xi\rangle^2 \left|\left(\mathcal{F}_{x\to\xi} u\right)(\xi)\right| \in L^1\left(\mathbb{R}^n_\xi\right)$, Eq. (3.10) implies that $\lim_{k\to\infty} s_k^{\tilde{r}} op\left(a_{s_k}\right) u\left(x\right) = 0$ by the dominated convergence theorem.

We now prove that

$$\lim_{k\to\infty} |\eta_k|^{\tilde{r}} \left\| R_k^{-1} op(a) R_k u \right\|_{L_p(\mathbb{R}^n)} = 0. \tag{3.11}$$

We only need to show that the functions $\left\{x \in \mathbb{R}^n \mapsto \left| |\eta_k|^{\tilde{r}} op\left(a_{s_k}\right) u\left(x\right)\right|^p\right\}_{k\in\mathbb{N}_0}$ are dominated by an integrable function.

By integration by parts, we have

$$\begin{aligned} & s_k^{\tilde{r}} x^\gamma \left(op\left(a_{s_k}\right) u\left(x\right)\right) \\ & = (-1)^{|\gamma|} \sum_{\sigma\le\gamma} \binom{\gamma}{\sigma} s_k^{\tilde{r}} \frac{1}{(2\pi)^n} \int_{\mathbb{R}^n} e^{ix\xi} D_\xi^\sigma \left(a\left(y_k + s_k^{-\tau} x, \eta_k + s_k^{\tau}\xi\right)\right) \\ & D_\xi^{\gamma-\sigma} \left(\mathcal{F}_{x\to\xi} u\right)(\xi)\, d\xi. \end{aligned} \tag{3.12}$$

For $\sigma \neq 0$, we obtain as in (3.8) that

$$s_k^{\tilde{r}} \left| D_\xi^\sigma \left(a\left(y_k + s_k^{-\tau} x, \eta_k + s_k^{\tau}\xi\right)\right)\right| \le C s_k^{\tilde{r}+2\tau|\sigma|-|\sigma|} \langle\xi\rangle^{|\sigma|}. \tag{3.13}$$

We note that $0 < \tilde{r} < \tau < \frac{1}{3}$ and $\tilde{r} + 2\tau|\sigma| - |\sigma| < 0$. Using Eqs. (3.10), (3.12), and (3.13), together with the fact that $\xi \in \mathbb{R}^n \mapsto \langle\xi\rangle^M D_\xi^\gamma \mathcal{F}_{x\to\xi} u\left(\xi\right)$ is integrable for all $M \in \mathbb{Z}$ and $\gamma \in \mathbb{N}_0^n$, we conclude that

$$s_k^{\tilde{r}} \left|op\left(a_{s_k}\right) u\left(x\right)\right| \le C_N \langle x\rangle^{-N},$$

for some constant $C_N > 0$ and the convergence of (3.11) follows from the dominated convergence theorem.

For the convergence in H_p^1 norm, we note that

$$\partial_{x_j} op(a_{s_k}) u = op\left(a_{s_k}\right) \partial_{x_j} u + s_k^{-\tau} op\left(\left(\partial_{x_j} a\right)_{s_k}\right) u. \tag{3.14}$$

Equation (3.11) implies that

$$\lim_{s\to\infty} s_k^{\tilde{r}} \left\|op\left(a_{s_k}\right) \partial_{x_j} u\right\|_{L_p(\mathbb{R}^n)} = 0. \tag{3.15}$$

Moreover, as $\tilde{r} < \tau$, the first item of the Lemma implies that

$$\lim_{s\to\infty} s_k^{\tilde{r}-\tau} \left\| op\left(\left(\partial_{x_j} a\right)_{s_k}\right) u \right\|_{L_p(\mathbb{R}^n)} = 0. \tag{3.16}$$

Equations (3.14)–(3.16) imply that

$$\lim_{s\to\infty} s_k^{\tilde{r}} \left\| \partial_{x_j} \left(op\left(a_{s_k}\right) u\right) \right\|_{L_p(\mathbb{R}^n)} = 0.$$

The result now follows easily.

3) We observe that

$$\begin{aligned}\partial_{x_j}\left(R_s u(x)\right) &= is\eta_j s^{\frac{\tau n}{p}} e^{isx\eta} u\left(s^\tau (x-y)\right) + s^{\frac{\tau n}{p}+\tau} e^{isx\eta}\left(\partial_{x_j} u\right)\left(s^\tau (x-y)\right) \\ &= is\eta_j R_s u(x) + s^\tau R_s\left(\partial_{x_j} u\right)(x).\end{aligned} \tag{3.17}$$

The above estimate together with the fact that $\|R_s u\|_{L_p(\mathbb{R}^n)} = \|u\|_{L_p(\mathbb{R}^n)}$ for all $u \in \mathcal{S}(\mathbb{R}^n)$ shows that $\|R_s u\|_{H^1_p(\mathbb{R}^n)} \le (1 + s\langle\eta\rangle)\|u\|_{H^1_p(\mathbb{R}^n)}$, for all $s \ge 1$. Corollary 10 implies that

$$\|R_s u\|_{B^0_{pq}(\mathbb{R}^n)} \le C_\theta \|R_s u\|^\theta_{H^1_p(\mathbb{R}^n)} \|R_s u\|^{1-\theta}_{L^p(\mathbb{R}^n)} \le C_\theta (1 + s\langle\eta\rangle)^\theta \|u\|_{H^1_p(\mathbb{R}^n)}.$$

That $\|R_s u\|_{F^0_{pq}(\mathbb{R}^n)} \le C_\theta (1 + s\langle\eta\rangle)^\theta \|u\|_{H^1_p(\mathbb{R}^n)}$ can be proved by precisely the same argument. We now choose $0 < \theta < \min\{r, \tau\}$ and conclude that

$$\begin{aligned}\|op(a)R_k u\|_{B^0_{pq}(\mathbb{R}^n)} &\le \left\| R_k\left(R_k^{-1} op(a) R_k u\right)\right\|_{B^0_{pq}(\mathbb{R}^n)} \\ &\le C_\theta \left(1 + |\eta_k| \left\langle \frac{\eta_k}{|\eta_k|}\right\rangle\right)^\theta \left\| R_k^{-1} op(a) R_k u\right\|_{H^1_p(\mathbb{R}^n)} \to 0.\end{aligned} \tag{3.18}$$

The same argument holds for $F^0_{pq}(\mathbb{R}^n)$. □

Theorem 21 *Let $a \in S^0(\mathbb{R}^n \times \mathbb{R}^n)$. If there is a sequence $\{(y_k, \eta_k)\}_{k\in\mathbb{N}_0}$ in $\mathbb{R}^n \times \mathbb{R}^n$ such that $\lim_{k\to\infty} |\eta_k| = \infty$ and $\lim_{k\to\infty} |\eta_k|^r a(y_k, \eta_k) = 0$, for some $r > 0$, then the operators $op(a) : B^0_{pq}(\mathbb{R}^n) \to B^0_{pq}(\mathbb{R}^n)$ and $op(a) : F^0_{pq}(\mathbb{R}^n) \to F^0_{pq}(\mathbb{R}^n)$ are not Fredholm operators for any $1 < p, q < \infty$.*

Proof Let $u \in \mathcal{S}(\mathbb{R}^n)$, $u \neq 0$, be such that $\operatorname{supp}(\mathcal{F}(u)) \subset \left\{\xi \in \mathbb{R}^n; \frac{1}{2} < |\xi| < 1\right\}$. Suppose that $A = op(a) : B^0_{pq}(\mathbb{R}^n) \to B^0_{pq}(\mathbb{R}^n)$ is Fredholm. Then there are operators B and K in $\mathcal{B}\left(B^0_{pq}(\mathbb{R}^n)\right)$ such that K is compact and

$$BA = I + K.$$

Let us define $R_k := R_{s_k}\left(y_k, \frac{\eta_k}{|\eta_k|}\right)$, where $s_k = |\eta_k|$. We assume without loss of generality that $|\eta_k| \geq 2^{\frac{1}{1-\tau}}$, $\forall k$. Then Lemma 17.1 implies that

$$\|u\|_{B^0_{pq}(\mathbb{R}^n)} \leq D_{pq} \|R_k u\|_{B^0_{pq}(\mathbb{R}^n)} = D_{pq} \|BAR_k u - KR_k u\|_{B^0_{pq}(\mathbb{R}^n)}$$

$$\leq D_{pq}\left(\|B\|_{\mathcal{B}\left(B^0_{pq}(\mathbb{R}^n)\right)} \|AR_k u\|_{B^0_{pq}(\mathbb{R}^n)} + \|KR_k u\|_{B^0_{pq}(\mathbb{R}^n)}\right).$$

However $\lim_{k\to\infty} \|AR_k u\|_{B^0_{pq}(\mathbb{R}^n)} = 0$ and $\lim_{k\to\infty} \|KR_k u\|_{B^0_{pq}(\mathbb{R}^n)} = 0$ by Lemmas 19.3 and 17.2, respectively. Therefore we conclude that $\|u\|_{B^0_{pq}(\mathbb{R}^n)} = 0$. As we have assumed $u \neq 0$, we obtain a contradiction.

Exactly the same argument can be used for $F^0_{pq}(\mathbb{R}^n)$. □

We finally can prove our main Theorem.

Proof (of Theorem 12) Suppose that $op(a) : B^s_{pq}(\mathbb{R}^n) \to B^{s-m}_{pq}(\mathbb{R}^n)$ is a Fredholm operator for some $s \in \mathbb{R}$ and let us denote by $\langle D\rangle^t$ the pseudodifferential operator with symbol $\langle\xi\rangle^t \in S^t(\mathbb{R}^n \times \mathbb{R}^n)$. Then

$$\langle D\rangle^{s-m} op(a) \langle D\rangle^{-s} : B^0_{pq}\left(\mathbb{R}^n\right) \to B^0_{pq}\left(\mathbb{R}^n\right)$$

is also Fredholm. Let us suppose also that there is a sequence $\{(y_k, \eta_k)\}_{k\in\mathbb{N}_0}$ such that $\lim_{k\to\infty} |\eta_k| = \infty$ and $\lim_{k\to\infty} |\eta_k|^{-m+r} a(y_k, \eta_k) = 0$, for some $0 < r < 1$. The symbol $c \in S^0(\mathbb{R}^n \times \mathbb{R}^n)$ of the operator $op(c) = \langle D\rangle^{s-m} op(a) \langle D\rangle^{-s}$ is equal to $\langle\xi\rangle^{-m} a(x,\xi) + q(x,\xi)$, where $q \in S^{-1}(\mathbb{R}^n \times \mathbb{R}^n)$. Hence

$$\lim_{k\to\infty} |\eta_k|^r c(y_k, \eta_k) = \lim_{k\to\infty} |\eta_k|^m \langle\eta_k\rangle^{-m} |\eta_k|^{-m+r} a(y_k, \eta_k) + |\eta_k|^r q(y_k, \eta_k) = 0.$$

This means that $op(c)$ is not Fredholm by Theorem 21, which is a contradiction. The proof for $F^0_{pq}(\mathbb{R}^n)$ is the same. □

Acknowledgements Pedro T. P. Lopes was partially supported by FAPESP (Processo número (Grant) 2016/07016-8).

References

1. H. Abels, C. Pfeuffer, Spectral invariance of non-smooth pseudo-differential operators. Integr. Equ. Oper. Theory **86**(1), 41–70 (2016). MR3557319
2. R.A.H.M. Cabral, S.T. Melo, Operators with analytic orbit for the torus action. Stud. Math. **243**, 243–250 (2018). https://doi.org/10.4064/sm8767-10-2017
3. A. Dasgupta, Ellipticity of Fredholm pseudo-differential operators on $L^p(\mathbb{R}^n)$, in *New Developments in Pseudo-Differential Operators*, Operator Theory: Advances and Applications, vol. 189 (Birkhäuser Verlag, Basel, 2009), pp. 107–116. MR2509095

4. A. Dasgupta, M.W. Wong, Spectral invariance of SG pseudo-differential operators on $L^p(\mathbb{R}^n)$, in *Pseudo-Differential Operators: Complex Analysis and Partial Differential Equations*, Operator Theory: Advances and Applications, vol. 205 (Birkhäuser Verlag, Basel, 2010), pp. 51–57. MR2664573
5. L. Hörmander, Pseudo-differential operators and hypoelliptic equations, in *Singular Integrals (Proc. Sympos. Pure Math., Vol. X, Chicago, Ill., 1966)* (American Mathematical Society, Providence, 1967), pp. 138–183. MR0383152
6. V.D. Kryakvin, Characterization of pseudodifferential operators in Hölder-Zygmund spaces. Differ. Equ. **49**(3), 306–312 (2013). Translation of Differ. Uravn. **49**(3), 318–324 (2013). MR3190875
7. V. Kryakvin, G. Omarova, Spectral invariance for pseudodifferential operators in Hölder-Zygmund spaces of the variable smoothness. J. Pseudo-Differ. Oper. Appl. **9**(1), 95–104 (2018). MR3766065
8. H.-G. Leopold, E. Schrohe, Spectral invariance for algebras of pseudodifferential operators on Besov-Triebel-Lizorkin spaces. Manuscripta Math. **78**(1), 99–110 (1993). MR1201764
9. P.T.P. Lopes, E. Schrohe, Spectral invariance of pseudodifferential boundary value problems on manifolds with conical singularities. J. Fourier Anal. Appl. (2018). https://doi.org/10.1007/s00041-018-9607-5.
10. A. Lunardi, Analytic semigroups and optimal regularity in parabolic problems, in *Modern Birkhäuser Classics* (Birkhäuser/Springer, Basel, 1995) [2013 reprint of the 1995 original] [MR1329547]. MR3012216
11. S. Molahajloo, M.W. Wong, Ellipticity, Fredholmness and spectral invariance of pseudo-differential operators on $\mathbb{S}^1$. J. Pseudo-Differ. Oper. Appl. **1**(2), 183–205 (2010). MR2679899
12. S. Rempel, B.-W. Schulze, *Index Theory of Elliptic Boundary Problems* (North Oxford Academic Publishing, London, 1985). Reprint of the 1982 edition. MR829709
13. E. Schrohe, J. Seiler, Ellipticity and invertibility in the cone algebra on L_p-Sobolev spaces. Integr. Equ. Oper. Theory **41**(1), 93–114 (2001). MR1844462
14. H. Triebel, Theory of function spaces. II, in *Monographs in Mathematics*, vol. 84 (Birkhäuser Verlag, Basel, 1992). MR1163193
15. H. Triebel, Theory of function spaces, in *Modern Birkhäuser Classics* (Birkhäuser/Springer, Basel, 2010). Reprint of 1983 edition [MR0730762]; Also published in 1983 by Birkhäuser Verlag [MR0781540]. MR3024598
16. M.W. Wong, An introduction to pseudo-differential operators, 3rd edn., in *Series on Analysis, Applications and Computation*, vol. 6 (World Scientific Publishing, Hackensack, 2014). MR3222682

Characterizations of Self-Adjointness, Normality, Invertibility, and Unitarity of Pseudo-Differential Operators on Compact and Hausdorff Groups

Majid Jamalpourbirgani and M. W. Wong

Abstract We give explicit formulas for the adjoint, product and inverse of a bounded pseudo-differential operator in terms of its symbol on a compact and Hausdorff group. As applications we give necessary and sufficient conditions to insure that a bounded pseudo-differential operator on a compact and Hausdorff group G is self-adjoint, normal, and unitary on $L^2(G)$, and invertible on $L^p(G)$ for $1 \leq p < \infty$.

Keywords Compact and Hausdorff groups · Symbols · Pseudo-differential operators · Self-adjoint · Normal · Invertible · Unitary

Mathematics Subject Classification (2000) Primary 47F05, 47G30; Secondary 35J70

1 Introduction

Let G be a compact and Hausdorff group on which the left (and right) Haar measure is denoted by μ. Let ξ be an irreducible and unitary representation of G on a complex and separable Hilbert space X_ξ. Since G is compact, it is well known that X_ξ is finite-dimensional. We let d_ξ be the dimension of X_ξ. The number d_ξ is also known as the degree of the representation ξ of G on X_ξ. Let $\widehat{G}$ be the set of all (equivalence classes) of irreducible and unitary representations of G, which is usually referred to as the *dual group* of G.

M. Jamalpourbirgani
School of Mathematics, Iran University of Science and Technology, Tehran, Tehran Province, Iran
e-mail: m_jamalpour@mathdep.iust.ac.ir

M. W. Wong (✉)
Department of Mathematics and Statistics, York University, Toronto, ON, Canada
e-mail: mwwong@mathstat.yorku.ca

S. Molahajloo, M. W. Wong (eds.), *Analysis of Pseudo-Differential Operators*, Trends in Mathematics, https://doi.org/10.1007/978-3-030-05168-6_4

Let $f \in L^p(G),\ p \geq 1$. Then we define the Fourier transform $\hat{f}$ of f by

$$\hat{f}(\xi) = \int_G f(x)\xi(x)^* dx, \quad \xi \in \widehat{G}.$$

It is also well known that the Fourier inversion formula states that for a good class of functions in $L^p(G),\ p \geq 1$,

$$f(x) = \sum_{\xi \in \widehat{G}} d_\xi \operatorname{tr}(\xi(x)\hat{f}(\xi)), \quad x \in G.$$

The Fourier inversion formula can be looked at as a formula for the identity operator on $L^p(G),\ p \geq 1$, and as such, is a perfect symmetry that gives us the identity operator on a suitable class of functions on G.

Good references for abstract harmonic analysis abound. See, for instance, [1, 4, 6] for abstract harmonic analysis in general and group representations, the dual group and the Fourier inversion formula in particular.

In order to obtain more interesting operators than the identity operator, we need to break the symmetry using *symbols* σ defined on the *phase space* $G \times \widehat{G}$. To wit, let σ be a *suitable* function defined on $G \times \widehat{G}$. Then for every point $(x, \xi) \in G \times \widehat{G}$, $\sigma(x, \xi)$ is a $d_\xi \times d_\xi$ matrix. For all $\xi \in \widehat{G}$, we denote by $M_{d_\xi}(\mathbb{C})$ the set of all $d_\xi \times d_\xi$ matrices with complex entries. A symbol σ on $G \times \widehat{G}$ in this paper is understood to be a mapping

$$G \times \widehat{G} \ni (x, \xi) \mapsto \sigma(x, \xi) \in M_{d_\xi}(\mathbb{C}).$$

We define the *pseudo-differential operator* T_σ on G with symbol σ by

$$(T_\sigma f)(x) = \sum_{\xi \in \widehat{G}} d_\xi \operatorname{tr}(\xi(x)\sigma(x, \xi)\hat{f}(\xi)), \quad x \in G.$$

The focus of this paper is on the functional analysis of bounded pseudo-differential operators on compact and Hausdorff groups. More explicitly, the overarching hypothesis is the boundedness of a pseudo-differential operator from $L^{p_1}(G)$ into $L^{p_2}(G)$, where $1 \leq p_1, p_2 < \infty$. Very general conditions on the boundedness of pseudo-differential operators can be found in [2] for compact Lie groups and in [8] for the unit circle centered at the origin. It should also be noted that the analysis of pseudo-differential operators on compact and Hausdorff groups can be found in [3, 5]. Results on operators related to pseudo-differential operators in the context of locally compact and Hausdorff groups can be found in [7]. Developing pseudo-differential operators at the level of topological groups rather than Lie groups manifests the many-faceted connections of these operators with mainstream areas of mathematics besides partial differential equations.

In Sect. 2 of the paper, we first give the result that every bounded linear operator $A : L^p(G) \to L^p(G)$, $1 \le p < \infty$, is a pseudo-differential operator of which the symbol can be uniquely determined. This immediately implies that the mapping of symbols to pseudo-differential operators on compact and Hausdorff groups is injective. We give in Sect. 3 a formula for the symbols of the adjoints of bounded pseudo-differential operators from $L^{p_1}(G)$ into $L^{p_2}(G)$, $1 \le p_1, p_2 < \infty$. In particular, we give a criterion for the self-adjointness, or equivalently, the non-self-adjointness of bounded pseudo-differential operators on $L^2(G)$. We give in Sect. 4 a formula for the product of two pseudo-differential operators on G. As an application, a criterion for a pseudo-differential operator on G to be normal is given. In Sect. 5, we give results on the invertibility of pseudo-differential operators on G. In particular, we give a necessary and sufficient condition for a pseudo-differential operator on G to be invertible. A criterion for the unitarity of pseudo-differential operators on G is also given.

2 Injectivity

We begin with the result that every bounded linear operator on $L^p(G)$, $1 \le p < \infty$, is a pseudo-differential operator from $L^p(G)$ into $L^p(G)$.

Theorem 2.1 *Let : $A : L^p(G) \to L^p(G)$ be a bounded linear operator, where $1 \le p < \infty$. Then $A : L^p(G) \to L^p(G)$ is a pseudo-differential operator T_σ : $L^p(G) \to L^p(G)$ such that*

$$(Af)(x) = (T_\sigma f)(x) = \sum_{\xi \in \widehat{G}} d_\xi \operatorname{tr}(\xi(x)\sigma(x,\xi)\hat{f}(\xi)), \quad x \in G,$$

where

$$\sigma(x,\xi) = \xi(x)^* a(x,\xi)$$

with

$$a(x,\xi)_{nm} = (A\xi_{nm})(x)$$

for all $(x,\xi) \in G \times \widehat{G}$ and all $1 \le n, m \le d_\xi$.

Proof Let $f \in C^\infty(G)$. Then for all $\eta \in \widehat{G}$ and all positive integers j and k with $1 \le j, k \le d_\eta$,

$$\begin{aligned}\widehat{(Af)}(\eta)_{kj} &= \int_G \overline{\eta(x)_{jk}}(Af)(x)\, d\mu(x) \\ &= \int_G \overline{(A^*\eta_{jk})(x)} f(x)\, d\mu(x)\end{aligned}$$

$$= \sum_{\xi \in \widehat{G}} \sum_{m,n=1}^{d_\xi} d_\xi \hat{f}(\xi)_{mn} \int_G \overline{(A^* \eta_{jk})(x)} \xi(x)_{nm} d\mu(x)$$

$$= \left[\sum_{\xi \in \widehat{G}} \sum_{m,n=1}^{d_\xi} d_\xi \hat{f}(\xi)_{mn} (A\xi_{nm})(\cdot) \right]^{\wedge} (\eta)_{kj}.$$

So, for all $x \in G$,

$$\begin{aligned}(Af)(x) &= \sum_{\xi \in \widehat{G}} \sum_{m,n=1}^{d_\xi} d_\xi \hat{f}(\xi)_{mn} (A\xi_{nm})(x) \\ &= \sum_{\xi \in \widehat{G}} d_\xi \operatorname{tr}(\xi(x)\xi(x)^* a(x,\xi) \hat{f}(\xi)) \\ &= (T_\sigma f)(x).\end{aligned}$$

Therefore $A = T_\sigma$ and the proof is complete. □

We can now show that the mapping of symbols on $G \times \widehat{G}$ into pseudo-differential operators on G is injective.

Corollary 2.2 *Let σ and τ be symbols on $G \times \widehat{G}$ such that $T_\sigma : L^p(G) \to L^p(G)$ and $T_\tau : L^p(G) \to L^p(G)$ are bounded linear operators, where $1 \le p < \infty$. Then $\sigma = \tau$.*

3 Adjoints

We need the following lemma.

Lemma 3.1 *Let σ be a function on $G \times \widehat{G}$ such that the corresponding pseudo-differential operator $T_\sigma : L^{p_1}(G) \to L^{p_2}(G)$ is a bounded linear operator, where $1 \le p_1, p_2 < \infty$. Then for all positive integers m and n with $1 \le m, n \le d_\xi$,*

$$(\xi(\cdot)\sigma(\cdot,\xi))_{mn} \in L^{p_2}(G).$$

Proof There exists a positive constant C such that

$$\|T_\sigma f\|_{L^{p_2}(G)} \le C \|f\|_{L^{p_1}(G)}, \quad f \in L^{p_1}(G).$$

Now, let $\xi \in \widehat{G}$ and let m and n be positive integers such that $1 \leq m, n \leq d_\xi$. Then we define the function f on G by

$$f(x) = \xi(x)_{mn}, \quad x \in G.$$

We note that

$$\begin{aligned} T_\sigma \xi(\cdot)_{mn} &= \int_G d_\eta \sum_{\eta \in \widehat{G}} \sum_{j,k=1}^{d_\eta} (\eta(\cdot)\sigma(\cdot, \eta))_{jk} \overline{\eta(y)}_{jk} \xi(y)_{mn} d\mu(y) \\ &= d_\xi (\xi(\cdot)\sigma(\cdot, \xi))_{mn}. \end{aligned}$$

So,

$$(\xi(\cdot)\sigma(\cdot, \xi))_{mn} = \frac{1}{d_\xi} T_\sigma \xi(\cdot)_{mn}. \tag{3.1}$$

Since $f = \xi(\cdot)_{mn} \in L^{p_1}(G)$, it follows that

$$\|\xi(\cdot)\sigma(\cdot, \xi)\|_{L^{p_2}(G)} = \frac{1}{d_\xi} \|T_\sigma \xi(\cdot)_{mn}\|_{L^{p_2}(G)} \leq \frac{1}{d_\xi} \|\xi(\cdot)_{mn}\|_{L^{p_1}(G)} < \infty$$

and this completes the proof. □

The following theorem gives a formula for the adjoint of a bounded pseudo-differential operator on G.

Theorem 3.2 *Let σ be a symbol such that the pseudo-differential operator T_σ : $L^{p_1}(G) \to L^{p_2}(G)$ is a bounded linear operator for $1 \leq p_1, p_2 < \infty$. Then its adjoint is the pseudo-differential operator $T_\tau : L^{p_2'}(G) \to L^{p_1'}(G)$, where*

$$\tau(x, \xi) = \xi(x)^* \sum_{\gamma \in \widehat{G}} \frac{d_\gamma^2}{d_\xi} (\mathrm{tr}[\gamma(x)(\gamma(y)\sigma(y, \gamma))^*])^\wedge(\xi^*), \quad (x, \xi) \in G \times \widehat{G}.$$

Proof Let γ and ξ be elements in $\widehat{G}$. Then for all positive integers t, m, n and l with $1 \leq t, m \leq d_\gamma$ and $1 \leq n, l \leq d_\xi$,

$$\begin{aligned} \int_G (\gamma(y)\sigma(y, \gamma))_{tm} \overline{\xi(y)}_{nl} d\mu(y) &= \int_G \frac{1}{d_\gamma} (T_\sigma \gamma_{tm})(y) \overline{\xi(y)}_{nl} d\mu(y) \\ &= \int_G \frac{1}{d_\gamma} \gamma(y)_{tm} \overline{(T_\tau \xi_{nl})(y)} d\mu(y) \\ &= \int_G \frac{d_\xi}{d_\gamma} \gamma(y)_{tm} \overline{(\xi(y)\tau(y, \xi))}_{nl} d\mu(y), \end{aligned} \tag{3.2}$$

and hence

$$\overline{\int_G (\gamma(y)\sigma(y,\gamma))_{tm}\overline{\xi(y)}_{nl}d\mu(y)} = \frac{d_\xi}{d_\gamma}\int_G (\xi(y)\tau(y,\xi))_{nl}\overline{\gamma(y)}_{tm}d\mu(y).$$

Now, using the Fourier transform on G and (3.1), we get

$$\overline{[((\gamma(\cdot)\sigma(\cdot,\gamma))_{tm})^\wedge(\xi)]_{ln}} = \frac{d_\xi}{d_\gamma}[((\xi(\cdot)\tau(\cdot,\xi))_{nl})^\wedge(\gamma)]_{mt}. \tag{3.3}$$

It follows from the Fourier inversion formula on G and (3.3) that for all $(x,\xi) \in G \times \widehat{G}$ and $1 \le n, l \le d_\xi$,

$$\begin{aligned}
((\xi(x)\tau(x,\xi))_{nl} &= \sum_{\gamma\in\widehat{G}} d_\gamma \mathrm{tr}(\gamma(x)((\xi(\cdot)\tau(\cdot,\xi))_{nl})^\wedge(\gamma)) \\
&= \sum_{\gamma\in\widehat{G}}\sum_{t,m=1}^{d_\gamma} d_\gamma\gamma(x)_{tm}[((\xi(\cdot)\tau(\cdot,\xi))_{nl})^\wedge(\gamma)]_{mt} \\
&= \sum_{\gamma\in\widehat{G}}\sum_{t,m=1}^{d_\gamma} \frac{d_\gamma^2}{d_\xi}\gamma(x)_{tm}\overline{[((\gamma(\cdot)\sigma(\cdot,\gamma))_{tm})^\wedge(\xi)]_{ln}}.
\end{aligned}$$

Using the Fourier transform on G, we get

$$\begin{aligned}
((\xi(x)\tau(x,\xi))_{nl} &= \sum_{\gamma\in\widehat{G}}\sum_{t,m=1}^{d_\gamma} \frac{d_\gamma^2}{d_\xi}\gamma(x)_{tm}\int_G \overline{((\gamma(y)\sigma(y,\gamma))_{tm}}\xi(y)_{nl}d\mu(y) \\
&= \sum_{\gamma\in\widehat{G}}\sum_{t,m=1}^{d_\gamma} \frac{d_\gamma^2}{d_\xi}\int_G \gamma(x)_{tm}[(\gamma(y)\sigma(y,\gamma))^*]_{mt}\xi(y)_{nl}d\mu(y) \\
&= \sum_{\gamma\in\widehat{G}} \frac{d_\gamma^2}{d_\xi}\int_G \mathrm{tr}[\gamma(x)(\gamma(y)\sigma(y,\gamma))^*]\xi(y)_{nl}d\mu(y) \\
&= \sum_{\gamma\in\widehat{G}} \frac{d_\gamma^2}{d_\xi}((\mathrm{tr}[\gamma(x)(\gamma(y)\sigma(y,\gamma))^*])^\wedge(\xi^*)_{nl}.
\end{aligned}$$

Therefore

$$\tau(x,\xi) = \xi(x)^*\sum_{\gamma\in\widehat{G}}\frac{d_\gamma^2}{d\xi}((\mathrm{tr}[\gamma(x)(\gamma(y)\sigma(y,\gamma))^*])^\wedge(\xi^*), \quad (x,\xi)\in G\times\widehat{G}.$$

□

A criterion for the self-adjointness of bounded pseudo-differential operators on G is provided by the following theorem.

Theorem 3.3 *Let σ be a symbol on $G \times \widehat{G}$. Then the pseudo-differential operator $T_\sigma : L^2(G) \to L^2(G)$ is self-adjoint if and only if for all γ and ξ in $\widehat{G}$ and all positive integers t, m, n, and l with $1 \le t, m \le d\gamma$ and $1 \le n, l \le d_\xi$,*

$$d_\gamma \int_G (\gamma(y)\sigma(y,\gamma))_{tm}\overline{\xi(y)_{nl}}\,d\mu(y) = d_\xi \int_G \gamma(y)_{tm}\overline{\xi(y)\sigma(y,\xi))_{nl}}\,d\mu(y).$$

Proof Suppose that $T_\sigma : L^2(G) \to L^2(G)$ is self-adjoint. Then for all $(y,\xi) \in G \times \widehat{G}$ and all positive integers n and l with $1 \le n, l \le d_\xi$,

$$(\xi(y)\sigma(y,\xi))_{nl} = (\xi(y)\tau(y,\xi))_{nl},$$

where τ is the symbol of the adjoint of $T_\sigma : L^2(G) \to L^2(G)$. By (3.2), we get for all γ and ξ in $\widehat{G}$ and all positive integers t, m, n, and l with $1 \le t, m \le d_\gamma$ and $1 \le n, l \le d_\xi$,

$$d_\gamma \int_G (\gamma(y)\sigma(y,\gamma))_{tm}\overline{\xi(y)_{nl}}\,d\mu(y) = d_\xi \int_G \gamma(y)_{tm}\overline{\xi(y)\sigma(y,\xi))_{nl}}\,d\mu(y).$$

Conversely, suppose that

$$d_\gamma \int_G (\gamma(y)\sigma(y,\gamma))_{tm}\overline{\xi(y)_{nl}}\,d\mu(y) = \int_G \gamma(y)_{tm}\overline{(\xi(y)\sigma(y,\xi))_{nl}}\,d\mu(y)$$

for all γ and ξ in $\widehat{G}$ and all positive integers t, m, n, and l with $1 \le t, m \le d_\gamma$ and $1 \le n, l \le d_\xi$. Then as in the proof of Theorem 3.2, we get

$$\tau(x,\xi) = \sigma(x,\xi)$$

for all $(x,\xi) \in G \times \widehat{G}$, where τ is the symbol of the adjoint of $T_\sigma : L^2(G) \to L^2(G)$. Therefore $T_\sigma : L^2(G) \to L^2(G)$ is self-adjoint. □

4 Products

The basic formula for the symbol of the product of two bounded pseudo-differential operators on G is the content of the following theorem.

Theorem 4.1 *Let σ and τ be symbols on $G \times \widehat{G}$ such that $T_\sigma : L^{p_1}(G) \to L^{p_2}(G)$ and $T_\tau : L^{p_2}(G) \to L^{p_3}(G)$ be bounded linear operators, where $1 \le p_1, p_2, p_3 < \infty$. Then $T_\sigma T_\tau : L^{p_1}(G) \to L^{p_3}(G)$ is the pseudo-differential operator $T_\lambda : L^{p_1}(G) \to L^{p_3}(G)$, where λ is the symbol on $G \times \widehat{G}$ given by*

$$\lambda(x, \xi) = \xi(x)^* \sum_{\omega \in \widehat{G}} d_\omega \int_G \mathrm{tr}[\omega(x)((\sigma^*(y, \omega))^* \omega(y)^*]\xi(y)\tau(y, \xi)\, d\mu(y)$$

for all $(x, \xi) \in G \times \widehat{G}$.

Proof By Theorem 2.1, we see that for all elements ξ and ω in $\widehat{G}$ and all positive integers n, m, k and l with $1 \le n, m \le d_\xi$ and $1 \le k, l \le d_\omega$, we have

$$\begin{aligned}
&\int_G (\xi(y)\lambda(y, \xi))_{mn} \overline{\omega(y)^*_{kl}} d\mu(y) \\
&= \int_G \frac{1}{d_\xi} (T_\sigma T_\tau \xi_{mn})(y) \overline{\omega(y)_{kl}} d\mu(y) \\
&= \int_G \frac{1}{d_\xi} (T_\tau \xi_{mn})(y) \overline{(T_\sigma^* \omega_{kl})(y)} d\mu(y) \\
&= \int_G d_\omega (\xi(y)\tau(y, \xi))_{mn} (((\sigma^*(y, \omega))^* \omega(y)^*)_{lk} d\mu(y).
\end{aligned}$$

So,

$$((\xi(\cdot)\lambda(\cdot, \xi))^\wedge_{mn}(\omega)_{lk} = \int_G d_\omega (\xi(y)\tau(y, \xi))_{mn} ((\sigma^*(y, \omega))^* \omega(y)^*)_{lk} d\mu(y).$$

Therefore for all $(x, \xi) \in G \times \widehat{G}$, we get by the Fourier inversion formula on G

$$\begin{aligned}
&((\xi(x)\lambda(x, \xi))_{mn} \\
&= \sum_{\omega \in \widehat{G}} d_\omega \mathrm{tr}[\omega(x)((\xi(\cdot)\lambda(\cdot, \xi))_{mn})^\wedge(\omega)] \\
&= \sum_{\omega \in \widehat{G}} \sum_{k,l=1}^{d_\omega} \omega(x)kl[((\xi(\cdot)\lambda(\cdot, \xi))_{mn})^\wedge(\omega)]_{lk} \\
&= \sum_{\omega \in \widehat{G}} \sum_{k,l=1}^{d_\xi} d_\omega \omega(x)_{kl} \int_G (\xi(y)\tau(y, \xi))_{mn} ((\sigma^*(y, \omega)^*) \omega(y)^*)_{lk} d\mu(y) \\
&= \sum_{\omega \in \widehat{G}} d_\omega \int_G \mathrm{tr}[\omega(x)(\sigma^*(y, \omega))^* \omega(y)^*](\xi(y)\tau(y, \xi))_{mn} d\mu(y).
\end{aligned}$$

Then for all $(x, \xi) \in G \times \widehat{G}$,

$$\xi(x)\lambda(x, \xi) = \sum_{\omega \in \widehat{G}} d_\omega \int_G \operatorname{tr}[\omega(x)(\sigma^*(y, \omega)^* \omega(y)^*]\xi(y)\tau(y, \xi)\, d\mu(y)$$

and hence

$$\lambda(x, \xi) = \xi(x)^* \sum_{\omega \in \widehat{G}} \int_G \operatorname{tr}[\omega(x)(\sigma^*(y, \omega)^*)\omega(y)^*](\xi(y)\tau(y, \xi))\, d\mu(y).$$

□

The following two theorems give, respectively, a characterization of the normality and unitarity of bounded pseudo-differential operators on G.

Theorem 4.2 *Let σ be a symbol on $G \times \widehat{G}$ such that the corresponding pseudo-differential operator $T_\sigma : L^2(G) \to L^2(G)$ is a bounded linear operator. Then $T_\sigma : L^2(G) \to L^2(G)$ is normal if and only if for all ξ and ω in $\widehat{G}$ with $1 \le m, n \le d_\xi$ and $1 \le l, k \le d_\omega$,*

$$\int_G (\xi(y)\sigma(y, \xi))_{mn} \overline{(\omega(y)\sigma(y, \omega))_{lk}} d\mu(y)$$
$$= \int_G (\xi(y)\sigma^*(y, \xi))_{mn} \overline{(\omega(y)(\sigma^*(y, \omega))_{lk}} d\mu(y).$$

Proof Suppose that $T_\sigma : L^2(G) \to L^2(G)$ is a normal operator. Then for all ξ and ω in $\widehat{G}$ with $1 \le m, n \le d_\xi$ and $1 \le l, k \le d_\omega$,

$$\int_G (\xi(y)\sigma(y, \xi))_{mn} \overline{(\omega(y)\sigma(y, \omega))_{lk}} d\mu(y)$$
$$= \frac{1}{d_\xi d_\omega} (T_\sigma \xi_{mn})(y) \overline{(T_\sigma \omega_{lk})(y)} d\mu(y)$$
$$= \int_G \frac{1}{d_\xi d_\omega} (T_\sigma^* T_\sigma \xi_{mn})(y) \overline{\omega_{lk}(y)} d\mu(y)$$
$$= \int_G \frac{1}{d_\xi d_\omega} (T_\sigma T_\sigma^* \xi_{mn})(y) \overline{\omega_{lk}(y)} d\mu(y)$$
$$= \int_G \frac{1}{d_\xi d_\omega} (T_\sigma^* \xi_{mn})(y) \overline{(T_\sigma^* \omega_{lk})(y)} d\mu(y)$$
$$= \int_G (\xi(y)\sigma^*(y, \xi))_{mn} \overline{(\omega(y)\sigma^*(y, \omega(y))_{lk}} d\mu(y).$$

Conversely, suppose that

$$\int_G (\xi(y)\sigma(y,\xi))_{mn}\overline{(\omega(y)\sigma(y,\omega))_{lk}}d\mu(y)$$
$$= \int_G (\xi(y)\sigma^*(y,\xi))_{mn}\overline{(\omega(y)\sigma^*(y,\omega))_{lk}}d\mu(y)$$

for all ξ and ω in $\widehat{G}$ with $1 \le m, n \le d_\xi$ and $1 \le l, k \le d_\omega$. Then for all $x \in G$,

$$\omega_{lk}(x) \int_G (\xi(y)\sigma(y,\xi))_{mn}\overline{(\omega(y)\sigma(y,\xi))_{lk}}d\mu(y)$$
$$= \omega_{lk}(x) \int_G (\xi(y)\sigma^*(y,\xi))_{mn}\overline{(\omega(y)\sigma^*(y,\omega))_{lk}}d\mu(y)$$

and so,

$$\int_G \mathrm{tr}[\omega(x)\sigma(y,\omega)^*\omega(y)^*](\xi(y)\sigma(y,\omega))_{mn}d\mu(y)$$
$$= \int_G \mathrm{tr}[\omega(x)(\sigma^*(y,\omega))^*\omega(y)^*](\xi(y)\sigma^*(y,\xi))_{mn}d\mu(y).$$

Therefore

$$\sum_{\omega\in\widehat{G}} d_\omega \int_G \mathrm{tr}[\omega(x)\sigma(y,\omega)^*\omega(y)^*](\xi(y)\sigma(y,\xi))_{mn}d\mu(y)$$
$$= \sum_{\omega\in\widehat{G}} d_\omega \int_G \mathrm{tr}[\omega(x)(\sigma^*(y,\omega))^*\omega(y)^*](\xi(y)\sigma^*(y,\xi))_{mn}d\mu(y)$$

and hence for all $x \in G$,

$$\xi(x)^* \sum_{\omega\in\widehat{G}} d_\omega \in_G \mathrm{tr}[\omega(x)\sigma(y,\omega)^*\omega(y)^*]\xi(y)\sigma(y,\xi)\, d\mu(y)$$
$$= \xi(x)^* \sum_{\omega\in\widehat{G}} d_\omega \int_G \mathrm{tr}[\omega(x)(\sigma^*(y,\omega))^*\omega(y)^*]\xi(y)\sigma^*(y,\xi)\, d\mu(y).$$

By Theorem 4.1, the symbol of $T_\sigma T_\sigma^*$ is equal to the symbol of $T_\sigma^* T_\sigma$. Therefore

$$T_\sigma T_\sigma^* = T_\sigma^* T_\sigma$$

and the proof is complete. □

Theorem 4.3 *Let σ be a symbol on $G \times \widehat{G}$ such that the corresponding pseudo-differential operator $T_\sigma : L^2(G) \to L^2(G)$ is a bounded linear operator. Then $T_\sigma : L^2(G) \to L^2(G)$ is unitary if and only if for all ξ and ω in $\widehat{G}$ with $1 \le m, n \le d_\xi$ and $1 \le l, k \le d_\omega$,*

$$\begin{aligned}
&\int_G (\xi(y)\sigma(y,\xi))_{mn}\overline{(\omega(y)\sigma(y,\omega))_{lk}}d\mu(y)\\
&= \int_G (\xi(y)\sigma^*(y,\xi))_{mn}\overline{(\omega(y)\sigma^*(y,\omega))_{lk}}d\mu(y)\\
&= \begin{cases} 0, & \xi \ne \omega, \\ 0, & m \ne k \text{ or } n \ne l, \\ \frac{1}{d_\xi}, & \xi = \omega,\ m = k,\ n = l. \end{cases}
\end{aligned}$$

Proof Suppose that $T_\sigma : L^2(G) \to L^2(G)$ is unitary. Then for all elements ξ and ω in $\widehat{G}$ with $1 \le m, n \le d_\xi$ and $1 \le l, k \le d_\omega$, we get by Theorem 4.2 and the Peter–Weyl theorem to the effect that $\{\sqrt{d_\xi}\xi_{mn} : 1 \le m, n \le d_\xi, \xi \in \widehat{G}\}$ is an orthonormal basis for $L^2(G)$,

$$\begin{aligned}
&\int_G (\xi(y)\sigma^*(y,\xi))_{mn}\overline{(\omega(y)\sigma^*(y,\xi))_{lk}}d\mu(y)\\
&= \int_G (\xi(y)\sigma(y,\xi))_{mn}\overline{(\omega(y)\sigma(y,\omega))_{lk}}d\mu(y)\\
&= \int_G \frac{1}{d_\xi d_\omega}(T_\sigma\xi_{mn})(y)\overline{(T_\sigma\omega_{lk})(y)}d\mu(y)\\
&= \int_G \frac{1}{d_\xi d_\omega}(T_\sigma^* T_\sigma\xi_{mn})(y)\overline{\omega_{lk}(y)}d\mu(y)\\
&= \int_G \xi_{mn}(y)\overline{\omega_{lk}(y)}d\mu(y)\\
&= \begin{cases} 0, & \xi \ne \omega, \\ 0, & m \ne k \text{ or } n \ne l, \\ \frac{1}{d_\xi}, & \xi = \omega,\ m = k,\ n = l. \end{cases}
\end{aligned}$$

For the converse, let $x \in G$. Then for all elements ξ and ω in $\widehat{G}$ with $1 \le m, n \le d_\xi$ and $1 \le l, k \le d_\omega$,

$$\begin{aligned}
&\omega_{lk}(x)\int_G (\xi(y)\sigma(y,\xi))_{mn}\overline{(\omega(y)\sigma(y,\omega))_{lk}}d\mu(y)\\
&= \omega_{lk}(x)\int_G\int_G (\xi(y)\sigma^*(y,\xi))_{mn}\overline{(\omega(y)\sigma^*(y,\omega))_{lk}}d\mu(y)\\
&= \omega_{lk}(x)\int_G \xi_{mn}(y)\overline{\omega_{lk}(y)}d\mu(y)
\end{aligned}$$

and so,

$$\begin{aligned}
&\int_G \operatorname{tr}[\omega(x)\sigma(y,\omega)^*\omega(y)^*](\xi(y)\sigma(y,\xi))_{mn}d\mu(y)\\
&= \int_G \operatorname{tr}[\omega(x)(\sigma^*(y,w))^*\omega(y)^*](\xi(y)\sigma^*(y,\xi))_{mn}d\mu(y)\\
&= \int_G \operatorname{tr}[\omega(x)\omega(y)^*]\xi_{mn}(y)\,d\mu(y).
\end{aligned}$$

Therefore

$$\begin{aligned}
&\sum_{\omega\in\widehat{G}} d_\omega \int_G \operatorname{tr}[\omega(x)\sigma(y,\omega)^*\omega(y)^*](\xi(y)\sigma(y,\xi))_{mn}d\mu(y)\\
&= \sum_{\omega\in\widehat{G}} d_\omega \int_G \operatorname{tr}[\omega(x)(\sigma^*(y,\omega))^*\omega(y)^*](\xi(y)\sigma^*(y,\xi))_{mn}d\mu(y)\\
&= \sum_{\omega\in\widehat{G}} d_\omega \int_G \operatorname{tr}[\omega(x)\omega(y)^*]\xi_{mn}(y)\,d\mu(y)
\end{aligned}$$

and hence

$$\begin{aligned}
&\xi(x)^* \sum_{\omega\in\widehat{G}} d_\omega \in_G \operatorname{tr} \int_G \operatorname{tr}[\omega(x)\sigma(y,\omega)^*\omega(y)^*]\xi(y)\sigma(y,\xi))\,d\mu(y)\\
&= \xi(x)^* \sum_{\omega\in\widehat{G}} d_\omega \int_G \operatorname{tr}[\omega(x)(\sigma^*(y,\omega))^*\omega(y)^*](\xi(y)\sigma^*(y,\xi))\,d\mu(y)\\
&= \xi(x)^* \sum_{\omega\in\widehat{G}} d_\omega \int_G \operatorname{tr}[\omega(x)\omega(y)^*]\xi(y)\,d\mu(y)\\
&= \xi(x)^* \int_G \delta(x\cdot y^{-1})\xi(y)\,d\mu(y)\\
&= \xi(x)^*\xi(x)\\
&= I,
\end{aligned}$$

where I is the identity matrix of order d_ξ. Thus, by Theorem 4.1,

$$T_\sigma T_\sigma^* = T_\sigma^* T_\sigma = I$$

and this proves that $T_\sigma : L^2(G) \to L^2(G)$ is unitary. □

Theorem 4.4 *Let σ be a symbol on $G \times \widehat{G}$. Then the corresponding pseudo-differential operator $T_\sigma : l^2(G) \to L^2(G)$ is unitary if and only if*

$$\{\sqrt{d_\xi}(\xi(\cdot)\sigma(\cdot,\xi))_{mn} : 1 \le m, n \le d_\xi, \xi \in \widehat{G}\}$$

and

$$\{\sqrt{d_\xi}(\xi(\cdot)\sigma^*(\cdot,\xi))_{m,n} : 1 \le m, n \le d_\xi, \xi \in \widehat{G}\}$$

are orthonormal bases for $L^2(G)$.

Proof Suppose that $T_\sigma : L^2(G) \to L^2(G)$ is unitary. Then $T_\sigma : L^2(G) \to L^2(G)$ is invertible and hence surjective. So, for all $f \in L^2(G)$, there exists a function $g \in L^2(G)$ such that

$$T_\sigma g = \sum_{\xi \in \widehat{G}} \sum_{m,n=1}^{d_\xi} d_\xi (\xi(\cdot)\sigma(\cdot,\xi))_{mn} \hat{g}(\xi)_{nm} = f.$$

By Theorem 4.3, we get for all elements ξ and η in $\widehat{G}$ and all positive integers m, n, k and l with $1 \le m, n \le d_\xi$ and $1 \le k, l \le d_\eta$,

$$\int_G \xi(x)_{mn}\overline{\eta(x)_{kl}}d\mu(x) = \int_G (\xi(x)\sigma(x,\xi))_{mn}\overline{(\eta(x)\sigma(x,\eta))_{kl}}d\mu(x).$$

We know by the Peter–Weyl theorem that $\{\sqrt{d_\omega}\omega_{mn} : 1 \le m, n \le d_\omega, \omega \in \widehat{G}\}$ is an orthonormal basis for $L^2(G)$. So, the set $\{\sqrt{d_\xi}(\xi(\cdot)\sigma(\cdot,\xi))_{mn} : 1 \le m, n \le d_\xi\}$ is orthonormal. Since $T_\sigma^* : L^2(G) \to L^2(G)$ is also unitary, it follows that $\{\sqrt{d_\xi}(\xi(\cdot)\sigma^*(\cdot,\xi))_{mn} : 1 \le m, n \le d_\xi, \xi \in \widehat{G}\}$ is an orthonormal basis for $L^2(G)$. The converse follows immediately from Theorem 4.3. □

5 Invertibility

A necessary and sufficient condition for a bounded pseudo-differential operator on G to be invertible is first given.

Theorem 5.1 *Let σ be a symbol on $G \times \widehat{G}$ such that the corresponding pseudo-differential operator $T_\sigma : L^p(G) \to L^p(G)$ is a bounded linear operator for $1 \le p < \infty$. Then $T_\sigma : L^p(G) \to L^p(G)$ is invertible if and only if there exists a symbol τ on $G \times \widehat{G}$ corresponding to a bounded pseudo-differential operator T_σ*

such that for all elements ξ and η in $\widehat{G}$ and all positive integers m, n, k and l with $1 \le m, n \le d_\xi$ and $1 \le k, l \le d_\eta$,

$$\begin{aligned}
&\int_G (\xi(x)\tau(x,\xi))_{mn}\overline{(\eta(x)\sigma^*(x,\eta))_{kl}}d\mu(x) \\
&= \int_G (\xi(x)\sigma(x,\xi))_{mn}\overline{(\eta(x)\tau^*(x,\eta))_{kl}}d\mu(x) \\
&= \begin{cases} 0, & \xi \neq \eta, \\ 0, & m \neq k \text{ or } m \neq l, \\ \frac{1}{d_\xi}, & \xi = \eta,\ m = k,\ n = l. \end{cases}
\end{aligned}$$

In this case, $T_\sigma^{-1} = T_\tau$.

Proof Suppose that $T_\sigma : L^p(G) \to L^p(G)$ is an invertible operator. Then for all $f \in L^p(G)$ and $g \in L^{p'}(G)$,

$$(f, g) = (T_\sigma T_\sigma^{-1} f, g) = (T_\sigma^{-1} f T_\sigma^* g)$$

and so,

$$\int_G f(x)\overline{g(x)}\,d\mu(x) = \int_G (T_\sigma^{-1} f)(x)\overline{(T_{\sigma*}G(x)}\,d\mu(x).$$

For all ξ and η in $\widehat{G}$ and all positive integers m, n, k, and l with $1 \le m, n \le d\xi$ and $1 \le k, l \le d_\eta$, let f and g be functions on G such that

$$f(x) = \xi(x)_{mn}, \quad x \in G$$

and

$$g(x) = \eta(x) =_{kl}, \quad x \in G.$$

Then letting $T_\sigma^{-1} - T_{\sigma^{-1}}$, we get

$$\int_G \xi(x)_{mn}\overline{\eta(x)_{kl}}d\mu(x) = d_\xi d_\eta (\xi(x)\sigma^{-1}(x,\xi))_{mn}\overline{(\eta(x)\sigma^*(x,\eta))_{kl}}d\mu(x).$$

Since $\{\sqrt{d_\xi}\omega_{mn} : 1 \le m, n \le d_\omega, \omega \in \widehat{G}\}$ is an orthonormal basis for $L^2(G)$, we have

$$\begin{aligned}
&\int_G d_\xi d_\eta (\xi(x)\sigma^{-1}(x,\xi))_{mn}\overline{(\eta(x)\sigma^*(x,\eta))_{kl}}d\mu(x) \\
&= \begin{cases} 0, & \xi \neq \eta, \\ 0, & m \neq k, \text{ or } n \neq l, \\ \frac{1}{d_\xi}, & \xi = \eta, m = k, 1, n = l. \end{cases}
\end{aligned}$$

Using the same proof and the fact that for all $f \in L^p(G)$ and $g \in L^{p'}(G)$,

$$(f, g) = (T_{\sigma^{-1}} T_\sigma f, g) = (T_\sigma f, T_{\sigma^{-1*}} g),$$

we get

$$\int_G \frac{1}{d_\xi} \frac{1}{d_\eta} (\xi(x)\sigma(x,\xi))_{mn} \overline{(\eta(x)(\sigma^{-1*}(x,\eta))_{kl}} d\mu(x)$$

$$= \begin{cases} 0, & \xi \neq \eta, \\ 0, & m \neq k, \text{ or } n \neq l, \\ \frac{1}{d_\xi}, & \xi = eta, m = k, n = l. \end{cases} \tag{5.1}$$

Conversely, suppose that there exists a symbol τ on $G \times \widehat{G}$ such that Theorem 5.1 is satisfied. Then by Theorem 4.1,

$$T_\sigma T_\tau = T_\tau T_\sigma = I$$

and the proof is complete. □

As a useful corollary, we give a necessary condition for the invertibility of a bounded pseudo-differential operator on G.

Theorem 5.2 *Let σ be a symbol on $G \times \widehat{G}$ such that the corresponding pseudo-differential operator $T_\sigma : L^p(G) \to L^p(G)$ is invertible, where $1 \leq p < \infty$. Then*

$$\int_G \text{tr}(\sigma^{-1}(x,\xi)(\sigma^*(x,\xi))^* \, d\mu(x) = \int_G \text{tr}(\sigma(x,\xi)(\sigma^{*-1}(x,\xi))^*) \, d\mu(x) = d_\xi.$$

Proof Let $\xi \in \widehat{G}$. Then by Theorem 5.1, we get for all positive integers m and n with $1 \leq m, n, \leq d_\xi$,

$$\int_G (\xi(x)\sigma^{-1}(x,\xi))_{mn} \overline{((\xi(x)\sigma^*(x,\xi))_{mn}} d\mu(x)$$

$$= \int_G (\xi(x)\sigma(x,\xi))_{mn} \overline{(\xi(x)\sigma^{-1*}(x,\xi))_{mn}} d\mu(x)$$

$$= \frac{1}{d_\xi}.$$

But for the first integral,

$$\begin{aligned}
&\int_G (\xi(x)\sigma^{-1}(x,\xi))_{mn}\overline{(\xi(x)\sigma^*(x,\xi))_{mn}}d\mu(x)\\
&+\int_G (\xi(x)\sigma^{-1}(x,\xi))_{mn}(\xi(x)\sigma^*(x,\xi))^*_{nm}d\mu(x)\\
&=\int_G (\xi(x)\sigma^{-1}(x,\xi))_{mn}((\sigma^*(x,\xi))^*\xi(x)^*)_{nm}d\mu(x)\\
&=\frac{1}{d_\xi}
\end{aligned}$$

and so,

$$\begin{aligned}
&\sum_{m,n=1}^{d_\xi}\int_G (\xi(x)\sigma^{-1}(x,\xi))_{mn}(\xi(x)^*((\sigma^*(x,\xi))^*)_{nm}d\mu(x)\\
&=\int_G \mathrm{tr}(\sigma^{-1}(x,\xi)(\sigma^*(x,\xi))^*)d\mu(x)\\
&=d_\xi.
\end{aligned}$$

Similarly, for the second integral,

$$\int_G \mathrm{tr}(\sigma(x,\xi)(\sigma^{*-1}(x,\xi))^*)\,d\mu(x)=d_\xi$$

and the proof is complete. □

Acknowledgements This work of the first-named author was supported by the International Visiting Research Traineeship (IVRT) sponsored by Professor M. W. Wong at York University in 2017. The research of the second-named author has been supported by the Natural Sciences and Engineering Research Council of Canada under Discovery Grant 0008562.

References

1. G.B. Folland, *A Course in Abstract Harmonic Analysis*, 2nd edn. (CRC Press, New York, 2016)
2. M.B. Ghaemi, M.J. Birgani, L^p-boundedness, compactness of pseudo-differential operators on compact Lie groups. J. Pseudo-Differ. Oper. Appl. **8**, 1–11 (2017)
3. M.B. Ghaemi, M. Jamalpourbirgani, M.W. Wong, Characterizations, adjoints and products of nuclear pseudo-differential operators on compact and Hausdorff groups. U.P.B. Sci. Bull. Ser. A **79**, 207–220 (2017)
4. L.H. Loomis, *Introduction to Abstract Harmonic Analysis* (Dover, Mineola, 2011)
5. S. Molahajloo, M. Pirhayati, Traces of pseudo-differential operators on compact and Hausdorff groups. J. Pseudo-Differ. Oper. Appl. **4**, 361–369 (2013)

6. W. Rudin, *Fourier Analysis on Groups*, Wiley Classics Library Edition (Wiley Inter-Science, New York, 1990)
7. M.W. Wong, *Wavelet Transforms and Localization Operators* (Birkhäuser, Basel, 2002)
8. M.W. Wong, *Discrete Fourier Analysis* (Birkhäuser, Basel, 2011)

Multilinear Commutators in Variable Lebesgue Spaces on Stratified Groups

Dongli Liu, Jian Tan, and Jiman Zhao

Abstract In this paper, we study the multilinear fractional integrals and Calderón–Zygmund singular integrals on stratified groups. We obtain the boundedness of the commutators of the multilinear fractional integrals and Calderón–Zygmund singular integrals in variable Lebesgue spaces.

Keywords Stratified groups · Multilinear fractional integrals · Multilinear Calderón–Zygmund singular integrals · Variable Lebesgue spaces

Mathematics Subject Classification (2000) 42B35, 43A80

1 Introduction

The study of function spaces is of great importance in harmonic analysis. Due to the applications to partial differential equations and the calculus of variations, more and more attention has been paid to the study of variable function spaces (see, for example, [12, 23, 28, 31, 34, 36]). Orlicz [29] firstly established the variable Lebesgue spaces in 1931. The variable Lebesgue spaces are a generalization of the classical Lebesgue spaces, replacing the constant exponent p with an exponent function $p(\cdot)$. In the variable Lebesgue spaces, Cruz-Uribe et al. [15] studied many classical operators on Euclidean spaces and proved the boundedness of these operators. Tan et al. [32] obtained characterizations of BMO in terms of commutators of multilinear fractional integrals and Calderón–Zygmund singular integrals. For other works about variable Lebesgue spaces, see, for example,

D. Liu · J. Zhao (✉)
School of Mathematical Sciences, Key Laboratory of Mathematics and Complex Systems, Ministry of Education, Beijing Normal University, Beijing, China
e-mail: dongliliu@bnu.edu.cn; jzhao@bnu.edu.cn

J. Tan
College of Science, Nanjing University of Posts and Telecommunications, Nanjing, China

S. Molahajloo, M. W. Wong (eds.), *Analysis of Pseudo-Differential Operators*, Trends in Mathematics, https://doi.org/10.1007/978-3-030-05168-6_5

[5, 10, 13, 21, 35]. Although variable Lebesgue spaces on Euclidean spaces have been well studied, there is still large space to study the variable Lebesgue spaces in various settings. Stratified groups appear in quantum physics and many parts of mathematics, including Fourier analysis, several complex variables, geometry, and topology. The geometry structure of stratified groups is so good that it inherits a lot of analysis properties from the Euclidean spaces. Apart from this, the difference between the geometry structures of Euclidean spaces and stratified groups makes the study of function spaces on them more complicated. However, the study of variable Lebesgue spaces on stratified groups is quite a few, which makes it deserve a further investigation.

On the other hand, the commutator is a class of important non-convolution-type operators in harmonic analysis. This kind of operator plays an important role in the study of partial differential equations, and its boundedness can be used to characterize certain function spaces. A classical result of Coifman et al. [9] studied the $L^p(R^n)$ boundedness of linear commutators generated by the Calderón–Zygmund singular integrals operator and BMO functions. Janson [22] and Uchiyama [33] established characterization of BMO by the commutators of singular integral operators. The research on the commutator has been paid much attention and has fruitful results (see, for example, [6–8, 24, 25]). In addition, many authors focus on the research of the commutator in various settings. Lu et al. [26] obtained the boundedness of commutators generated by linear operators and BMO functions in the weighted Lebesgue spaces on homogeneous groups. Guliyev et al. [19] studied the fractional integral operator I_α on stratified groups in the weighted Lebesgue spaces and obtained the boundedness of the fractional integral operator I_α. In [37], the authors studied the sharp estimates for the multilinear commutators related to the singular integral operator on the spaces of homogenous type. For other works about the commutator in various settings, see, for example, [2–4, 11, 17, 20].

In this paper, on stratified groups we use the technique of [13, 15, 27] and [32] to obtain the boundedness of commutators of the multilinear fractional integrals and multilinear Calderón–Zygmund singular integrals in variable Lebesgue spaces.

2 The Preliminaries

Firstly, we recall some preliminaries concerning stratified groups. We refer the reader to [16]. A Lie group G is called stratified if it is nilpotent, connected, and simple connected, and its Lie algebra g is endowed with a vector space decomposition $g = \bigoplus_{i=1}^{m} V_k$ such that $[V_1, V_k] = V_{k+1}$, for $1 \le k < m$ and $[V_1, V_m] = 0$. As usual G is identified with its Lie algebra g through the exponential map. And the exponential map is a diffeomorphism from g to G and the bi-invariant Hear measure of G is induced by the Lebesgue measure of its Lie algebra g. Let $X_1 \in V_1, X_2 \in V_2, \cdots, X_m \in V_m$. If G is stratified, then its Lie algebra g admits a family of dilations, namely $\delta_r\left(\sum_{i=1}^{m} X_i\right) = \sum_{i=1}^{m} r^i X_i$, where $r > 0$. In this paper, we use Q to denote the homogenous dimension of G, y^{-1} is the inverse of y, $y^{-1}x$ denotes the group multiplication of y^{-1} by x and the group identity of G will be referred to as the origin denote by e. A homogenous norm on G is a continuous function $x \to \rho(x)$ from G to $[0, \infty)$, which is C^∞ on $G\backslash\{0\}$ and satisfies $\rho(x^{-1}) = \rho(x)$, $\rho(\delta_t x) = t\rho(x)$ for all $x \in G$, $t > 0$, and $\rho(e) = 0$. Moreover, there exists a constant $c_0 \geq 1$ such that $\rho(xy) \leq c_0(\rho(x) + \rho(y))$ for all $x, y \in G$. With this norm, we define the G ball centered at x with radius r by $B(x,r) = \{y \in G : \rho(y^{-1}x) < r\}$, let $B_r = B(e,r) = \{y \in G : \rho(y) < r\}$ be the open ball centered at e with radius r.

Secondly, we recall the definition of variable Lebesgue spaces on stratified groups. The function $p(\cdot) : G \to (0, \infty)$ is called the variable exponent. For a measurable subset $E \subset G$, let $p^+(E) = \sup_{x\in E} p(x)$, $p^-(E) = \inf_{x\in E} p(x)$. For conciseness, we abbreviate $p^+(G)$ and $p^-(G)$ to p^+ and p^-. Let $\mathscr{P}_0(G)$ be the set of measurable function $p(\cdot) : G \to (0, \infty)$ such that $0 < p^- \leq p^+ < \infty$. Let $\mathscr{P}_1(G)$ be the set of measurable function $p(\cdot) : G \to [1, \infty)$ such that $1 < p^- \leq p^+ < \infty$. For a measurable function $f : G \to R$, we define the norm

$$\|f\|_{p(\cdot)} = \inf\left\{\lambda > 0 : \int_G \left(\frac{|f(x)|}{\lambda}\right)^{p(x)} dx \leq 1\right\}$$

The variable exponent Lebesgue spaces $L^{p(\cdot)}$ consist of those measurable functions $f : G \to R$ for which $\|f\|_{p(\cdot)} < \infty$. According to [1], when $p^- \geq 1$, $L^{p(\cdot)}$ is a Banach space and by the unit ball property we know that $\|f\|_{p(\cdot)} \leq 1$ if and only if $\int_G |f(x)|^{p(x)} dx \leq 1$.

Next we define the conditions on the exponent [1]. Let $\Omega \subset G$. We say that $p : \Omega \to R$ is locally log-Hölder continuous in Ω if there exists $c_1 > 0$ such that

$$|p(x) - p(y)| \leq \frac{c_1}{\log(e + 1/\rho(x^{-1}y))}$$

for all $x, y \in G$. We say that p satisfies the log-Hölder decay condition with basepoint $x_0 \in G$ if there exist $p_\infty \in R$ and a constant $c_2 > 0$ such that

$$|p(x) - p_\infty| \leq \frac{c_2}{\log(e + \rho(x_0^{-1}x))}$$

for all $x \in G$. We say that p is log-Hölder continuous in G if both conditions are satisfied. We define a class of exponent p whose reciprocal is log-Hölder continuous:

$$P_d^{\log}(\Omega) = \left\{p : \Omega \to [1, \infty) \mid \frac{1}{p} \text{ is log-Hölder continuous}\right\}$$

Then we give some notations. Here and hereafter p' will always denote the conjugate of p. And $d\vec{y}$ will denote $\prod_{i=1}^m dy_i$. $\mathscr{D}$ denotes the set of all C^∞ functions with compact support. Let $\mathscr{B}(G)$ be the set of $p(\cdot) \in \mathscr{P}_1(G)$ such that the maximal operator M is bounded on $L^{p(\cdot)}(G)$.

1. $\mathbf{A_p}$ Weight

A weight function $\omega(x) > 0$ belongs to the class $A_p(G)(1 < p < \infty)$ [18], if

$$\sup \left(\frac{1}{|B|}\int_B \omega(x)dx\right)\left(\frac{1}{|B|}\int_B \omega(x)^{1-p'}dx\right)^{p-1} < \infty$$

where the supremum is taken over all balls $B \subset G$. $\omega \in A_1(G)$ if there exists $c > 0$ such that

$$\frac{1}{|B|}\int_B \omega(x)dx \leq c \inf_{x\in B} ess\ \omega(x)$$

for every ball $B \subset G$. In fact, if $\omega \in A_p(G)$ for some $p \in (1, \infty)$, then there is $\epsilon > 0$ such that $\omega \in A_{p-\epsilon}(G)$ and $\omega \in A_{p_1}(G)$ for any $p_1 \geq p \geq 1$. Let $A_\infty(G) = \cup_{1\leq p<\infty} A_p(G)$.

2. About Maximal Function

Let $0 \leq \alpha < Q$ and $f : G \to R$ is a locally integrable function. The fractional maximal function [18] is defined by

$$M_\alpha f(x) = \sup_{x\in B} \frac{1}{|B|^{1-\alpha/Q}}\int_B |f(y)|dy$$

where the supremum is taken over all $B \subset G$.
In fact, $M_0 f(x)$ is the classical Hardy-Littlewood maximal function Mf. We consider the closely related sharp maximal function operator $M^\sharp$ defined by

$$M^\sharp f(x) = \sup_{x\in B} \frac{1}{|B|}\int_B |f - f_B|dx$$

where $f_B = \frac{1}{|B|}\int_B f(x)dx$. For $\delta > 0$, we define the δ sharp function $M_\delta^\sharp$ as $M_\delta^\sharp f = (M^\sharp(|f|^\delta))^{1/\delta}$ and $M_\delta(f) = (M(|f|^\delta))^{1/\delta}$. Further, let $\omega \in A_\infty$, then for all $1 < p < \infty$ [18], the following inequality holds:

$$\int_G (Mf(x))^p\omega(x)dx \leq C\int_G (M^\sharp f(x))^p\omega(x)dx$$

3. BMO Space [16]

Suppose that f is a locally integrable function on G and B is a ball, we set $f_B = \frac{1}{|B|}\int_B f(y)dy$. Define

$$BMO(G) = \{f \in L^1_{loc}(G) : \|f\|_* < \infty\}$$

where

$$\|f\|_* = \sup_B \frac{1}{|B|}\int_B |f(x) - f_B|dx$$

We now formulate some remarks about $BMO(G)$ [4, 16, 18].

(i) There are constants $C_1, C_2 > 0$ such that for every $f \in BMO(G)$, every ball B, and every $\alpha > 0$

$$|\{x \in B : |f(x) - f_B| > \alpha\}| \le C_1|B|\exp^{-C_2\alpha/\|f\|_*}$$

(ii) If $f \in BMO(G)$, then for $1 < p < \infty$, the following inequality holds:

$$\left(\frac{1}{|B|}\int_B |f(x) - f_B|^p dx\right)^{1/p} \le C\|f\|_*$$

(iii) By the definition of BMO and the sharp maximal function, if $f \in L^1_{loc}(G)$, then

$$f \in BMO(G) \Longleftrightarrow M^\sharp f \in L^\infty(G)$$

(iv) If $f \in BMO(G)$, then

$$\|f\|_* \sim \sup_{x\in B}\inf_{c\in C}\frac{1}{|B|}\int_B |f(y) - c|dy$$

And then, we give the definitions of multilinear fractional integral operator, multilinear Calderón–Zygmund operator and their commutators. For any $1 \le j \le m$, we define the commutator of multilinear integral operator by

$$[b, T]_j(\vec{f})(x) = bT(\vec{f})(x) - T(f_1, \cdots, bf_j, \cdots, f_m)(x)$$

where b is a locally integral function and T is an m-linear integral operator. According to the definition of the fractional integral operator I_α on G [19] and the classical multilinear fractional integral operator, we can define the multilinear fractional integral operator on stratified groups by

$$I_\alpha(\vec{f})(x) = \int_{G^m} \frac{\prod_{i=1}^m f_i(y_i)}{(\sum_{i=1}^m \rho(y_i^{-1}x))^{Qm-\alpha}} d\vec{y}$$

Then $[b, I_\alpha]_j(\vec{f})$ is defined by

$$[b, I_\alpha]_j(\vec{f})(x) = \int_{G^m} \frac{(b(x) - b(y_j))\prod_{i=1}^m f_i(y_i)}{(\sum_{i=1}^m \rho(y_i^{-1}x))^{Qm-\alpha}} d\vec{y}$$

We say that T is a Calderón–Zygmund operator on G [4] if the following conditions are satisfied:

(i) $T : L^p(G) \to L^p(G)$ is linear and continuous for every $p \in (1, \infty)$;
(ii) there exists a measurable function $K : G \times G \to R$ such that for $f \in \mathscr{D}$, for a.e $x \notin supp(f)$

$$T(f)(x) = \int_G K(x, y) f(y) dy$$

(iii) the kernel K satisfies the following pointwise Hörmander condition, namely there exists positive constant $C > 0$, $\beta > 0$ and $M > 1$ such that

$$|K(x_0, y) - K(x, y)| \leq C \frac{\rho(x_0^{-1}x)^\beta}{|B(x_0, 2\rho(x_0^{-1}y))|\rho(x_0^{-1}y)^\beta}$$

holds for every $x_0 \in G, r > 0, x \in B(x_0, r), y \in G\backslash B(x_0, Mr)$.
(iv) the kernel K also satisfies the inequality

$$|K(x, y)| \leq \frac{C}{|B(x, \rho(x^{-1}y))|}$$

Similarly, we define the multilinear Calderón–Zygmund operator on G. We say that T is an m-linear Calderón–Zygmund operator on G, if $T : L^{p_1}(G) \times \cdots \times L^{p_m}(G) \longrightarrow L^p(G)$ for some $1 < p_1, \cdots, p_m < \infty$ and $\frac{1}{p} = \sum_{i=1}^m \frac{1}{p_i}$, and for all $f_i \in \mathscr{D}$ and all $x \notin \cap_{i=1}^m supp(f_i)$, we have

$$T(\vec{f})(x) = \int_{G^m} K(x, y_1, \cdots, y_m) \prod_{i=1}^m f_i(y_i) d\vec{y}$$

where K is a locally integral function defined on $(G \times G^m)\backslash\{(x, y_1, \cdots, y_m) : x = y_1 = \cdots = y_m\}$ and satisfies the following properties:

$$|K(y_0, y_1, \cdots, y_m)| \leq \frac{C}{(\sum_{k,l=0}^m \rho(y_k^{-1}y_l))^{Qm}}$$

and

$$|K(y_0, \cdots, y_j, \cdots, y_m) - K(y_0, \cdots, y_j', \cdots, y_m)| \leq \frac{C\rho(y_j^{-1}y_j')^\beta}{(\sum_{k,l=0}^m \rho(y_k^{-1}y_l))^{Qm+\beta}}$$

where $C > 0, \beta > 0, 0 \leq j \leq m$ and $\rho(y_j^{-1}y_j') \leq \frac{1}{2}\max_{0\leq k\leq m} \rho(y_j^{-1}y_k)$.

Finally, we recall the definition of the space $Osc_{\exp L^r}(G)$ and the definition of multilinear commutator of the singular integral operator [37]. For $r \geq 1$, let $\|b\|_{osc_{\exp L^r}}(G) = \sup_B \|b - b_B\|_{\exp L^r, B}$, where

$$\|b\|_{\exp L^r, B} = \inf_B \left\{ \lambda > 0 : \frac{1}{|B|} \int_B (\exp(|b(x)|/\lambda)^r - 1) dx \leq 1 \right\}$$

The space $Osc_{\exp L^r}(G)$ is defined by

$$Osc_{\exp L^r}(G) = \{b \in L^1_{loc} : \|b\|_{osc_{\exp L^r}} < \infty\}$$

It is obvious that $Osc_{\exp L^r}(G)$ coincides with the $BMO(G)$ space if $r = 1$. For $r_j > 0, b_j \in Osc_{\exp L^{r_j}}(G)$ for $j = 1, 2, \cdots m$, let $\|\vec{b}\| = \prod_{j=1}^{m} \|b_j\|_{Osc_{\exp L^{r_j}}}$.

The multilinear commutator of the singular integral operator is defined by

$$T_{\vec{b}}(f)(x) = \int_G \prod_{i=1}^{m} (b_j(x) - b_j(y)) K(x, y) f(y) dy$$

In this section, we give some basic properties of variable Lebesgue spaces on stratified groups and some important lemmas we need in the paper.

Lemma 2.1 *Suppose $q(\cdot), r(\cdot) \geq 1$ and for all $x \in G$, define $p(\cdot)$ by*

$$\frac{1}{p(x)} = \frac{1}{q(x)} + \frac{1}{r(x)}$$

Then for all $f \in L^{q(\cdot)}$ and $g \in L^{r(\cdot)}$, the following inequality holds:

$$\|fg\|_{p(\cdot)} \leq C \|f\|_{q(\cdot)} \|g\|_{r(\cdot)}$$

Proof Since the proof is similar to the corresponding one in [10], we omit the proof. □

Corollary 2.2 *Given exponent functions $p_i(\cdot) \in \mathscr{P}_1(G), i = 1, 2, \cdots m$, and for all $x \in G$, define $p(\cdot)$ by*

$$\frac{1}{p(x)} = \sum_{i=1}^{m} \frac{1}{p_i(x)}$$

Then for all $f_i \in L^{p_i(\cdot)}(G), i = 1, 2, \cdots m$, the following inequality holds:

$$\left\| \prod_{i=1}^{m} f_i \right\|_{p(\cdot)} \leq C \prod_{i=1}^{m} \|f_i\|_{p_i(\cdot)}$$

Lemma 2.3 ([1]) *If $p \in P_d^{\log}(G)$ with $p^- > 1$, then there exists $C > 0$ depending on p such that for all $f \in L^{p(\cdot)}(G)$, the following inequality holds:*

$$\|Mf\|_{p(\cdot)} \le C\|f\|_{p(\cdot)}$$

Corollary 2.4 *If $p(\cdot)$ is log-Hölder and $p \in \mathscr{P}_1(G)$, then there exists $C > 0$ depending on p such that for all $f \in L^{p(\cdot)}(G)$, the following inequality holds:*

$$\|Mf\|_{p(\cdot)} \le C\|f\|_{p(\cdot)}$$

Proof Corollary 2.4 follows immediately from the fact that $p(\cdot)$ is log-Hölder continuous is equivalent to $\frac{1}{p(\cdot)}$ is log-Hölder continuous and Lemma 2.3. □

Lemma 2.5 ([11]) *Given $0 \le \alpha < Q$, let $p(\cdot) \in \mathscr{P}_1(G)$ and be log-Hölder continuous, for each $x \in G$, define $q(\cdot)$ pointwise by*

$$\frac{1}{p(x)} - \frac{1}{q(x)} = \frac{\alpha}{Q}$$

Then there exists a constant C such that for all $f \in L^{p(\cdot)}(G)$, the following inequality holds:

$$\|M_\alpha f\|_{q(\cdot)} \le C\|f\|_{p(\cdot)}$$

Hereafter $\mathscr{F}$ will denote a family of ordered pairs of non-negative, measurable function (f, g).

Lemma 2.6 *Given a family $\mathscr{F}$ and an open set $\Omega \subset G$, assume that for some p_0 and q_0, $0 < p_0 \le q_0 < \infty$ and every weight $\omega \in A_1$*

$$\left(\int_\Omega f(x)^{q_0}\omega(x)dx\right)^{\frac{1}{q_0}} \le C\left(\int_\Omega g(x)^{p_0}\omega(x)^{p_0/q_0}dx\right)^{\frac{1}{p_0}} \qquad (f, g) \in \mathscr{F}$$

Given $p(\cdot) \in \mathscr{P}_0(\Omega)$ such that $p_0 < p^- \le p^+ < p_0q_0/(q_0 - p_0)$. For $\forall x \in \Omega$, define the function $q(\cdot)$ by

$$\frac{1}{p(x)} - \frac{1}{q(x)} = \frac{1}{p_0} - \frac{1}{q_0}$$

If $(q(x)/q_0)' \in \mathscr{B}(\Omega)$, then for all $(f, g) \in \mathscr{F}$ such that $f \in L^{q(\cdot)}(\Omega)$

$$\|f\|_{q(\cdot),\Omega} \le C\|g\|_{p(\cdot),\Omega}$$

Proof Fix $p(\cdot) \in \mathscr{P}_0(\Omega)$ such that $p^- > p_0$ and let $\bar{p}(x) = \frac{p(x)}{p_0}$. Let $\bar{q}(x) = \frac{q(x)}{q_0}$, since $\bar{q}' \in \mathscr{B}(\Omega)$, then there exists a constant $A > 0$ such that $\|Mf\|_{\bar{q}'(\cdot),\Omega} \leq A\|f\|_{\bar{q}'(\cdot),\Omega}$. Define a new operator $\mathfrak{R}$ on $L^{\bar{q}'(\cdot)}(\Omega)$ by

$$\mathfrak{R}h(x) = \sum_{k=0}^{\infty} \frac{M^k h(x)}{2^k A^k}$$

where for $k \geq 1$, $M^k = M \circ M \circ \cdots \circ M$ denotes k interactions of the maximal operator, and M^0 is the identity operator. From the definition we have:

(i) if h is non-negative, then $h(x) \leq \mathfrak{R}h(x)$;
(ii) $\|\mathfrak{R}h\|_{\bar{q}'(\cdot),\Omega} \leq 2\|h\|_{\bar{q}'(\cdot),\Omega}$;
(iii) for every $x \in \Omega$, $M(\mathfrak{R}h)(x) \leq 2A\mathfrak{R}h(x)$, so $\mathfrak{R}h \in A_1$ with a constant that does not depend on h.

Before giving the following proof, we state two facts about variable Lebesgue spaces.

First, if $p(\cdot), q(\cdot) \in \mathscr{P}_0(\Omega)$ and $\frac{p(x)}{q(x)} = r$, then it follows from the definition of the norm that

$$\|f\|^r_{p(\cdot),\Omega} = \||f|^r\|_{q(\cdot)}$$

Second, we have the generalized Hölder's inequality

$$\int_\Omega |f(x)g(x)|dx \leq C\|f\|_{p(\cdot),\Omega}\|g\|_{p'(\cdot),\Omega}$$

and

$$\|f\|_{p(\cdot),\Omega} \leq \sup_g \left| \int_\Omega f(x)g(x)dx \right| \leq C\|f\|_{p(\cdot),\Omega}$$

where the supremum is taken over all $g \in L^{p'}(\Omega)$ such that $\|g\|_{p'(\cdot),\Omega} = 1$.

Thus we have

$$\|f\|^{q_0}_{q(\cdot),\Omega} = \|f^{q_0}\|_{\bar{q}(\cdot),\Omega} \leq \sup \int_\Omega f(x)^{q_0} h(x)dx$$

where the sup is taken over all non-negative $h \in L^{\bar{q}'(\cdot)}$ such that $\|h\|_{\bar{q}'(\cdot)} = 1$.

In order to complete the proof, it will be sufficient to show that

$$\int_\Omega f(x)^{q_0} h(x)dx \leq C\|g\|^{q_0}_{p(\cdot)}$$

Applying (i) above and the hypothesis, we have

$$\begin{aligned}\int_\Omega f(x)^{q_0}h(x)dx &\le \int_\Omega f(x)^{q_0}(\Re h)(x)dx\\ &\le C\left(\int_\Omega g(x)^{p_0}(\Re h)(x)^{p_0/q_0}dx\right)^{\frac{q_0}{p_0}}\\ &\le C\|g^{p_0}\|_{\bar{p}(\cdot)}^{q_0/p_0}\|(\Re h)^{p_0/q_0}\|_{\bar{p}'(\cdot)}^{q_0/p_0}\\ &\le C\|g\|_{p(\cdot)}^{q_0}\|(\Re h)^{p_0/q_0}\|_{\bar{p}'(\cdot)}^{q_0/p_0}\end{aligned}$$

So in order to complete the proof, we only need to show that $\|(\Re h)^{p_0/q_0}\|_{\bar{p}'(\cdot)}^{\frac{q_0}{p_0}}$ is bounded and independent of h. We know that $\frac{1}{p(x)}-\frac{1}{q(x)}=\frac{1}{p_0}-\frac{1}{q_0}$, it is easy to verify that $\bar{p}'(x)=\frac{p(x)}{p(x)-p_0}=\frac{q_0}{p_0}\frac{q(x)}{q(x)-q_0}=\frac{q_0}{p_0}\bar{q}'(x)$. Therefore

$$\|(\Re h)^{p_0/q_0}\|_{\bar{p}'(\cdot)}^{q_0/p_0}=\|\Re h\|_{\bar{q}'(\cdot)}\le C\|h\|_{\bar{q}'(\cdot)}\le C$$

□

Corollary 2.7 *Given a family $\mathscr{F}$ and an open set $\Omega\subset G$, suppose that for some p_0, $0<p_0<\infty$ and for every $\omega\in A_1$*

$$\int_\Omega f(x)^{p_0}\omega(x)dx\le C\int_\Omega g(x)^{p_0}\omega(x)dx \qquad (f,g)\in\mathscr{F}$$

Let $p(\cdot)\in\mathscr{P}_0(\Omega)$ be such that $p_0<p^-$, and $(p(\cdot)/p_0)'\in\mathscr{B}(\Omega)$. Then for all $(f,g)\in\mathscr{F}$ such that $f\in L^{p(\cdot)}(\Omega)$

$$\|f\|_{p(\cdot),\Omega}\le C\|g\|_{p(\cdot),\Omega}$$

Since the proof of the following two lemmas is similar to the corresponding ones on Euclidean spaces [14, 15], we omit it.

Lemma 2.8 *Given a family $\mathscr{F}$ and an open set $\Omega\subset G$, assume that for some p_0, $0<p_0<\infty$, for every $\omega\in A_\infty$ and for all $(f,g)\in\mathscr{F}$, the following inequality holds:*

$$\int_\Omega f(x)^{p_0}\omega(x)dx\le C\int_\Omega g(x)^{p_0}\omega(x)dx \tag{2.1}$$

Then for all $0<p<\infty$, $\omega\in A_\infty$ and $(f,g)\in\mathscr{F}$, we have

$$\int_\Omega f(x)^{p}\omega(x)dx\le C\int_\Omega g(x)^{p}\omega(x)dx \tag{2.2}$$

Lemma 2.9 *Given a family $\mathscr{F}$ and an open set $\Omega \subset G$, assume that for some p_0, $1 < p_0 < \infty$, for every $\omega \in A_{p_0}$ and for all $(f, g) \in \mathscr{F}$, (2.1) holds. Then for all $1 < p < \infty$, $\omega \in A_p$ and $(f, g) \in \mathscr{F}$, (2.2) holds.*

Lemma 2.10 *Given a family $\mathscr{F}$ and an open set $\Omega \subset G$, assume that for some $p_0, 0 < p_0 < \infty$ and for every $\omega \in A_\infty$, (2.1) holds. Let $p(\cdot) \in \mathscr{P}_0(\Omega)$ be such that there exists $0 < p_1 < p^-$ with $(p(\cdot)/p_1)' \in \mathscr{B}(\Omega)$. Then for all $(f, g) \in \mathscr{F}$, we have*

$$\|f\|_{p(\cdot),\Omega} \le C\|g\|_{p(\cdot),\Omega}$$

Proof Since (2.1) holds for some p_0 and every $\omega \in A_\infty$, then by Lemma 2.8, for all $0 < p_1 < \infty$, $\omega \in A_\infty$ and $(f, g) \in \mathscr{F}$, we have

$$\int_\Omega f(x)^{p_1}\omega(x)dx \le C\int_\Omega g(x)^{p_1}\omega(x)dx$$

Then applying Corollary 2.7 with p_1 in place of p_0, we complete the proof. □

Corollary 2.11 *Given a family $\mathscr{F}$ and an open set $\Omega \subset G$, assume that for some $p_0, 1 < p_0 < \infty$ and for every $\omega \in A_{p_0}$, (2.1) holds. Let $p(\cdot) \in \mathscr{P}_1(\Omega)$ be such that there exists $1 < p_1 < p^-$ with $(p(\cdot)/p_1)' \in \mathscr{B}(\Omega)$. Then for all $(f, g) \in \mathscr{F}$, we have*

$$\|f\|_{p(\cdot),\Omega} \le C\|g\|_{p(\cdot),\Omega}$$

By Lemma 2.10 with the pairs $(Mf, M^\sharp f)$, we can obtain

Corollary 2.12 *Let $p(\cdot) \in \mathscr{P}_0(\Omega)$ be such that there exists $0 < p_1 < p^-$ with $(p(\cdot)/p_1)' \in \mathscr{B}(\Omega)$. Then for all $(Mf, M^\sharp f) \in \mathscr{F}$*

$$\|Mf\|_{p(\cdot),\Omega} \le C\|M^\sharp f\|_{p(\cdot),\Omega}$$

Lemma 2.13 ([37]) *If $1 < p < \infty$ and $\omega \in A_p$, then*

$$\|T_{\vec{b}}(f)\|_{L^p(\omega)} \le C\|\vec{b}\|\|f\|_{L^p(\omega)}$$

3 The Main Results and Proofs

Theorem 3.1 *Suppose that $b \in L^1_{loc}(G)$, $0 < \alpha < Qm$, $p_i(\cdot) \in \mathscr{P}_1$ and is log-Hölder continuous, $i = 1, 2, \cdots, m$. Let $q(\cdot)$ satisfy*

$$\sum_{i=1}^m \frac{1}{p_i(x)} - \frac{\alpha}{Q} = \frac{1}{q(x)} < 1$$

For any $1 \leq j \leq m$, *then* $[b, I_\alpha]_j$ *is bounded from* $L^{p_1(\cdot)}(G) \times L^{p_2(\cdot)}(G) \times \cdots \times L^{p_m(\cdot)}(G) \longrightarrow L^{q(\cdot)}(G)$ *if* $b \in BMO$, *namely*

$$\|[b, I_\alpha]_j(\vec{f})\|_{q(\cdot)} \leq C\|b\|_* \prod_{i=1}^{m} \|f_i\|_{p_i(\cdot)}$$

Theorem 3.2 *Suppose that* $b \in L^1_{loc}(G)$, $p_i(\cdot) \in \mathscr{P}_1$ *and is log-Hölder continuous,* $i = 1, 2, \cdots, m$. T *is an m-linear Calderón–Zygmund operator and* $T : L^1 \times \cdots \times L^1 \to L^{1/m,\infty}$. *Let* $p(\cdot)$ *satisfy*

$$\sum_{i=1}^{m} \frac{1}{p_i(x)} = \frac{1}{p(x)} < 1$$

Then for all $1 \leq j \leq m$, $[b, T]_j$ *is bounded from* $L^{p_1(\cdot)}(G) \times L^{p_2(\cdot)}(G) \times \cdots \times L^{p_m(\cdot)}(G) \longrightarrow L^{p(\cdot)}(G)$ *if* $b \in BMO$, *namely*

$$\|[b, T]_j(\vec{f})\|_{p(\cdot)} \leq C\|b\|_* \prod_{i=1}^{m} \|f_i\|_{p_i(\cdot)}$$

Theorem 3.3 *Let* $b_i \in Osc_{\exp L^{r_i}}$ *and* $r_i \geq 1$, $1 \leq i \leq m$. *Then for* $p(\cdot) \in \mathscr{P}_1$ *and is log-Hölder continuous and* $\omega \in A_p$, *we have*

$$\|T_{\vec{b}}(f)\|_{p(\cdot)} \leq C \prod_{j=1}^{m} \|b_j\|_{Osc_{\exp L^{r_j}}} \|f\|_{p(\cdot)}$$

Firstly, we give the proof of Theorem 3.1.

Proof First of all, we prove the following sharp maximal estimate

$$M^\sharp([b, I_\alpha]_j(\vec{f}))(x) \leq C\|b\|_* \left[(M(|I_\alpha(\vec{f})|^r)(x))^{1/r} + \prod_{i=1}^{m} (M_{\alpha_i s_i}(|f_i|^{s_i})(x))^{1/s_i} \right]$$

where $\sum_{i=1}^{m} \alpha_i = \alpha, 0 < \alpha_i < Q, 1 < s_i < p_i^-, 1 < r < q^-$.

Fix a ball B, we put $f_j = f_j^0 + f_j^\infty$, where $f_j^0 = f_j \chi_{2c_0B}$. Then we divide $[b, I_\alpha]_j(\vec{f})(x)$ into four parts, namely

$$\begin{aligned}
[b, I_\alpha]_j(\vec{f})(x) &= \int_{G^m} \frac{(b(x) - b_B + b_B - b(y_j)) \prod_{i=1}^m f_i(y_i)}{(\sum_{i=1}^m \rho(y_i^{-1}x))^{Qm-\alpha}} d\vec{y} \\
&= (b(x) - b_B) I_\alpha(\vec{f})(x) - I_\alpha(f_1^0, \cdots, (b - b_B) f_j^0, \cdots, f_m^0)(x) \\
&- I_\alpha(f_1^\infty, \cdots, (b - b_B) f_j^\infty, \cdots, f_m^\infty)(x) \\
&- \sum I_\alpha(f_1^{r_1}, \cdots, (b - b_B) f_j^{r_j}, \cdots, f_m^{r_m})(x) \\
&:= A_1(x) - A_2(x) - A_3(x) - A_4(x)
\end{aligned}$$

where in the last sum each $r_i = 0$ or ∞ and in each term there is at least one $r_j = 0$ and $r_l = \infty$.

Firstly, we consider A_1. By Hölder's inequality, we get

$$\begin{aligned}
\frac{1}{|B|} \int_B |A_1(z)| dz &\leq \left(\frac{1}{|B|} \int_B |b(z) - b_B|^{r'} dz \right)^{1/r'} \left(\frac{1}{|B|} \int_B |I_\alpha(\vec{f})(z)|^r dz \right)^{\frac{1}{r}} \\
&\leq C \|b\|_* (M(|I_\alpha(\vec{f})|^r)(x))^{1/r}
\end{aligned}$$

According to [19], we know the fact that: if $1 < p < \frac{Q}{\alpha}, 0 < \alpha < Q$, then condition $\frac{1}{p} - \frac{1}{q} = \frac{\alpha}{Q}$ is necessary and sufficient for the boundedness of I_α from $L^p(G)$ to $L^q(G)$. We choose $r, \beta_i > 1$ such that $r\beta_i = s_i$. Thus $1 < \beta_i < p_i^-$. So there exists $u > 1$ such that $\frac{1}{u} = \sum_{i=1}^m \frac{1}{\beta_i} - \frac{\alpha}{Q}$. Let $\frac{1}{u} = \sum_{i=1}^m \frac{1}{u_i}$, then $\frac{1}{\beta_i} - \frac{1}{u_i} = \frac{\alpha_i}{Q}$. Then by Hölder's inequality and the above fact, we have

$$\begin{aligned}
\frac{1}{|B|} \int_B |A_2(z)| dz &\leq \left(\frac{1}{|B|} \int_B |I_\alpha(f_1^0, \cdots, (b - b_B) f_j^0, \cdots, f_m^0)(z)|^u dz \right)^{1/u} \\
&\leq \frac{1}{|B|^{\frac{1}{u}}} \prod_{i \neq j} \left(\int_B [I_{\alpha_i}(|f_i^0|)(z)]^{u_i} dz \right)^{\frac{1}{u_i}} \\
&\times \left(\int_B [I_{\alpha_j}(|(b - b_B) f_j^0|)(z)]^{u_j} dz \right)^{\frac{1}{u_j}} \\
&\leq \frac{C}{|B|^{\frac{1}{u}}} \prod_{i \neq j} \left(\int_B |f_i^0|^{\beta_i} dz \right)^{\frac{1}{\beta_i}} \left(\int_B |(b - b_B) f_j^0|^{\beta_j} dz \right)^{\frac{1}{\beta_j}} \\
&\leq C \|b\|_* \prod_{i=1}^m (M_{\alpha_i s_i}(|f_i|^{s_i})(x))^{1/s_i}
\end{aligned}$$

Let x_0 be the center of $B = B(x_0, r_0)$, then for $x \in B$ and $y_i \in (2c_0B)^c$, we have $\rho(y_i^{-1}x) \sim \rho(y_i^{-1}x_0)$. Applying the Hölder's inequality with exponent s_j and its conjugate s'_j, we get

$$
\begin{aligned}
&|I_\alpha(f_1^\infty, \cdots, (b-b_B)f_j^\infty, \cdots, f_m^\infty)(x) - I_\alpha(f_1^\infty, \cdots, (b-b_B)f_j^\infty, \cdots, f_m^\infty)(x_0)| \\
&\le C\int_{(G\setminus(2c_0B))^m} \frac{e(x_0^{-1}x)}{(\sum_{i=1}^m e(y_i^{-1}x_0))^{Qm-\alpha+1}}|b(y_j)-b_B|\prod_{i=1}^m |f_i(y_i)|d\vec{y} \\
&\le C\left(\int_{G\setminus(2c_0B)} \frac{r_0^{1/m}|b(y_j)-b_B|^{s_j'}}{e(y_j^{-1}x_0)^{Q+1/m}}dy_j\right)^{\frac{1}{s_j'}} \\
&\times \left(\int_{G\setminus(2c_0B)} \frac{r_0^{1/m}|f_j(y_j)|^{s_j}}{e(y_j^{-1}x_0)^{Q+1/m-\alpha_j s_j}}dy_j\right)^{\frac{1}{s_j}} \\
&\times \int_{(G\setminus(2c_0B))^{m-1}} \frac{r_0^{(m-1)/m}\prod_{i\neq j}|f_i(y_i)|}{(\sum_{i\neq j} e(y_i^{-1}x_0))^{Q(m-1)-\sum_{i\neq j}\alpha_i+(m-1)/m}}\prod_{i\neq j} dy_i \\
&\le C\left(\sum_{i=1}^\infty \int_{2^{i+1}c_0B} \frac{r_0^{1/m}|b(y_j)-b_B|^{s_j'}}{(2^i c_0 r_0)^{Q+1/m}}dy_j\right)^{1/s_j'} \\
&\times \left(\sum_{i=1}^\infty \int_{2^{i+1}c_0B} \frac{r_0^{1/m}|f_j(y_j)|^{s_j}}{(2^i c_0 r_0)^{Q+1/m-\alpha_j s_j}}dy_j\right)^{1/s_j} \\
&\times \int_{(G\setminus(2c_0B))^{m-1}} \frac{r_0^{(m-1)/m}\prod_{i\neq j}|f_i(y_i)|}{(\sum_{i\neq j} e(y_i^{-1}x_0))^{Q(m-1)-\sum_{i\neq j}\alpha_i+(m-1)/m}}\prod_{i\neq j} dy_i \\
&\le C\|b\|_*\prod_{i=1}^m (M_{\alpha_i s_i}(|f_i|^{s_i})(x))^{1/s_i}
\end{aligned}
$$

Now we consider A_4. Without loss of generality, we can assume that $r_1 = r_2 = \cdots = r_d = 0$ and $r_{d+1} = \cdots = r_m = \infty$. When $d+1 \le j \le m$, applying the Hölder's inequality, we can obtain

$$
\begin{aligned}
&|I_\alpha(f_1^{r_1}, \cdots, (b-b_B)f_j^{r_j}, \cdots, f_m^{r_m})(x) - I_\alpha(f_1^{r_1}, \cdots, (b-b_B)f_j^{r_j}, \cdots, f_m^{r_m})(x_0)| \\
&\le C\int_{G^m} \frac{r_0}{(\sum_{i=1}^m e(y_i^{-1}x_0))^{Qm-\alpha+1}}|b(y_j)-b_B|\prod_{i=1}^m |f_i^{r_i}|d\vec{y}
\end{aligned}
$$

$$\leq C\prod_{i=1}^{d}\int_{2c_0B}|f_i|dy_i\int_{(G\backslash(2c_0B))^{m-d}}\frac{r_0|b(y_j)-b_B|\prod_{i=d+1}^{m}|f_i|}{(\sum_{i=d+1}^{m}e(y_i^{-1}x_0))^{Qm-\alpha+1}}\prod_{i=d+1}^{m}dy_i$$

$$\leq C\|b\|_*\prod_{i=1}^{m}(M_{\alpha_i s_i}(|f_i|^{s_i})(x))^{1/s_i}$$

Repeating the same steps as we deal with the case when $d+1\leq j\leq m$, we also obtain that when $1\leq j\leq d$,

$$|I_\alpha(f_1^{r_1},\cdots,(b-b_B)f_j^{r_j},\cdots,f_m^{r_m})(x)-I_\alpha(f_1^{r_1},\cdots,(b-b_B)f_j^{r_j},\cdots,f_m^{r_m})(x_0)|$$

$$\leq C\|b\|_*\prod_{i=1}^{m}(M_{\alpha_i s_i}(|f_i|^{s_i})(x))^{1/s_i}$$

Then we can easily get the sharp maximal estimate.

Since $p_i(\cdot)$ is log-Hölder continuous and the fact that $p(\cdot)$ is log-Hölder continuous is equivalent to $\frac{1}{p(\cdot)}$ is log-Hölder continuous, it is easy to see that $q(\cdot)$ is log-Hölder continuous. Note that $1<r<q^-$, then M is of $(\frac{q(\cdot)}{r},\frac{q(\cdot)}{r})$. Thus if $b\in BMO$, we have

$$\|[b,I_\alpha]_j(\vec{f})\|_{q(\cdot)}\leq\|M[b,I_\alpha]_j(\vec{f})\|_{q(\cdot)}\leq\|M^\sharp([b,I_\alpha]_j(\vec{f}))\|_{q(\cdot)}$$
$$\leq C\|b\|_*\left(\|(M(|I_\alpha(\vec{f})|^r))^{1/r}\|_{q(\cdot)}+\left\|\prod_{i=1}^{m}(M_{\alpha_i s_i}(|f_i|^{s_i}))^{1/s_i}\right\|_{q(\cdot)}\right)$$

Now we are in a position to estimate I_α, we consider two cases.

$$I_\alpha(\vec{f})(x)=\int_{(2c_0B)^m}\frac{\prod_{i=1}^{m}f_i(y_i)}{(\sum_{i=1}^{m}\rho(y_i{}^{-1}x))^{Qm-\alpha}}d\vec{y}$$
$$+\int_{G^m\backslash(2c_0B)^m}\frac{\prod_{i=1}^{m}f_i(y_i)}{(\sum_{i=1}^{m}\rho(y_i{}^{-1}x))^{Qm-\alpha}}d\vec{y}$$
$$:=I+II$$

Now we want to obtain the estimate of I_α. Fix ϵ such that $\epsilon = \sum_{i=1}^m \epsilon_i$ and $0 < \epsilon_i < \min\{\alpha_i, Q - \alpha_i\}$. For I, we have

$$
\begin{aligned}
|I| &\le \int_{(2c_0B)^m} \frac{\prod_{i=1}^m |f_i(y_i)|}{(\sum_{i=1}^m \rho(y_i{}^{-1}x))^{Qm-\alpha}} d\vec{y} \\
&= \sum_{j=0}^{\infty} \int_{(2^{-j+1}c_0B)^m \setminus (2^{-j}c_0B)^m} \frac{\prod_{i=1}^m |f_i(y_i)|}{(\sum_{i=1}^m \rho(y_i{}^{-1}x))^{Qm-\alpha}} d\vec{y} \\
&\le C \sum_{j=0}^{\infty} \int_{(2^{-j+1}c_0B)^m} \frac{\prod_{i=1}^m |f_i(y_i)|}{(2^{-j}c_0r_0)^{Qm-\alpha}} d\vec{y} \\
&\le C|c_0B|^{\frac{\epsilon}{Q}} \prod_{i=1}^m M_{\alpha_i-\epsilon_i} f_i(x)
\end{aligned}
$$

Similarly, we get

$$
\begin{aligned}
|II| &\le \int_{G^m \setminus (2c_0B)^m} \frac{\prod_{i=1}^m |f_i(y_i)|}{(\sum_{i=1}^m \rho(y_i{}^{-1}x))^{Qm-\alpha}} d\vec{y} \\
&= \sum_{j=1}^{\infty} \int_{(2^{j+1}c_0B)^m \setminus (2^{j}c_0B)^m} \frac{\prod_{i=1}^m |f_i(y_i)|}{(\sum_{i=1}^m \rho(y_i{}^{-1}x))^{Qm-\alpha}} d\vec{y} \\
&\le C \sum_{j=1}^{\infty} \int_{(2^{j+1}c_0B)^m} \frac{\prod_{i=1}^m |f_i(y_i)|}{(2^{j}c_0r_0)^{Qm-\alpha}} d\vec{y} \\
&\le C|c_0B|^{\frac{-\epsilon}{Q}} \prod_{i=1}^m M_{\alpha_i+\epsilon_i} f_i(x)
\end{aligned}
$$

Let $|c_0B|^{\frac{2\epsilon}{Q}} = \frac{\prod_{i=1}^m M_{\alpha_i+\epsilon_i} f_i(x)}{\prod_{i=1}^m M_{\alpha_i-\epsilon_i} f_i(x)}$, then we obtain that

$$
|I_\alpha(\vec{f})(x)| \le C \left(\prod_{i=1}^m M_{\alpha_i+\epsilon_i} f_i(x) \right)^{1/2} \left(\prod_{i=1}^m M_{\alpha_i-\epsilon_i} f_i(x) \right)^{1/2}
$$

Let us now prove that

$$
\|I_\alpha(\vec{f})\|_{q(\cdot)} \le C \prod_{i=1}^m \|f_i\|_{p_i(\cdot)}
$$

when $\frac{1}{q(x)} = \sum_{i=1}^{m} \frac{1}{p_i(x)} - \frac{\alpha}{Q}$. Without loss of generality, we can assume $\|f_i\|_{p_i(\cdot)} = 1$. We recall that $\|f\|_{q(\cdot)} \leq C$ if and only if $\int_G |f(x)|^{q(x)}dx \leq C$. Since $q^+ < \infty$, in order to prove $\|I_\alpha(\vec{f})\|_{q(\cdot)} \leq C$, it will be sufficient to prove $\int_G |I_\alpha(\vec{f})(x)|^{q(x)}dx \leq C$. Define $r(\cdot) : G \to [1, \infty)$ by $r(x) = \frac{2}{\epsilon q(x)/Q+1}$. Then for all $x \in G$, we have

$$\sum_{i=1}^{m} \frac{1}{p_i(x)} - \frac{1}{r(x)q(x)/2} = \frac{\alpha - \epsilon}{Q}$$

$$\sum_{i=1}^{m} \frac{1}{p_i(x)} - \frac{1}{r(x)'q(x)/2} = \frac{\alpha + \epsilon}{Q}$$

By Hölder's inequality for variable L^p, we have

$$\begin{aligned} \int_G |I_\alpha(\vec{f})(x)|^{q(x)}dx &\leq C \int_G \left[\prod_{i=1}^{m} M_{\alpha_i+\epsilon_i} f_i(x)\right]^{\frac{q(x)}{2}} \left[\prod_{i=1}^{m} M_{\alpha_i-\epsilon_i} f_i(x)\right]^{\frac{q(x)}{2}} dx \\ &\leq C \left\| \left[\prod_{i=1}^{m} M_{\alpha_i+\epsilon_i} f_i\right]^{\frac{q(\cdot)}{2}} \right\|_{r'(\cdot)} \left\| \left[\prod_{i=1}^{m} M_{\alpha_i-\epsilon_i} f_i\right]^{\frac{q(\cdot)}{2}} \right\|_{r(\cdot)} \\ &:= CI_1 \times II_1 \end{aligned}$$

Now we will estimate I_1 and II_1. We can assume that each is greater than 1, since otherwise nothing need to be proved. It's easy to verify that $(\frac{r'(\cdot)q(\cdot)}{2})^- > 1$, then we can choose exponent function $s_i(x) \geq s_i^- > 1$ such that $\sum_{i=1}^{m} \frac{1}{s_i(x)} = \frac{1}{\frac{r'(\cdot)q(\cdot)}{2}}$ and $\frac{1}{p_i(x)} - \frac{1}{s_i(x)} = \frac{\alpha_i+\epsilon_i}{Q}$, where $i = 1, \cdots, m$ and $x \in G$.

$$\begin{aligned} I_1 &= \inf\left\{\lambda > 0 : \int_G \left[\frac{(\prod_{i=1}^{m} M_{\alpha_i+\epsilon_i} f_i)^{\frac{q(x)}{2}}}{\lambda}\right]^{r'(x)} dx \leq 1\right\} \\ &\leq \inf\left\{\lambda > 0 : \int_G \left[\frac{(\prod_{i=1}^{m} M_{\alpha_i+\epsilon_i} f_i)}{\lambda^{\frac{2}{q^+}}}\right]^{\frac{q(x)r'(x)}{2}} dx \leq 1\right\} \\ &\leq C \prod_{i=1}^{m} \|M_{\alpha_i+\epsilon_i} f_i\|_{s_i(\cdot)}^{q^+/2} \leq C \prod_{i=1}^{m} \|f_i\|_{p_i(\cdot)}^{q^+/2} \leq C \end{aligned}$$

Then repeating the same step as we deal with I_1, we can get $II_1 \leq C$. Thus

$$\|I_\alpha(\vec{f})\|_{q(\cdot)} \leq C\prod_{i=1}^{m}\|f_i\|_{p_i(\cdot)}$$

Therefore we obtain

$$\left\|\left[M(|I_\alpha(\vec{f})|^r)\right]^{1/r}\right\|_{q(\cdot)} \leq C\||I_\alpha(\vec{f})|^r\|_{\frac{q(\cdot)}{r}}^{1/r} \leq C\|I_\alpha(\vec{f})\|_{q(\cdot)} \leq C\prod_{i=1}^{m}\|f_i\|_{p_i(\cdot)}$$

Then we choose $q_i(x) > 1$ such that $\frac{1}{q(x)} = \sum_{i=1}^{m}\frac{1}{q_i(x)}$, and $\frac{1}{q_i(x)} = \frac{1}{p_i(x)} - \frac{\alpha_i}{Q}$. At the same time, we know $\frac{1}{q_i(x)/s_i} = \frac{1}{p_i(x)/s_i} - \frac{\alpha_i s_i}{Q}$, where $1 \leq i \leq m$. So we easily get

$$\left\|\prod_{i=1}^{m}(M_{\alpha_i s_i}(|f_i|^{s_i}))^{1/s_i}\right\|_{q(\cdot)} \leq C\prod_{i=1}^{m}\||f_i|^{s_i}\|_{p_i(\cdot)/s_i}^{1/s_i} \leq C\prod_{i=1}^{m}\|f_i\|_{p_i(\cdot)}$$

Thus we obtain

$$\|[b, I_\alpha]_j(\vec{f})\|_{q(\cdot)} \leq C\|b\|_*\prod_{i=1}^{m}\|f_i\|_{p_i(\cdot)}$$

□

Secondly, we give the proof of Theorem 3.2.

Proof Firstly we can obtain the following sharp maximal estimate

$$M^\sharp([b, T]_j(\vec{f}))(x) \leq C\|b\|_*\left[(M(|T(\vec{f})|^s)(x))^{1/s} + \prod_{i=1}^{m}(M(|f_i|^{s_i})(x))^{1/s_i}\right]$$

where $\frac{1}{s} = \sum_{i=1}^{m}\frac{1}{s_i} < 1$ and $1 < s_i < p_i^-$ for all $1 \leq i \leq m$. Fix a ball B, for $1 \leq j \leq m$, let $f_j = f_j^0 + f_j^\infty$, where $f_j^0 = f_j\chi_{4\sqrt{Q}c_0B}$, then

$$\begin{aligned}
[b, T]_j(\vec{f})(x) &= [b(x) - b_B]T(\vec{f})(x) - T(f_1, \cdots, (b - b_B)f_j, \cdots, f_m)(x) \\
&= (b - b_B)T(\vec{f})(x) - T(f_1^0, \cdots, (b - b_B)f_j^0, \cdots, f_m^0)(x) \\
&- T(f_1^\infty, \cdots, (b - b_B)f_j^\infty, \cdots, f_m^\infty)(x) \\
&- \sum T(f_1^{r_1}, \cdots, (b - b_B)f_j^{r_j}, \cdots, f_m^{r_m})(x) \\
&:= B_1 - B_2 - B_3 - B_4
\end{aligned}$$

where in the last sum each $r_i = 0$ or ∞ and in each term there is at least one $r_l = 0$ and $r_k = \infty$. According to the Hölder's inequality, we get

$$\frac{1}{|B|}\int_B |B_1(z)|dz \le \left(\frac{1}{|B|}\int_B |b(z)-b_B|^{s'}dz\right)^{1/s'}\left(\frac{1}{|B|}\int_B |T(\vec{f})(z)|^s dz\right)^{1/s}$$
$$\le C\|b\|_*\left(M(|T(\vec{f})|^s)(x)\right)^{1/s}$$

Similarly, we choose $1 < u, q$, and $q_i < \infty$ such that $uq_i = s_i$ and $\frac{1}{q} = \sum_{i=1}^{m}\frac{1}{q_i}$. Then by applying the boundedness of T and the Hölder's inequality, we obtain

$$\frac{1}{|B|}\int_B |B_2(z)|dz \le \left(\frac{1}{|B|}\int_B |T(f_1^0,\cdots,(b-b_B)f_j^0,\cdots,f_m^0)|^q dz\right)^{1/q}$$
$$\le C|B|^{-1/q}\prod_{i\neq j}\|f_i\chi_{4\sqrt{Q}c_0B}\|_{q_i}\|(b-b_B)\|_{q_ju'}\|f_j\chi_{4\sqrt{Q}c_0B}\|_{q_ju}$$
$$\le C\|b\|_*\prod_{i=1}^{m}\left(M(|f_i|^{s_i})(x)\right)^{\frac{1}{s_i}}$$

Let $B = B(x_0, r_0)$. If $y_i \notin 4\sqrt{Q}c_0B$, then $\rho(x_0^{-1}x) \le \frac{1}{2}\max_{1\le i\le m}\rho(y_i^{-1}x)$ for any $x \in B$. Then we have

$$|B_3(x_0) - B_3(x)| \le \int_{G^m}|K(x,\vec{y}) - K(x_0,\vec{y})||b(y_j)-b_B|\prod_{i=1}^{m}|f_i^\infty(y_i)|d\vec{y}$$
$$\le \int_{G\setminus(4\sqrt{Q}c_0B)}\frac{Cr_0^{\frac{\beta}{m}}|b(y_j)-b_B||f_j(y_j)|}{\rho(y_j^{-1}x)^{Q+\frac{\beta}{m}}}dy_j$$
$$\times\prod_{i\neq j}\int_{G\setminus(4\sqrt{Q}c_0B)}\frac{r_0^{\frac{\beta}{m}}|f_i(y_i)|}{\rho(y_i^{-1}x)^{Q+\frac{\beta}{m}}}dy_i$$
$$\le C\|b\|_*\prod_{i=1}^{m}\left(M(|f_i|^{s_i})(x)\right)^{\frac{1}{s_i}}$$

Finally we consider B_4. We abbreviate $(4\sqrt{Q}c_0B)^l \times (G\setminus(4\sqrt{Q}c_0B))^{m-l}$ to G_l. Without loss of generality, we can assume $r_1 = \cdots = r_l = 0$ and $r_{l+1} = \cdots =$

$r_m = \infty$. When $1 \le j \le l$, we can get

$$\frac{1}{|B|}\int_B |T(f_1^0, \cdots, (b-b_B)f_j^0, \cdots, f_l^0, f_{l+1}^\infty, \cdots, f_m^\infty)(z)|dz$$

$$\le \frac{C}{|B|}\int_B\int_{G_l} \frac{|f_1^0(y_1)\cdots(b-b_B)f_j^0(y_j)\cdots f_l^0(y_l)f_{l+1}^\infty(y_{l+1})\cdots f_m^\infty(y_m)|}{(\sum_{i=1}^m \rho(y_i^{-1}z))^{Qm}} d\vec{y}dz$$

$$\le \frac{C}{|B|}\int_B \Big(\int_{4\sqrt{Q}c_0B} \frac{|b(y_j)-b_B||f_j(y_j)|}{|4\sqrt{Q}c_0B|}dy_j \prod_{i=1, i\ne j}^{l} \int_{4\sqrt{Q}c_0B} \frac{|f_i(y_i)|}{|4\sqrt{Q}c_0B|}dy_i$$

$$\times |4\sqrt{Q}c_0B|^l \prod_{k=l+1}^{m} \int_{G\setminus(4\sqrt{Q}c_0B)} \frac{|f_k(y_k)|}{\rho(y_k^{-1}z)^{Q+\frac{Ql}{m-l}}}dy_k \Big) dz$$

$$\le C\|b\|_* \prod_{i=1}^m \big(M(|f_i|^{s_i})(x)\big)^{\frac{1}{s_i}}$$

In fact, when $l+1 \le j \le m$, we also obtain

$$\frac{1}{|B|}\int_B |T(f_1^0, \cdots, f_l^0, f_{l+1}^\infty, \cdots, (b-b_B)f_j^\infty, \cdots, f_m^\infty)(z)|dz$$

$$\le C\|b\|_* \prod_{i=1}^m \big(M(|f_i|^{s_i})(x)\big)^{\frac{1}{s_i}}$$

In order to complete our proof, we need to prove the following inequality holds for $0 < \delta < \frac{1}{m}$

$$M_\delta^\sharp(T(\vec{f}))(x) \le C\prod_{i=1}^m Mf_i(x)$$

Let $B = B(x_0, r)$ be an arbitrary ball containing x. Since $0 < \delta < 1$ implies that $||a|^\delta - |c|^\delta| \le |a-c|^\delta$ for $a, c \in R$, it is enough to show that for some constant c_B, there exists constant C such that

$$\left(\frac{1}{|B|}\int_B |T(\vec{f})(z) - c_B|^\delta dz\right)^{1/\delta} \le C\prod_{i=1}^m Mf_i(x)$$

Let $f_i = f_i^0 + f_i^\infty$, where $f_i^0 = f_i\chi_{4\sqrt{Q}c_0B}$. Let $c_B = T(f_1^\infty, \cdots, f_m^\infty)(x_0)$. Then

$$T(\vec{f})(z) - c_B = T(f_1^\infty, \cdots, f_m^\infty)(z) - T(f_1^\infty, \cdots, f_m^\infty)(x_0) + \sum T(f_1^{k_1}, \cdots, f_m^{k_m})(z)$$

where in the last sum each $k_i = 0$ or ∞ and in each term there is at least one $k_j = 0$.

Using the Hölder's inequality and the property of kernel K, we have

$$\begin{aligned}
&\left(\frac{1}{|B|}\int_B |T(f_1^\infty,\cdots,f_m^\infty)(z) - T(f_1^\infty,\cdots,f_m^\infty)(x_0)|^\delta dz\right)^{1/\delta}\\
&\le \frac{1}{|B|}\int_B |T(f_1^\infty,\cdots,f_m^\infty)(z) - T(f_1^\infty,\cdots,f_m^\infty)(x_0)| dz\\
&\le \frac{C}{|B|}\int_B (|B|^{\frac{\beta}{Qm}})^m \prod_{i=1}^m \int_{G\backslash(4\sqrt{Q}c_0B)} \frac{|f_i|}{\rho(y_i^{-1}x_0)^{Q+\beta/m}} dy_i dz\\
&\le C\prod_{i=1}^m Mf_i(x)
\end{aligned}$$

Here we note $(G\backslash(4\sqrt{Q}c_0B))^h \times (4\sqrt{Q}c_0B)^{m-h} = G^h$, then we get

$$\begin{aligned}
&\left(\frac{1}{|B|}\int_B |T(f_1^\infty,\cdots,f_h^\infty,f_{h+1}^0,\cdots,f_m^0)(z)|^\delta dz\right)^{1/\delta}\\
&\le \frac{1}{|B|}\int_B |T(f_1^\infty,\cdots,f_h^\infty,f_{h+1}^0,\cdots,f_m^0)(z)| dz\\
&\le \frac{C}{|B|}\int_B \left(\prod_{i=h+1}^m \int_{4\sqrt{Q}c_0B} |f_i(y_i)| \prod_{k=1}^h \int_{G\backslash(4\sqrt{Q}c_0B)} \frac{|f_k(y_k)|}{\rho(y_k^{-1}z)^{Qm/h}} dy_k\right) dz\\
&\le C\prod_{i=1}^m M(f_i)(x)
\end{aligned}$$

Using the Kolmogorov's estimate [30] and the fact $T : L^1 \times \cdots \times L^1 \to L^{1/m,\infty}$, we can obtain

$$\begin{aligned}
\left(\frac{1}{|B|}\int_B |T(f_1^0,\cdots,f_m^0)(z)|^\delta dz\right)^{1/\delta} &\le C\|T(f_1^0,\cdots,f_m^0)(z)\|_{L^{1/m,\infty}(B,\frac{dz}{B})}\\
&\le C\prod_{i=1}^m \frac{1}{|B|}\int_B |f_i(z)| dz\\
&\le C\prod_{i=1}^m M(f_i)(x)
\end{aligned}$$

Thus for $0 < \delta < \frac{1}{m}$, we have

$$M_\delta^\sharp(T(\vec{f}))(x) \le C\prod_{i=1}^m Mf_i(x)$$

Therefore

$$\begin{aligned}\|T(\vec{f})\|_{p(\cdot)} &\le \|M_\delta(T(\vec{f}))\|_{p(\cdot)} \le \|M_\delta^\sharp(T(\vec{f}))\|_{p(\cdot)} \\ &\le C\|\prod_{i=1}^m Mf_i\|_{p(\cdot)} \le C\prod_{i=1}^m \|Mf_i\|_{p_i(\cdot)} \le C\prod_{i=1}^m \|f_i\|_{p_i(\cdot)}\end{aligned}$$

Knowing that $\frac{1}{s} = \sum_{i=1}^m \frac{1}{s_i} > \sum_{i=1}^m \frac{1}{p_i^-} = \frac{1}{p^-}$, we can get

$$\begin{aligned}\|[b,T]_j(\vec{f})\|_{p(\cdot)} &\le \|M[b,T]_j(\vec{f})\|_{p(\cdot)} \le C\|M^\sharp[b,T]_j(\vec{f})\|_{p(\cdot)} \\ &\le C\|b\|_*\left(\left\|(M(|T(\vec{f})|^s))^{1/s}\right\|_{p(\cdot)} + \left\|\prod_{i=1}^m (M(|f_i|^{s_i}))^{1/s_i}\right\|_{p(\cdot)}\right) \\ &\le C\|b\|_*\left(\left\|T(\vec{f})\right\|_{p(\cdot)} + \prod_{i=1}^m \||f_i|^{s_i}\|_{\frac{p_i(\cdot)}{s_i}}^{1/s_i}\right) \le C\|b\|_*\prod_{i=1}^m \|f_i\|_{p_i(\cdot)}\end{aligned}$$

Therefore we obtain

$$\|[b,T]_j(\vec{f})\|_{p(\cdot)} \le C\|b\|_*\prod_{i=1}^m \|f_i\|_{p_i(\cdot)}$$

□

Finally, we give the proof of Theorem 3.3.

Proof It follows immediately from Corollary 2.11 and Lemma 2.13. □

Acknowledgements The authors would like to express great gratitude to the referees for their valuable remarks which improve the presentation of this article. The author "J. Zhao" was supported by National Natural Science Foundation of China (Grant Nos. 11471040 and 11761131002).

References

1. T. Adamowicz, P. Harjulehto, P. Hästö, Maximal operator in variable exponent Lebesgue spaces on unbounded quasimetric measure spaces. Math. Scand. **116**(1), 5–22 (2015)
2. M. Bramanti, Commutators of integral operators with positive kernels. Le Matematiche **49**(1), 149–168 (1994)

3. M. Bramanti, M.C. Cerutti, Commutators of singular integrals and fractional integrals on homogeneous spaces. Contemp. Math. **189**, 81–94 (1995)
4. M. Bramanti, M.C. Cerutti, Commutators of singular integrals on homogeneous spaces. Boll. Un. Mat. Ital. **10**(4), 843–883 (1996)
5. C. Capone, D. Cruz-Uribe, A. Fiorenza, The fractional maximal operators and fractional integrals on variable L^p spaces. Rev. Mat. Iberoam. **23**(23), 743–770 (2007)
6. L. Chaffee, Commutators of multilinear singular integral operators with pointwise multiplication. Thesis (Ph.D.). University of Kansas, 2015
7. S. Chanillo, A note on commutators. Indiana Univ. Math. J. **31**(1), 7–16 (1982)
8. X. Chen, Q. Xue, Weighted estimates for a class of multilinear fractional type operators. J. Math. Anal. Appl. **362**(2), 355–373 (2010)
9. R.R. Coifman, R. Rochberg, G. Weiss, Factorization theorems for Hardy spaces in several variables. Ann. Math. **103**(2), 611–635 (1976)
10. D. Cruz-Uribe, A. Fiorenza, *Variable Lebesgue Spaces: Foundations and Harmonic Analysis*, 1st edn. (Birkhäuser, Basel, 2013)
11. D. Cruz-Uribe, P. Shukla, The boundedness of fractional maximal operators on variable Lebesgue spaces over spaces of homogeneous type. IEEE Int. Conf. Fuzzy Syst. **5**(3), 183–196 (2015)
12. D. Cruz-Uribe, L. Wang, Variable Hardy spaces. Indiana Univ. Math. J. **63**(2), 447–493 (2014)
13. D. Cruz-Uribe, L. Wang, Extrapolation and weighted norm inequalities in the variable Lebesgue spaces. Trans. Am. Math. Soc. **369**(2), 1205–1235 (2017)
14. D. Cruz-Uribe, J.M. Martell, C. Pérez, Extrapolation from A_∞ weights and applications. J. Funct. Anal. **213**(2), 412–439 (2004)
15. D. Cruz-Uribe, A. Fiorenza, J.M. Martell, C. Pérez, The boundedness of classical operators on variable L^p spaces. Ann. Acad. Sci. Fenn. Math. **31**(1), 239–264 (2006)
16. G.B. Folland, E.M. Stein, *Hardy Spaces on Homogeneous Groups*, 1st edn. (Princeton University Press, Princeton, 1982)
17. X. Fu, D. Yang, W. Yuan, Generalized fractional integrals and their commutators over non-homogeneous metric measure spaces. Taiwan. J. Math. **18**(2), 509–557 (2014)
18. I. Genebashvili, A. Gogatishvili, V. Kokilashvili, M. Krbec, *Weight Theory for Integral Transforms on Spaces of Homogeneous Type*, 1st edn. (Longman, Harlow, 1998)
19. V.S. Guliyev, R.C. Mustafayev, A. Serbetci, Stein-Weiss inequalities for the fractional integral operators in Carnot groups and applications. Complex Var. Elliptic Equ. **55**(8–10), 847–863 (2010)
20. G. Hu, Y. Meng, D. Yang, Multilinear commutators of singular integrals with non-doubling measures. Integr. Equ. Oper. Theory **51**(2), 235–255 (2005)
21. A. Huang, J. Xu, Multilinear singular integrals and commutators in variable exponent Lebesgue spaces. Appl. Math. J. Chinese Univ. **25**(1), 69–77 (2010)
22. S. Janson, Mean oscillation and commutators of singular integral operators. Ark. Mat. **16**(1–2), 263–270 (1978)
23. O. Kováčik, J. Rákosnik, On spaces $L^{p(x)}$ and $W^{k,p(x)}$. Czechoslov. Math. J. **41**(4), 592–618 (1991)
24. Y. Liang, L.D. Ky, D. Yang, Weighted endpoint estimates for commutators of Calderón-Zygmund operators. Proc. Am. Math. Soc. **144**(2), 5171–5181 (2016)
25. Z. Liu, S. Lu, Two-weight weak-type norm inequalities for the commutators of fractional integral. Integr. Equ. Oper. Theory **48**(3), 397–409 (2004)
26. G. Lu, S. Lu, D. Yang, Singular integrals and commutators on homogeneous groups. Anal. Math. **28**(2), 103–134 (2002)
27. S. Lu, Y. Ding, D. Yan, *Singular Integral and Related Topics*, 1st edn. (World Scientific, Singapore, 2011)
28. E. Nakai, Y. Sawano, Hardy spaces with variable exponents and generalized Campanato spaces. J. Funct. Anal. **262**(9), 3665–3748 (2012)
29. W. Orlicz, Über konjugierte Exponentenfolgen. Stud. Math. **3**(1), 200–211 (1931)

30. C. Pérez, R. Torres, Sharp maximal function estimates for multilinear singular integrals. Contemp. Math. **46**(2), 229–274 (2002)
31. J. Tan, J. Zhao, Rough fractional integrals and its commutators on variable Morrey spaces. C. R. Math. Acad. Sci. Paris **353**(12), 1117–1122 (2015)
32. J. Tan, Z. Liu, J. Zhao, On some multilinear commutators in variable Lebesgue spaces. J. Math. Inequal. **11**(3), 715–734 (2017)
33. A. Uchiyama, On the compactness of operators of Hankel type. Tohoku Math. J. **30**(1), 163–171 (2010)
34. H. Wang, Z. Liu, The wavelet characterization of Herz-type Hardy spaces with variable exponent. Ann. Funct. Anal. **3**(1), 128–141 (2012)
35. J. Xu, The boundedness of multilinear commutators of singular integrals on Lebesgue spaces with variable exponent. Czechoslov. Math. J. **57**(1), 13–27 (2010)
36. D. Yang, C. Zhou, W. Yuan, Triebel-Lizorkin-Type spaces with variable exponents. Banach J. Math. Anal. **9**(4), 146–202 (2015)
37. Y. Zhu, L. Liu, Weighted sharp boundedness for multilinear commutators of singular integral on spaces of homogeneous type. Vietnam J. Math. **39**(4), 381–390 (2011)

Volterra Operators with Asymptotics on Manifolds with Edge

M. Hedayat Mahmoudi and B.-W. Schulze

Abstract We study Volterra property and parabolicity of a class of anisotropic pseudo-differential operators on a manifold with edge. This exposition belongs to a more comprehensive approach. In the present consideration we focus on asymptotic aspects of parametrices or inverses in the subalgebra of anisotropic operators of Mellin plus Green type. In the zero order case we also add the identity map. The resulting space constitutes a necessary step for constructing Volterra parametrices in general.

Keywords Volterra operators of Mellin plus Green type · Anisotropic edge operators · Invertibility over finite time intervals

Mathematics Subject Classification (2000) Primary 35S35, Secondary 35J70

1 Introduction

The analysis on manifolds with singularities has been applied to elliptic problems in many variants, e.g., Fredholm theory of boundary value problems (BVPs) in weighted spaces of distributions and subspaces with asymptotics. In the present exposition we refer to the pseudo-differential algebra on manifolds with edge, containing parametrices within the calculus, see [26] or [27, 28]. A similar program for $\mathbb{R} \times X$ and a closed smooth manifold X is well-known for Volterra operators and parabolic equations, cf. the work of Piriou [21], with time-variable t. Its covariable τ is involved with anisotropy l. There is an operator algebra in the thesis of Buchholz [4] containing an anisotropic Volterra edge calculus on the level of pairs of Hilbert spaces with strongly continuous group actions. Another investigation of Mikayelyan [20] has developed a formalism, partly motivated by Piriou [21]

M. Hedayat Mahmoudi · B.-W. Schulze (✉)
Institute of Mathematics, University of Potsdam, Potsdam, Germany
e-mail: hedayatm@uni-potsdam.de; schulze@math.uni-potsdam.de

S. Molahajloo, M. W. Wong (eds.), *Analysis of Pseudo-Differential Operators*, Trends in Mathematics, https://doi.org/10.1007/978-3-030-05168-6_6

and the framework of Boutet de Monvel [3] as well as of the edge calculus, i.e., more general X, including operators of trace, potential or Green type with respect to Y. Our final goal will be a modified picture of parabolicity, using a new Volterra quantization of smoothing Mellin plus Green symbols which is contained in the present paper and then in another chapter an adequate Volterra quantization which admits non-smoothing contributions of the edge calculus. In other words we investigate a necessary aspect for a manifold B with edge Y, concerning the case of anisotropic Mellin plus Green edge operators which remained open in [4]. Parabolic problems in a specific singular context have been studied by Krainer [16, 17]. Those are not really touched by our models. Differential operators on finite time-intervals are more related to the work of Piriou [22] and Agranovich, Vishik [1] when the respective manifolds are smooth, though with boundary in these papers. Analogously as in the isotropic edge theory, in L^2-based weighted Sobolev spaces we apply here an anisotropic variant of such a calculus. The new element compared with [4] is a Volterra quantization of smoothing Mellin plus Green symbols in this context. Note that Volterra quantization for non-smoothing contributions requires more tools from the edge calculus. This will be studied in another part of the program. Other approaches in L^p-spaces have been developed by Grubb [11], or in a completely different framework by Amann [2], see also the bibliography there.

Note that there are more investigations on parabolicity in the pseudo-differential set-up, e.g., Cho Čan and Eskin [6] in Vishik-Eskin's technique, and in terms of methods from isotropic degenerate operators, see, in particular, Buchholz [4], Kapanadze et al. [15], Krainer [17], Hedayat Mahmoudi and Schulze [12] using method from Egorov and Schulze [8], cf. also Gil et al. [9, 10], Hirschmann [13] and more general aspects on pseudo-differential operators from Hörmander [14], Kumano-go [18], Lyu and Schulze [19], Rempel and Schulze [23, 24], Schulze [25], Schulze and Seiler [29], Seiler [31], Shubin [32] and Sternin [33].

2 Manifolds with Edge

In this section we give a brief account of notions on manifolds with edge. Those will play the role of spatial configurations of anisotropic manifolds with edge, containing an extra component, the time.

A manifold B with edge Y is a topological space such that $B \setminus Y$ and Y are both smooth manifolds of dimension b and $q > 0$, respectively, and there is a neighbourhood W of Y which is locally close to Y modeled on

$$X^\Delta \times \mathbb{R}^q \tag{2.1}$$

for

$$X^\Delta = (\overline{\mathbb{R}}_+ \times X)/(\{0\} \times X) \tag{2.2}$$

where (2.2) is called the model cone of the wedge (2.1). "Locally modeled" means that (2.1) admits homeomorphisms

$$\chi : N \to X^{\Delta} \times \mathbb{R}^q \tag{2.3}$$

for relatively open subsets $N \subseteq W$ which form a covering of W, and (2.3) restricts to diffeomorphisms (i.e., charts on Y)

$$\chi_{\text{edge}} : N \cap Y \to \mathbb{R}^q$$

where the sets $U := N \cap Y$ form an open covering of Y and (2.3) restricts to diffeomorphisms

$$\chi_{\text{int}} : N \setminus Y \to X^{\wedge} \times \mathbb{R}^q. \tag{2.4}$$

Concerning transition maps under different wedge charts (2.3) we assume that they induce an X^{Δ}-bundle over Y. The situation is similar to smooth manifolds B with boundary Y. In this case W corresponds to a collar neighbourhood of Y in B and it may be identified with the normal bundle $[0, 1) \times Y$ when we fixed a Riemannian metric on W which induces a Riemannian metric on Y. Then (2.3) corresponds to a map $N := [0, 1) \times U \to \overline{\mathbb{R}}_+ \times \mathbb{R}^q$ for a coordinate neighbourhood $U \subset Y$. It is well-known in this case that W represents a trivial $\overline{\mathbb{R}}_+$-bundle while the above-mentioned X^{Δ}-bundle is not necessarily trivial when $n := \dim X > 0$.

There is another equivalent definition of a manifold B with edge. In this case we replace local wedges $X^{\Delta} \times \mathbb{R}^q$ by

$$\overline{\mathbb{R}}_+ \times X \times \mathbb{R}^q, \tag{2.5}$$

called stretched wedges. Analysis will take place on open stretched wedges

$$X^{\wedge} \times \mathbb{R}^q \tag{2.6}$$

for corresponding open stretched cones $X^{\wedge} := \mathbb{R}_+ \times X$ in variables (r, x). Local models $\overline{\mathbb{R}}_+ \times X \times \mathbb{R}^q$ allow us in an invariant manner to form a so-called stretched manifold $\mathbb{B}$ which has a smooth boundary $\partial\mathbb{B}$ which is an X-bundle over Y. For $B = X^{\Delta} \times \mathbb{R}^q$ we simply have

$$\mathbb{B} = \overline{\mathbb{R}}_+ \times X \times \mathbb{R}^q, \ \partial\mathbb{B} = \{0\} \times X \times \mathbb{R}^q. \tag{2.7}$$

We first assume that Y is a smooth manifold with standard charts $U \to \mathbb{R}^q$ and develop ideas of the edge calculus for such an isotropic edge. Later on we replace Y by $\mathbb{R} \times Y$ with the time axis $\mathbb{R} \ni t$, and then some part of the analysis will be formulated in anisotropic form.

Note that for stretched wedges $\mathbb{B}$ in (2.5) there are also the doubled spaces

$$2\mathbb{B} = \mathbb{R} \times X \times \mathbb{R}^q$$

which are obtained by gluing together two copies of $\mathbb{B}$ along the common boundary to a smooth manifold with empty boundary. A similar construction makes sense for the original manifold B with edge, i.e., we obtain $\mathbb{B}$ in an invariant way. There is then a canonical surjective map

$$\mathbb{B} \to B,$$

locally corresponding to $\overline{\mathbb{R}}_+ \times X \times \mathbb{R}^q \to X^\Delta \times \mathbb{R}^q$ for the quotient map $\overline{\mathbb{R}}_+ \times X \to X^\Delta$. It will be sufficient for our considerations to impose some conditions on the choice for the above-mentioned X-bundle $\partial\mathbb{B}$ over Y. Let us assume that it is trivial and that transition maps for the associated $\overline{\mathbb{R}}_+ \times X$ bundle which models $\mathbb{B}$ close to Y are independent of r for $0 \leq r < \varepsilon$ for some $\varepsilon > 0$. In order to organize a pseudo-differential calculus on B we first focus on open stretched wedges (2.6). Suitable coordinate invariance properties, see Dorschfeldt [7] or [28], will lead to edge operators globally on B.

The compact manifold B with edge Y of dimension $q > 0$ will be regarded as the spatial configuration of a corresponding space with an extra time-variable $t \in \mathbb{R}$. Then Y is replaced by $\mathbb{R} \times Y$ which is the edge of $\mathbb{R} \times B$. In this case the calculus of pseudo-differential operators is formulated in anisotropic terms, locally in variables $(t, y) \in \mathbb{R}^{1+q}$ and covariables. The space $B \setminus Y$ itself has the variables $\boldsymbol{x}$ with covariables $\boldsymbol{\xi}$ which split close to Y to variables $(y, r, x) \in \mathbb{R}^q \times X^\wedge$ and corresponding covariables $(\eta, \rho, \xi) \in \mathbb{R}^{q+1+n}$ for $n = \dim X$. We systematically refer to the isotropic edge calculus on B and recall basics from several articles and textbooks, see [26–28, 30] and then we add variables t and covariables τ in anisotropic form, motivated by the heat operator $\partial_t - \Delta$ for the Laplacian in $\mathbb{R}^n$ with the anisotropic homogeneous principal symbol

$$i\tau + |\xi|^2.$$

The anisotropic extension of the edge pseudo-differential calculus to anisotropic covariables (τ, η) is more or less straightforward and topic of Buchholz [4], cf. [5] and works of Krainer [16], in a different context.

3 Weighted Cone Spaces

Let B be a manifold with edge Y. If B is locally modeled on $X^\Delta \times \mathbb{R}^q$ for a closed smooth manifold X and local coordinates $\mathbb{R}^q$ on Y, we formulate spaces of distributions locally on stretched wedges $X^\wedge \times \mathbb{R}^q$ in variables $((r, x), y) \in X^\wedge \times \mathbb{R}^q$. Together with the time axis $\mathbb{R}$ in the variable t we look at the anisotropic wedge $\mathbb{R} \times B$, where $\mathbb{R} \times (B \setminus Y)$ is locally identified with $X^\wedge \times \mathbb{R} \times \mathbb{R}^q$ in the variables $((r, x), (t, y))$. In this section we define Green operators with asymptotic types $\mathcal{P}, \mathcal{Q}$.

From the isotropic edge calculus over B we first recall the definition of Kegel spaces $\mathcal{K}^{s,\gamma}(X^\wedge)$ and subspaces $\mathcal{K}^{s,\gamma}_{\mathcal{P}}(X^\wedge)$ with asymptotics of type $\mathcal{P}$. Those spaces are formulated in terms of the Mellin transform

$$Mu(w) := \int_0^\infty r^w u(r) \frac{dr}{r}, \tag{3.1}$$

first defined for $u \in C_0^\infty(\mathbb{R}_+)$, where $w \in \mathbb{C}$ and often restricted to

$$\Gamma_\beta := \{w \in \mathbb{C} : \operatorname{Re} w = \beta\} \tag{3.2}$$

for any real β. Then a simple property is that

$$M_\gamma u := Mu|_{\Gamma_{\frac{1}{2}-\gamma}} \in \mathcal{S}(\Gamma_{\frac{1}{2}-\gamma})$$

extends by continuity to an isomorphism

$$M_\gamma : r^\gamma L^2(\mathbb{R}_+) \to L^2(\Gamma_{\frac{1}{2}-\gamma}) \tag{3.3}$$

for every weight γ with the inverse

$$(M_\gamma^{-1} g)(r) = \int_{\Gamma_{\frac{1}{2}-\gamma}} r^{-w} g(w) đw \tag{3.4}$$

for $đw := (2\pi)^{-1} dw$ with integration over the respective weight line from $\operatorname{Im} w = -\infty$ to $\operatorname{Im} w = +\infty$. We have weighted Mellin Sobolev spaces on the half-axis

$$\mathcal{H}^{s,\gamma}(\mathbb{R}_+) \tag{3.5}$$

of smoothness $s \in \mathbb{R}$ and weight $\gamma \in \mathbb{R}$, defined as the completion of $C_0^\infty(\mathbb{R}_+)$ with respect to the norm

$$\|u\|_{\mathcal{H}^{s,\gamma}(\mathbb{R}_+)} := \left\{ \int_{\Gamma_{\frac{1}{2}-\gamma}} \langle w \rangle^{2s} |Mu(w)|^2 đw \right\}^{1/2}. \tag{3.6}$$

Then

$$\mathcal{K}^{s,\gamma}(\mathbb{R}_+) : \{\omega u_0 + (1-\omega) u_\infty : u_0 \in \mathcal{H}^{s,\gamma}(\mathbb{R}_+), u_\infty \in H^s(\mathbb{R}_+)\} \tag{3.7}$$

for some cut-off function ω with respect to $r = 0$ (i.e., $\omega \in C^\infty(\mathbb{R}_+)$ is strictly positive, and $\omega \equiv 1$ for $r < \varepsilon_0$, $\omega \equiv 0$ for $r > \varepsilon_1$ for some $0 < \varepsilon_0 < \varepsilon_1 < \infty$) and $H^s(\mathbb{R}_+) = H^s(\mathbb{R})|_{r>0}$.

Similar notation will be used for spaces over $X^\wedge = \mathbb{R}_+ \times X$ for some closed smooth manifold X of dimension $n > 0$. The space

$$\mathcal{H}^{s,\gamma}(\mathbb{R}_+ \times \mathbb{R}^n) \tag{3.8}$$

for $s,\ \gamma \in \mathbb{R}$ is defined as the completion of $C_0^\infty(\mathbb{R}_+ \times \mathbb{R}^n)$ with respect to the norm

$$\|u\|_{\mathcal{H}^{s,\gamma}(\mathbb{R}_+\times\mathbb{R}^n)} := \Big\{ \int_{\Gamma_{\frac{1}{2}-\gamma}} \int_{\mathbb{R}^n} \langle w, \xi\rangle^{2s} |MFu(w,\xi)|^2 đw đ\xi \Big\}^{1/2}. \tag{3.9}$$

with $F = F_{x\to\xi}$ being the Fourier transform in $\mathbb{R}^n$ and $đ\xi := (2\pi)^{-n} d\xi$. Let us fix on X a finite covering by coordinate neighbourhoods $U_1, \dots, U_N$, charts $\chi_j : U_j \to \mathbb{R}^n$, $j = 1, \dots, N$, a subordinate partition of unity $\varphi_1, \dots \varphi_N$, and define the space

$$\mathcal{H}^{s,\gamma}(X^\wedge) \tag{3.10}$$

as the completion of $C_0^\infty(\mathbb{R}_+ \times X)$ with respect to the norm

$$\|u\|_{\mathcal{H}^{s,\gamma}(X^\wedge)} := \Big\{ \sum_{j=1}^{N} \|(\mathrm{id}_{\mathbb{R}_+} \times \chi_j^{-1})_*)(\varphi_j u)\|^2_{\mathcal{H}^{s,\gamma}(\mathbb{R}_+\times\mathbb{R}^n)} \Big\}^{1/2}. \tag{3.11}$$

Recall that we produce equivalent norms if we change the charts or the subordinate partition of unity. Then we set

$$\mathcal{K}^{s,\gamma}(X^\wedge) : \big\{\omega u_0 + (1-\omega)u_\infty : u_0 \in \mathcal{H}^{s,\gamma}(X^\wedge), u_\infty \in H^s_{\mathrm{cone}}(X^\wedge)\big\}, \tag{3.12}$$

where $H^s_{\mathrm{cone}}(X^\wedge)$ may be interpreted as a modification of $H^s(\mathbb{R}^{1+n}_{\tilde{x}})$ for $\tilde{x} \neq 0$ with respect to the behaviour of Sobolev distributions when $|\tilde{x}| \to \infty$. The precise definition is as follows:

Let $(U_1, \dots, U_N)$ be an open covering of X by coordinate neighbourhoods, choose an open covering of S^n by coordinate neighbourhoods $(V_1, \dots, V_N)$, and set

$$\Gamma_{V_j} := \big\{\tilde{x} \in \mathbb{R}^{1+n} : \tilde{x}/|\tilde{x}| \in V_j\big\}.$$

Denote points in U_j by x, choose a diffeomorphism $\kappa_j : U_j \to V_j$, $\kappa_j(x) =: x_j$, and define diffeomorphisms

$$\mathrm{cone}_j : \mathbb{R}_+ \times U_j \to \Gamma_{V_j} \tag{3.13}$$

by $\mathrm{cone}_j(r, x) := r\kappa_j(x)$. Let $(\sigma_1, \dots, \sigma_N)$ be a partition of unity on S^n subordinate to $(V_1, \dots, V_N)$. Set $H^s(\Gamma_{V_j}) := H^s(\mathbb{R}^{1+n})|_{\Gamma_{V_j}}$, and define the space

$$H^s_{\mathrm{cone}}(X^\wedge) := \omega H^s(\mathbb{R} \times X)|_{\mathbb{R}_+\times X} + (1-\omega)H^s_{\mathrm{cone}}(X^\wedge) \tag{3.14}$$

with $(1-\omega)H^s_{\text{cone}}(X^\wedge)$ being the set of all functions

$$\Big\{(1-\omega)\sum_{j=1}^{N}(\sigma_j v_j)\circ \text{cone}_j : v_j \in H^s(\Gamma_{V_j}),\, j=1,\dots,N\Big\}. \tag{3.15}$$

It may be convenient to refer to another equivalent definition of spaces $H^s_{\text{cone}}(X^\wedge)$ by replacing V_j by an open n-dimensional unit ball B in $\mathbb{R}^n$, identified with $\{\tilde{x} = ((\tilde{x}_0, \tilde{x}') \in \mathbb{R}^{1+n} : \tilde{x}_0 = 1, \tilde{x}' := (\tilde{x}_1, \dots, \tilde{x}_n) \in B\}$. Then replacing (3.13) by diffeomorphisms

$$\text{cone}^\wedge_j : \mathbb{R}_+ \times U_j \to B^\wedge, \tag{3.16}$$

where $\text{cone}^\wedge_j(r,x) := r\kappa_j(x)$ with another diffeomorphism $\kappa_j : U_j \to B$ and $B^\wedge$ identified with the set $\{(r, r\tilde{x}') : r > 0\}$ we have the space $H^s(B^\wedge) := H^s(\mathbb{R}^{1+n})|_{B^\wedge}$, and we can employ formula (3.14) where the second space has the meaning

$$\Big\{(1-\omega)\sum_{j=1}^{N}\sigma_j v_j \circ \text{cone}^\wedge_j : v_j \in H^s(B^\wedge),\, j=1,\dots,N\Big\}. \tag{3.17}$$

In these considerations ω is a cut-off function on the r half-axis. Both $H^s_{\text{cone}}(X^\wedge)$ and $\mathcal{K}^{s,\gamma}(X^\wedge)$ are Hilbert spaces with scalar products coming from the respective non-direct sums. For $s = \gamma = 0$ we normalize the scalar products by taking the ones from L^2-spaces, e.g., we identify $\mathcal{K}^{0,0}(X^\wedge)$ with $r^{-n/2}L^2(\mathbb{R}_+ \times X)$.

In order to introduce Green operators we now recall a few notions from the isotropic edge calculus, namely, asymptotic types. Those make sense in different variants, e.g., discrete or continuous ones, outlined in [27] or [28]. We focus here on constant (with respect to edge variables) asymptotic types, indicating subspaces of $\mathcal{K}^{s,\gamma}(X^\wedge)$ of elements with such asymptotics. In this context we fix weight data (γ, Θ), where $\gamma \in \mathbb{R}$ is some weight and $\Theta = (\vartheta, 0]$, $\vartheta < 0$ a weight interval where we control asymptotics or flatness with respect to γ. Let us set

$$\mathcal{K}^{s,\gamma}_\Theta(X^\wedge) := \varprojlim_{\varepsilon>0} \mathcal{K}^{s,\gamma-\vartheta-\varepsilon}(X^\wedge) \tag{3.18}$$

in its natural Fréchet topology, indicating flatness of distributions of order Θ with respect to γ near $r = 0$. In addition we define spaces of singular functions belonging to discrete asymptotic types $\mathcal{P}$ associated with (γ, Θ). Such a $\mathcal{P}$ is defined by a sequence

$$\mathcal{P} := \{(p_j, m_j)\}_{j=1,\dots,N} \subset \mathbb{C} \times \mathbb{N} \tag{3.19}$$

for some $N \in \mathbb{N}$, such that $\pi_{\mathbb{C}}\mathcal{P} := \{p_j\}_{j=1,\dots,N}$ is finite when N is finite, otherwise, for infinite N it satisfies $\operatorname{Re} p_j \to -\infty$ for $j \to \infty$, and we ask

$$\pi_{\mathbb{C}}\mathcal{P} \subset \{w \in \mathbb{C} : \frac{n+1}{2} - \gamma + \vartheta < \operatorname{Re} w < \frac{n+1}{2} - \gamma\}.$$

Singular functions with asymptotics of type $\mathcal{P}$ for finite $\pi_{\mathbb{C}}\mathcal{P}$ are elements of spaces

$$\mathcal{E}_{\mathcal{P}}(X^\wedge) := \{\omega \sum_{j=0}^{N} \sum_{k=0}^{m_j} c_{jk} r^{-p_j} \log^k r : c_{jk} \in C^\infty(X)\} \tag{3.20}$$

for some cut-off function ω. Those are Fréchet and contained in $\mathcal{K}^{\infty,\gamma}(X^\wedge)$, and they intersect (3.18) only in $\{0\}$. Thus

$$\mathcal{K}^{s,\gamma}_{\mathcal{P}}(X^\wedge) := \mathcal{E}_{\mathcal{P}}(X^\wedge) \oplus \mathcal{K}^{s,\gamma}_{\Theta}(X^\wedge) \tag{3.21}$$

is Fréchet in the topology of the direct sum. For an infinite discrete asymptotic type $\mathcal{P}$ we form $\mathcal{P}_b := \{(p,m) \in \mathcal{P} : \operatorname{Re} p > (n+1)/2 - \gamma - b\}$ for some $b \in \mathbb{N}$. According to (3.21) we have the spaces $\mathcal{K}^{s,\gamma}_{\mathcal{P}_b}(X^\wedge)$ for every b with continuous embeddings

$$\mathcal{K}^{s,\gamma}_{\mathcal{P}_{b+1}}(X^\wedge) \hookrightarrow \mathcal{K}^{s,\gamma}_{\mathcal{P}_b}(X^\wedge)$$

for every b, and then we define

$$\mathcal{K}^{s,\gamma}_{\mathcal{P}}(X^\wedge) := \varprojlim_{b\in\mathbb{N}} \mathcal{K}^{s,\gamma}_{\mathcal{P}_b}(X^\wedge). \tag{3.22}$$

Throughout this exposition we assume that the occurring asymptotic types $\mathcal{P}$ satisfy the shadow condition, concerning the meaning, see [28]. In the following considerations we need some modifications of the behaviour of spaces for $r \to \infty$, namely,

$$\mathcal{K}^{s,\gamma;e}(X^\wedge) := [r]^{-e}\mathcal{K}^{s,\gamma}(X^\wedge)$$

for any $e \in \mathbb{R}$ where $r \to [r]$ is any smooth strictly positive function in $r \in \mathbb{R}$ which is $\equiv 1$ for $0 < r < \varepsilon_1$ and equal to r for $\varepsilon_2 < r$ for some $0 < \varepsilon_1 < \varepsilon_2 < \infty$. Analogously we set

$$\mathcal{K}^{s,\gamma;e}_{\mathcal{P}}(X^\wedge) := [r]^{-e}\mathcal{K}^{s,\gamma}_{\mathcal{P}}(X^\wedge).$$

Another notation in this context is:

$$\mathcal{S}^{\gamma}_{\mathcal{P}}(X^\wedge) := \varprojlim_{s,e\in\mathbb{R}} \mathcal{K}^{s,\gamma;e}_{\mathcal{P}}(X^\wedge).$$

4 Volterra Symbols and Edge Spaces

In our calculus we specify notions and results of [4, 5] concerning anisotropic symbols taking values in Hilbert (and later on also Fréchet) spaces with group action. Let us first recall some general definitions.

Consider a (separable) Hilbert space H with group action $\kappa = \{\kappa_\delta\}_{\delta\in\mathbb{R}_+}$, i.e., where

$$\kappa_\delta : H \to H,\ \delta \in \mathbb{R}_+,$$

is a family of isomorphisms such that $\kappa_\delta\kappa_{\delta'} = \kappa_{\delta\delta'}$ for all $\delta, \delta' \in \mathbb{R}_+$, $\kappa_1 = \mathrm{id}_H$, and $\delta \to \kappa_\delta h$ determines an element in $C(\mathbb{R}_+, H)$ for every $h \in H$.

Proposition 4.1 *There are constants $C, M > 0$ such that*

$$\|\kappa_\delta\|_{\mathcal{L}(H)} \le C \max\{\delta, \delta^{-1}\}^M.$$

This result is standard. Incidentally we write $M := M(\kappa)$. In this section we formulate results for the case

$$H := \mathcal{K}^{s,\gamma}(X^\wedge) \tag{4.1}$$

with $\kappa = \{\kappa_\delta\}_{\delta\in\mathbb{R}_+}$, defined by

$$(\kappa_\delta u)(r, x) := \delta^{(n+1)/2} u(\delta r, x). \tag{4.2}$$

Other choices of Hilbert spaces with the same group action are

$$\tilde{H} := \mathcal{K}^{s-\mu,\gamma-\mu}(X^\wedge). \tag{4.3}$$

It will be also interesting to admit (Fréchet) subspaces with asymptotics, cf. Sect. 3, which admit the action of (4.2). In order to formulate anisotropic analogues of Sobolev spaces and operator-valued symbols with twisted symbolic estimates we first recall that we fixed an anisotropy $l \in \mathbb{N} \setminus \{0\}$, and for $(\tau, \eta) \in \mathbb{R}^{1+q}$ we set

$$|\tau, \eta|_l := (|\tau|^2 + |\eta|^{2l})^{1/2l}$$

and

$$[\tau, \eta]_l := \omega(\tau, \eta) + (1 - \omega(\tau, \eta))|\tau, \eta|_l$$

for any fixed real-valued $\omega \in C_0^\infty(\mathbb{R}^{1+q})$ such that

$$\omega(\tau, \eta) = \begin{cases} 1 & \text{for } 0 \le |\tau, \eta|_l \le \varepsilon_0 \\ 0 & \text{for } \qquad |\tau, \eta|_l \ge \varepsilon_1 \end{cases} \tag{4.4}$$

for some $0 < \varepsilon_0 < \varepsilon_1$. Moreover, we set

$$\langle \tau, \eta \rangle_l := (1 + |\tau|^2 + |\eta|^{2l})^{1/2l}.$$

Similar expressions will be used for $v \in \mathbb{C}$ in place of $\tau \in \mathbb{R}$, i.e.,

$$|v, \eta|_l, [v, \eta]_l, \text{ and } \langle v, \eta \rangle_l.$$

Note that for some constants $c, c_1, c_2 > 0$ and all $(v, \eta) \in \mathbb{C} \times \mathbb{R}^q$ we have an analogue of Peetre's inequality

$$[v, \eta]_l^s \leq c^{|s|}[v - v', \eta - \eta']_l^{|s|}[v', \eta']_l^s$$

for every $s \in \mathbb{R}$. Note that

$$\|\kappa_{\frac{[v,\eta]_l}{[v',\eta']_l}}\|_{\mathcal{L}(H)} \leq c[v - v', \eta - \eta']_l^M$$

for constants $c, M > 0$. Moreover, we have

$$c_1[v, \eta]_l \leq (1 + |v|^2 + |\eta|^{2l})^{1/2l} \leq c_2[v, \eta]_l,$$

$$c_1[v, \eta]_l \leq [v, \eta]_1 \leq c_2[v, \eta]_l^l, \tag{4.5}$$

for suitable constants $c_1, c_2 > 0$.

We are now in the position to define abstract anisotropic edge symbols, referring to pairs of (separable) Hilbert spaces

$$H \quad \text{and} \quad \tilde{H},$$

equipped with strongly continuous groups

$$\kappa = \{\kappa_\delta\}_{\delta>0} \quad \text{and} \quad \tilde{\kappa} = \{\tilde{\kappa}_\delta\}_{\delta>0},$$

respectively. The space of such symbols

$$S^{\mu,l}(\Omega \times \mathbb{R}^{1+q}; H, \tilde{H}) \tag{4.6}$$

for order μ and anisotropy l is defined as the set of all $a(t, y, \tau, \eta) \in C^\infty(\Omega \times \mathbb{R}^{1+q}, \mathcal{L}(H, \tilde{H}))$ for

$$\Omega := (t_0, t_1) \times \Sigma, \tag{4.7}$$

$\Sigma \subseteq \mathbb{R}^q$ open, such that

$$\|\tilde{\kappa}(\tau,\eta)_l^{-1}\{D^{\alpha}_{t,y}D^{\beta}_{\tau,\eta}a(t,y,\tau,\eta)\}\kappa(\tau,\eta)_l\|_{\mathcal{L}(H,\tilde{H})} \leq c\,\langle\tau,\eta\rangle_l^{\mu-|\beta|_l} \tag{4.8}$$

for all $\alpha \in \mathbb{N}^{1+q}$, $\beta \in \mathbb{N}^{1+q}$ and all $t \in (t_0', t_1')$, $[t_0', t_1'] \Subset (t_0, t_1)$ and $y \in \Sigma'$ for $\Sigma' \Subset \Sigma$, for constants $c = c((t_0', t_1') \times \Sigma')$ and all $(\tau, \eta) \in \mathbb{R}^{1+q}$. Here

$$\kappa(\tau,\eta)_l := \kappa_{\langle\tau,\eta\rangle_l},$$

etc. In this formalism we also have the case that the involved spaces are Fréchet spaces, written as projective limit of Hilbert spaces with group action. In our case the spaces of symbols refer to

$$H := \mathcal{K}_{\mathcal{P}}^{\infty,\gamma;\infty}(X^{\wedge}),$$

where

$$(\kappa_\delta u)(r,x) = \delta^{(n+1)/2}u(\delta r, x),$$

and for $\tilde{H}$ we may have spaces of the form

$$\mathcal{S}_{\mathcal{P}}^{\gamma}(X^{\wedge})$$

for suitable weights and asymptotic types $\mathcal{P}$ with the same expressions for κ_δ in the scales of spaces which are involved in the projective limits. Recall that we also have subspaces of classical anisotropic symbols

$$S^{\mu,l}_{\mathrm{cl}}(\Omega \times \mathbb{R}^{1+q}; H, \tilde{H}) \subset S^{\mu,l}(\Omega \times \mathbb{R}^{1+q}; H, \tilde{H}), \tag{4.9}$$

characterized by anisotropic homogeneous components $a_{(\mu-j)}(t, y, \tau, \eta)$, satisfy the homogeneity relations

$$a_{(\mu-j)}(t,y,\delta^l\tau,\delta\eta) = \delta^{\mu-j}\tilde{\kappa}(\tau,\eta)_l\, a_{(\mu-j)}(t,y,\tau,\eta)\kappa(\tau,\eta)_l^{-1}$$

for all $\delta > 0$ and $(\tau, \eta) \neq 0$.

Definition 4.2 The space of Volterra symbols

$$S^{\mu,l}_{V}(\Omega \times \mathbb{R}^{1+q}; H, \tilde{H}) \subset S^{\mu,l}(\Omega \times \mathbb{R}^{1+q}; H, \tilde{H}), \tag{4.10}$$

is defined as the set of all $a(t, y, \tau, \eta) \in C^\infty(\Omega \times \mathbb{R}^{1+q}; H, \tilde{H})$ which admit an analytic extension $a(t, y, v, \eta)$ with respect to v into the lower complex v-half-plane

$\mathbb{C}^- := \{v = \tau + i\sigma : \sigma < 0\}$, such that

$$a(t, y, v, \eta) \in \mathcal{A}(\mathbb{C}^-, C^\infty(\Omega \times \mathbb{R}^q, \mathcal{L}(H, \tilde{H}))) \cap C^\infty(\overline{\mathbb{C}^-}, C^\infty(\Omega \times \mathbb{R}^q, \mathcal{L}(H, \tilde{H})))$$

satisfying the symbolic estimates

$$\|\tilde{\kappa}(v, \eta)_l^{-1}\{D_{t,y}^\alpha D_{v,\eta}^\beta a(t, y, v, \eta)\}\kappa(v, \eta)_l\|_{\mathcal{L}(H,\tilde{H})} \leq c\,\langle v, \eta\rangle_l^{\mu-|\beta|_l} \tag{4.11}$$

for all $\alpha \in \mathbb{N}^{1+q}$, $\beta \in \mathbb{N}^{1+q}$ and all $t \in (t_0', t_1')$ for $[t_0', t_1'] \Subset (t_0, t_1)$ and $y \in \Sigma'$ for $\Sigma' \Subset \Sigma$, for constants $c = c((t_0', t_1') \times \Sigma')$ and all $(v, \eta) \in \overline{\mathbb{C}^-} \times \mathbb{R}^q$, $\kappa(v, \eta)_l := \kappa_{\langle v,\eta\rangle_l}$.

Similarly as before we also admit the case that the involved spaces are Fréchet spaces, written as projective limit of Hilbert spaces with group action. In our case the spaces of symbols may refer to $\tilde{H} := \mathcal{K}_{\mathcal{P}}^{\infty,\gamma;\infty}(X^\wedge)$.

Definition 4.3 The space of classical Volterra symbols

$$S_{\mathrm{cl},V}^{\mu,l}(\Omega \times \mathbb{R}^{1+q}; H, \tilde{H}) \tag{4.12}$$

is defined to be the set of all $a(t, y, \tau, \eta) \in S_V^{\mu,l}(\Omega \times \mathbb{R}^{1+q}; H, \tilde{H})$ such that there is a sequence of homogeneous components

$$a_{(\mu-j)}(t, y, v, \eta) \in S^{(\mu-j),l}(\Omega \times ((\overline{\mathbb{C}^-} \times \mathbb{R}^q) \setminus \{0\}); H, \tilde{H}),$$

$j \in \mathbb{N}$, such that for every $N \in \mathbb{N}$ there is an $M \in \mathbb{N}$ satisfying the relation

$$a(t, y, v, \eta) - \sum_{j=0}^{M} \chi(v, \eta) a_{(\mu-j)}(t, y, v, \eta) \in S^{\mu-N,l}(\Omega \times (\overline{\mathbb{C}^-} \times \mathbb{R}^q); H, \tilde{H})$$

for any excision function $\chi(v, \eta)$.

Notation for spaces of homogeneous functions is used here in a similar manner as for real covariables:

$$S^{(\nu),l}(\Omega \times ((\overline{\mathbb{C}^-} \times \mathbb{R}^q) \setminus \{0\}); H, \tilde{H}),$$

for some real ν denotes the space of all $f_{(\nu)}(t, y, v, \eta) \in C^\infty(\Omega \times ((\overline{\mathbb{C}^-} \times \mathbb{R}^q) \setminus \{0\}), \mathcal{L}(H, \tilde{H}))$ satisfying relation

$$f_{(\nu)}(t, y, \delta^l v, \delta\eta) = \delta^\nu \tilde{\kappa}_{|v,\eta|_l} f_{(\nu)}(t, y, v, \eta)\kappa_{|v,\eta|_l}^{-1}$$

for all $\delta > 0$ and all $(t, y, v, \eta) \in \Omega \times ((\overline{\mathbb{C}^-} \times \mathbb{R}^q) \setminus \{0\})$.

5 Anisotropic Green Symbols

We now turn to Green symbols and Green operators in the anisotropic edge calculus.

Definition 5.1 An element

$$g(t, y, \tau, \eta) \in S_{\mathrm{cl}}^{\nu, l}(\Omega \times \mathbb{R}^{1+q}; H, \tilde{H})$$

for $H := \mathcal{K}^{s,\gamma;e}(X^\wedge)$, $\tilde{H} := \mathcal{K}^{\infty,\gamma-\mu;e}(X^\wedge)$ for some $s, e \in \mathbb{R}$ is called a Green symbol of order ν, belonging to the weight data $\boldsymbol{g} = (\gamma, \gamma - \mu, \Theta)$ if it has the properties

$$g(t, y, \tau, \eta) \in S_{\mathrm{cl}}^{\nu, l}(\Omega \times \mathbb{R}^{1+q}; \mathcal{K}^{s,\gamma;e}(X^\wedge), \mathcal{K}_{\mathcal{P}}^{\infty,\gamma-\mu;\infty}(X^\wedge)) \tag{5.1}$$

and

$$g^*(t, y, \tau, \eta) \in S_{\mathrm{cl}}^{\nu, l}(\Omega \times \mathbb{R}^{1+q}; \mathcal{K}^{s,-\gamma+\mu;e}(X^\wedge), \mathcal{K}_{\mathcal{Q}}^{\infty,-\gamma;\infty}(X^\wedge)) \tag{5.2}$$

for all $s, e \in \mathbb{R}$, where g^* denotes the point wise formal adjoint of g with respect to the scalar product of $\mathcal{K}^{0,0;0}(X^\wedge)$ and g-dependent asymptotic types $\mathcal{P}, \mathcal{Q}$.

Let

$$R_G^{\nu,l}(\Omega \times \mathbb{R}^{1+q}, \boldsymbol{g}) \tag{5.3}$$

denote the space of all Green symbols of that kind. If we want to indicate subspaces for fixed $\mathcal{P}, \mathcal{Q}$ we write

$$R_{\mathrm{G}}^{\nu,l}(\Omega \times \mathbb{R}^{1+q}, \boldsymbol{g})_{\mathcal{P},\mathcal{Q}}.$$

Applying Definition 4.2 we obtain the space of Volterra Green symbols

$$R_{\mathrm{G,V}}^{\nu,l}(\Omega \times \mathbb{R}^{1+q}, \boldsymbol{g}). \tag{5.4}$$

6 Mellin Operators with Asymptotics

Another necessary class are Volterra symbols of Mellin type. Let us first formulate some ingredients of the anisotropic edge calculus. First we need smoothing Mellin symbols with asymptotics. By a Mellin asymptotic type we understand a sequence

$$\mathcal{R} := \{(p_j, n_j)\}_{j \in \mathbb{I}} \subset \mathbb{C} \times \mathbb{N} \tag{6.1}$$

for some index set $\mathbb{I} \subseteq \mathbb{Z}$ such that $\pi_{\mathbb{C}}\mathcal{R} := \{p_j\}_{j\in\mathbb{I}}$ intersects $\{w \in \mathbb{C} : |\mathrm{Re}\, w| \leq N\}$ in a finite set for every $N \in \mathbb{N}$.

The space $M_{\mathcal{R}}^{-\infty}(X)$ of smoothing Mellin symbols with asymptotics of type $\mathcal{R}$ is defined to be the space of all $f(w) \in \mathcal{A}(\mathbb{C}\setminus\pi_{\mathbb{C}}\mathcal{R}, L^{-\infty}(X))$ which are meromorphic with poles at the points $p_j \in \pi_{\mathbb{C}}\mathcal{R}$ of multiplicity $n_j + 1$ for all j. Moreover, for every $\pi_{\mathbb{C}}\mathcal{R}$-excision function χ, (i.e., $\chi \in C^\infty(\mathbb{C})$ such that $\chi(w) \equiv 0$ for $\mathrm{dist}\,(w, \pi_{\mathbb{C}}\mathcal{R}) < \varepsilon_0$, $\chi(w) \equiv 1$ for $\mathrm{dist}\,(w, \pi_{\mathbb{C}}\mathcal{R}) > \varepsilon_1$ for some $0 < \varepsilon_0 < \varepsilon_1 < \infty$), we ask

$$\chi(w)f(w)|_{\Gamma_\beta} \in \mathcal{S}(\Gamma_\beta, L^{-\infty}(X))$$

for every real β, uniformly in compact β-intervals, and moreover, the Laurent coefficients of $f(w)$ at the points $w = p_j$ are operators of finite rank in $L^{-\infty}(X)$ for all j.

We now form operator functions

$$m(t, y, \tau, \eta) = \omega_{(\tau,\eta)} \sum_{j=0}^{N} r^j \sum_{|\alpha|_l \leq j} \mathrm{op}_M^{\gamma_{j\alpha}-n/2}(f_{j\alpha})(t, y) p_\alpha(\tau, \eta)\omega'_{(\tau,\eta)} \tag{6.2}$$

for

$$\omega_{(\tau,\eta)}(r) := \omega(r)e^{-r^l(i\tau+|\eta|^l)}$$

and, analogously, $\omega'_{(\tau,\eta)}$, for cut-off functions $\omega(r)$, $\omega'(r)$. Note that

$$e^{-r^l(i(\tau+i\sigma)+|\eta|^l)}$$

behaves like a Schwartz function for $r \to +\infty$ and $\sigma \leq 0$. In addition, in the region $\{\mathrm{Im}\,\tau \leq 0\}$ this function is homogeneous in the sense

$$e^{-r^l(i\delta^l(\tau+i\sigma)+|\delta\eta|^l)} = e^{-(\delta r)^l(i(\tau+i\sigma)+|\eta|^l)}$$

for $\delta > 0$.

In (6.2) we assume $f_{j\alpha}(t, y, w) \in C^\infty(\Omega, M_{\mathcal{R}_{j\alpha}}^{-\infty}(X))$ such that

$$\gamma - j \leq \gamma_{j\alpha} \leq \gamma, \quad \pi_{\mathbb{C}}\mathcal{R}_{j\alpha} \cap \Gamma_{(n+1)/2-\gamma_{j\alpha}} = \emptyset$$

for all j, α. The functions $p_\alpha(\tau, \eta)$ in (6.2) are polynomials in (τ, η) of anisotropic homogeneity $|\alpha|_l$.

We form the space of anisotropic smoothing Mellin plus Green symbols

$$R_{\mathrm{M+G}}^{\mu,l}(\Omega \times \mathbb{R}^{1+q}, \boldsymbol{g}) \tag{6.3}$$

for $\boldsymbol{g} = (\gamma, \gamma - \mu, \Theta)$ as the set of all $m(t, y, \tau, \eta) + g(t, y, \tau, \eta)$ such that $m(t, y, \tau, \eta)$ is of the form (6.2) and $g(t, y, \tau, \eta) \in R^{\mu,l}(\Omega \times \mathbb{R}^{1+q}, \boldsymbol{g})$. We have

$$R^{\mu,l}_{\mathrm{M+G}}(\Omega \times \mathbb{R}^{1+q}, \boldsymbol{g}) \subset S^{\mu,l}_{\mathrm{cl}}(\Omega \times \mathbb{R}^{1+q}; H, \tilde{H})$$

for

$$H = \mathcal{K}^{s,\gamma}(X^\wedge), \;\; \tilde{H} = \mathcal{K}^{\infty,\gamma-\mu}(X^\wedge), \tag{6.4}$$

and

$$H = \mathcal{K}^{s,\gamma}_{\mathcal{P}}(X^\wedge), \;\; \tilde{H} = \mathcal{K}^{\infty,\gamma-\mu}_{\mathcal{Q}}(X^\wedge), \tag{6.5}$$

for any asymptotic type $\mathcal{P}$ and some resulting $\mathcal{Q}$.

The space of Volterra Mellin plus Green symbols

$$R^{\mu,l}_{\mathrm{M+G},V}(\Omega \times \mathbb{R}^{1+q}, \boldsymbol{g}) \tag{6.6}$$

for $\boldsymbol{g} = (\gamma, \gamma - \mu, \Theta)$ is defined to be the set of all families of operators $m(t, y, \tau, \eta) + g(t, y, \tau, \eta)$ for $m(t, y, \tau, \eta)$ of the form (6.2) and $g(t, y, \tau, \eta) \in R^{\mu,l}_{\mathrm{G},V}(\Omega \times \mathbb{R}^{1+q}, \boldsymbol{g})$.

Proposition 6.1 *We have*

$$R^{\mu,l}_{\mathrm{M+G},V}(\Omega \times \mathbb{R}^{1+q}, \boldsymbol{g}) \subset S^{\mu,l}_{\mathrm{cl},V}(\Omega \times \mathbb{R}^{1+q}; H, \tilde{H}) \tag{6.7}$$

for (6.4), (6.5).

Let us set

$$L^{-\infty}(\Omega, \boldsymbol{g}) = \{\mathrm{Op}_{t,y}(c) : c(t, y, t', y', \tau, \eta) \in R^{-\infty}_{\mathrm{G}}(\Omega \times \Omega \times \mathbb{R}^{1+q}, \boldsymbol{g})\}$$

for $R^{-\infty}_{\mathrm{G}}(\Omega \times \Omega \times \mathbb{R}^{1+q}, \boldsymbol{g}) := \bigcap_{\nu \in \mathbb{R}} R^{\nu,l}_{\mathrm{G}}(\Omega \times \Omega \times \mathbb{R}^{1+q}, \boldsymbol{g})$, where $R^{\nu,l}_{\mathrm{G}}(\Omega \times \Omega \times \mathbb{R}^{1+q}, \boldsymbol{g})$ is a natural generalization of (5.3) to double symbols.

Moreover, $L^{-\infty}_V(\Omega, \boldsymbol{g})$ is defined to be the space of integral operators with kernels $c(t, y, t', y') \in C^\infty(\Omega \times \Omega, \mathcal{L}(\mathcal{K}^{s,\gamma;e}(X^\wedge), \mathcal{K}^{\infty,\gamma;\infty}_{\mathcal{P}}(X^\wedge)))$ for some asymptotic type $\mathcal{P}$ such that $c(t, y, t', y') \equiv 0$ for $t < t'$.

Definition 6.2

(i) Let

$$L^{\mu,l}_{\mathrm{M+G}}(\Omega, \boldsymbol{g}) \tag{6.8}$$

for $\boldsymbol{g} = (\gamma, \gamma - \mu, \Theta)$ defined to be the set of operators

$$A = \mathrm{Op}_{t,y}(a) + C \tag{6.9}$$

for arbitrary

$$a(t, y, \tau, \eta) \in R^{\mu,l}_{\mathrm{M+G}}(\Omega \times \mathbb{R}^{1+q}, \boldsymbol{g}),$$

and $C \in L^{-\infty}(\Omega, \boldsymbol{g})$. The class of those operators (6.9) such that $a(t, y, \tau, \eta) \in R^{\nu,l}_{\mathrm{G}}(\Omega \times \mathbb{R}^{1+q}, \boldsymbol{g})$ will be denoted by

$$L^{\nu,l}_{\mathrm{G}}(\Omega, \boldsymbol{g}).$$

(ii) By

$$L^{\mu,l}_{\mathrm{M+G},V}(\Omega, \boldsymbol{g}) \tag{6.10}$$

we denote the set of operators (6.9) for arbitrary

$$a(t, y, \tau, \eta) \in R^{\mu,l}_{\mathrm{M+G},V}(\Omega \times \mathbb{R}^{1+q}, \boldsymbol{g}), \tag{6.11}$$

and

$$C \in L^{-\infty}_{V}(\Omega, \boldsymbol{g}). \tag{6.12}$$

The class of those operators (6.9) such that $a(t, y, \tau, \eta) \in R^{\nu,l}_{\mathrm{G},V}(\Omega \times \mathbb{R}^{1+q}, \boldsymbol{g})$ and (6.12) will be denoted by

$$L^{\nu,l}_{\mathrm{G},V}(\Omega, \boldsymbol{g}). \tag{6.13}$$

7 Anisotropic Weighted Spaces

Definition 7.1 We define the spaces $\mathcal{W}^{s,l}(\mathbb{R} \times \mathbb{R}^q, H)$ for a Hilbert space H with group action $\kappa = \{\kappa_\delta\}_{\delta\in\mathbb{R}_+}$ as closure of $\mathcal{S}(\mathbb{R} \times \mathbb{R}^q, H)$ with respect to the norm

$$\|u\|_{\mathcal{W}^{s,l}(\mathbb{R}\times\mathbb{R}^q,H)} = \Big\{ \iint \langle\tau, \eta\rangle_l^{2s} \|\kappa^{-1}_{\langle\tau,\eta\rangle_l}(Fu)(\tau, \eta)\|^2_H đ\tau đ\eta\Big\}^{1/2}, \tag{7.1}$$

with $F = F_{t\to\tau,y\to\eta}$ being the Fourier transform in $\mathbb{R} \times \mathbb{R}^q$, also indicated by "hat".

Remark 7.2 It will be convenient to replace (7.1) by

$$\|u\|_{\mathcal{W}^{s,l}(\mathbb{R}\times\mathbb{R}^q,H)} = \Big\{ \iint [\tau, \eta]_l^{2s} \|\kappa^{-1}_{[\tau,\eta]_l}(Fu)(\tau, \eta)\|^2_H đ\tau đ\eta\Big\}^{1/2}, \tag{7.2}$$

which defines an equivalent norm.

If we point out that $\mathcal{W}^{s,l}(\mathbb{R} \times \mathbb{R}^q, H)$ is a Hilbert space, then the corresponding scalar product will depend on the choice of the norm, in the case (7.2) we can set

$$(f, g)_{\mathcal{W}^{s,l}(\mathbb{R}\times\mathbb{R}^q,H)} = \iint [\tau, \eta]_l^{2s} (\kappa_{[\tau,\eta]_l}^{-1} \hat{f}, \kappa_{[\tau,\eta]_l}^{-1} \hat{g})_H đ\tau đ\eta.$$

Example 7.3 The space

$$H = H^s(\mathbb{R}_{\tilde{x}}^{1+n})$$

for $s \in \mathbb{R}$ admits the group action

$$(\kappa_\delta u)(\tilde{x}) := \delta^{\frac{n+1}{2}} u(\delta \tilde{x}),\ \delta \in \mathbb{R}_+.$$

Then we have

$$\mathcal{W}^{s,l}(\mathbb{R}^{1+q}, H^s(\mathbb{R}^{1+n})) = H^{s,l}(\mathbb{R}^{1+q+1+n}),$$

where the space on the right-hand side just coincides with

$$\mathcal{W}^{s,l}(\mathbb{R}^{1+(q+1+n)}, \mathbb{C})$$

for the complex plane, endowed with the action $\mathrm{id}_{\mathbb{C}}$ for all $\delta \in \mathbb{R}_+$.

Proposition 7.4 *There is a canonical embedding* $\mathcal{W}^{s,l}(\mathbb{R}^{1+q}, H) \hookrightarrow \mathcal{S}'(\mathbb{R}^{1+q}, H)$, *defined by*

$$\langle u, \varphi \rangle = \lim_{k\to\infty} \int \varphi(t, y) u_k(t, y) dt dy \tag{7.3}$$

for any $\varphi \in \mathcal{S}(\mathbb{R}^{1+q})$ *and a sequence* $(u_k)_{k=1,\ldots,\infty} \subset \mathcal{S}(\mathbb{R}^{1+q}, H)$ *converging to* u *in the space* $\mathcal{W}^{s,l}(\mathbb{R}^{1+q}, H)$.

Those observations allow us to introduce the space

$$\mathcal{W}_K^{s,l}(\Omega, H) := \{u \in \mathcal{W}^{s,l}(\mathbb{R}^{1+q}, H) : \operatorname{supp} u \subseteq K\}$$

for any open $\Omega \subseteq \mathbb{R}^{1+q}$, $K \Subset \Omega$ which gives us $\mathcal{W}_{\mathrm{comp}}^{s,l}(\Omega, H) = \varinjlim_{K\Subset\Omega} \mathcal{W}_K^{s,l}(\Omega, H)$ which is equal to $\bigcup_{K\Subset\Omega} \mathcal{W}_K^{s,l}(\Omega, H)$, and

$$\mathcal{W}_{\mathrm{loc}}^{s,l}(\Omega, H) := \{u \in \mathcal{S}'(\mathbb{R}^{1+q}, H) : \varphi u \in \mathcal{W}^{s,l}(\mathbb{R}^{1+q}, H) \text{ for every } \varphi \in C_0^\infty(\Omega)\}.$$

Remark 7.5 The spaces

$$\mathcal{W}^{\infty}_{\mathrm{comp/loc}}(\Omega, H) = \bigcap_{s\in\mathbb{R}} \mathcal{W}^{s,l}_{\mathrm{comp/loc}}(\Omega, H)$$

as well as

$$\mathcal{W}^{\infty}(\mathbb{R}^{1+q}, H) = \bigcap_{s\in\mathbb{R}} \mathcal{W}^{s,l}(\mathbb{R}^{1+q}, H)$$

are independent of the involved group action and also of l, cf. the estimates (4.5).

Proposition 7.6 *The operator $\mathcal{M}_\varphi$ of multiplication by $\varphi(t, y) \in \mathcal{S}(\mathbb{R}^{1+q})$, first acting in $C_0^\infty(\mathbb{R}^{1+q}, H)$ extends to a continuous operator*

$$\mathcal{M}_\varphi : \mathcal{W}^{s,l}(\mathbb{R}^{1+q}, H) \to \mathcal{W}^{s,l}(\mathbb{R}^{1+q}, H). \tag{7.4}$$

In other words the operator norm $\|\mathcal{M}_\varphi\|$ is finite and $\varphi \to \mathcal{M}_\varphi$ represents a continuous operator

$$\mathcal{S}(\mathbb{R}^{1+q}) \to \mathcal{L}(\mathcal{W}^{s,l}(\mathbb{R}^{1+q}, H)) \tag{7.5}$$

for any $s \in \mathbb{R}$.

This gives us (7.4) on the anisotropic abstract edge spaces. Moreover the above relations show $C_\varphi \to 0$ as $\varphi \to 0$ in $\mathcal{S}(\mathbb{R}^{1+q})$ which implies the continuity of (7.5).

Let $t_0 \in \mathbb{R}$ and set

$$\mathcal{W}_0^{s,l}([t_0, \infty) \times \mathbb{R}^q, H) := \{u \in \mathcal{W}^{s,l}(\mathbb{R} \times \mathbb{R}^q, H) : \operatorname{supp} u \subseteq [t_0, \infty) \times \mathbb{R}^q\}. \tag{7.6}$$

Note that the space $C_0^\infty((t_0, \infty) \times \mathbb{R}^q, H)$ is dense in (7.6) for every $s \in \mathbb{R}$. Clearly $\mathcal{W}_0^{s,l}([t_0, \infty) \times \mathbb{R}^q, H)$ is a closed subspace of $\mathcal{W}^{s,l}(\mathbb{R} \times \mathbb{R}^q, H)$. Analogously we can form the space $\mathcal{W}_0^{s,l}((-\infty, t_0] \times \mathbb{R}^q, H)$. By virtue of $\mathcal{W}^{s,l}(\mathbb{R}^{1+q}, H) \subset \mathcal{S}'(\mathbb{R}^{1+q}, H)$ it makes sense to pass to restrictions $\mathcal{W}^{s,l}(\mathbb{R}^{1+q}, H)|_{(t_0,\infty)\times\mathbb{R}^q} =: \mathcal{W}^{s,l}((t_0, \infty) \times \mathbb{R}^q, H)$, $\mathcal{W}^{s,l}(\mathbb{R}^{1+q}, H)|_{(-\infty,t_0)\times\mathbb{R}^q} =: \mathcal{W}^{s,l}((-\infty, t_0) \times \mathbb{R}^q, H)$ in the space of H-valued distributions over the respective open sets. This gives rise to linear maps

$$\mathrm{r}_+(t_0) : \mathcal{W}^{s,l}(\mathbb{R}^{1+q}, H) \to \mathcal{W}^{s,l}((t_0, \infty) \times \mathbb{R}^q, H), \tag{7.7}$$

$$\mathrm{r}_-(t_0) : \mathcal{W}^{s,l}(\mathbb{R}^{1+q}, H) \to \mathcal{W}^{s,l}((-\infty, t_0) \times \mathbb{R}^q, H). \tag{7.8}$$

Observe that there are canonical isomorphisms

$$\mathcal{W}^{s,l}((t_0, \infty) \times \mathbb{R}^q, H) \cong \mathcal{W}^{s,l}(\mathbb{R}^{1+q}, H)/\mathcal{W}_0^{s,l}((-\infty, t_0] \times \mathbb{R}^q, H) \tag{7.9}$$

and

$$\mathcal{W}^{s,l}((-\infty, t_0) \times \mathbb{R}^q, H) \cong \mathcal{W}^{s,l}(\mathbb{R}^{1+q}, H)/\mathcal{W}_0^{s,l}([t_0, \infty) \times \mathbb{R}^q, H) \tag{7.10}$$

for every $s \in \mathbb{R}$.

Lemma 7.7 *The space* $C_0^\infty((t_0, \infty) \times \mathbb{R}^q, H)$ *is dense in* $\mathcal{W}_0^{s,l}([t_0, \infty) \times \mathbb{R}^q, H)$.

For the following considerations it makes sense to employ translation operators in the time variable t, defined by

$$(T_c u)(t, y) := u(t + c, y).$$

Clearly those define isomorphisms

$$T_c : \mathcal{W}^{s,l}(\mathbb{R}^{1+q}, H) \to \mathcal{W}^{s,l}(\mathbb{R}^{1+q}, H)$$

for any real s. Combined with restriction operators we obtain isomorphisms

$$T_c : \mathcal{W}_0^{s,l}([t_0, \infty) \times \mathbb{R}^q, H) \to \mathcal{W}_0^{s,l}([t_0 - c, \infty) \times \mathbb{R}^q, H)$$

and

$$T_c : \mathcal{W}_0^{s,l}((-\infty, t_0] \times \mathbb{R}^q, H) \to \mathcal{W}_0^{s,l}((-\infty, t_0 - c] \times \mathbb{R}^q, H)$$

and also by factorization

$$T_c : \mathcal{W}^{s,l}((t_0, \infty) \times \mathbb{R}^q, H) \to \mathcal{W}^{s,l}((t_0 - c, \infty) \times \mathbb{R}^q, H)$$

and

$$T_c : \mathcal{W}^{s,l}((-\infty, t_0) \times \mathbb{R}^q, H) \to \mathcal{W}^{s,l}((-\infty, t_0 - c) \times \mathbb{R}^q, H).$$

Remark 7.8 For any $a(\tau, \eta) \in S^{\mu,l}(\mathbb{R}^{1+q}; H, \tilde{H})$ and $\mathrm{Op}(a) : \mathcal{W}^{s,l}(\mathbb{R}^{1+q}, H) \to \mathcal{W}^{s-\mu,l}(\mathbb{R}^{1+q}, \tilde{H})$ we have $T_{-c}\mathrm{Op}(a)T_c = \mathrm{Op}(a)$ for every $c \in \mathbb{R}$.

Proposition 7.9 *For every* $s \in \mathbb{R}$ *and any* $t_0 \in \mathbb{R}$ *there exist continuous operators*

$$\mathrm{e}_-^s(t_0) : \mathcal{W}^{s,l}((t_0, \infty) \times \mathbb{R}^q, H) \to \mathcal{W}^{s,l}(\mathbb{R}^{1+q}, H)$$

and

$$\mathrm{e}_+^s(t_0) : \mathcal{W}^{s,l}((-\infty, t_0) \times \mathbb{R}^q, H) \to \mathcal{W}^{s,l}(\mathbb{R}^{1+q}, H)$$

such that

$$\mathrm{r}_+(t_0)\mathrm{e}_-^s(t_0) = \mathrm{id} \text{ and } \mathrm{r}_-(t_0)\mathrm{e}_+^s(t_0) = \mathrm{id},$$

respectively. In that way $\mathrm{e}^s_\pm(t_0)$ *have the meaning of corresponding extension operators.*

Proof Let H and L be Hilbert spaces and $T : H \to L$ a continuous surjective operator. Then its adjoint $T^* : L \to H$ is injective and $TT^* : L \to L$ is an isomorphism. Thus we can form $T^*(TT^*)^{-1} : L \to H$ and we have $T(T^*(TT^*)^{-1}) = \mathrm{id}_L$, i.e., T has a right inverse. Now for

$$T = \mathrm{r}_+(t_0),\, H = \mathcal{W}^{s,l}(\mathbb{R}^{1+q}, H),$$

$$L = \mathcal{W}^{s,l}((t_0, \infty) \times \mathbb{R}^q, H)$$

we can set $\mathrm{e}^s_-(t_0) = T^*(TT^*)^{-1}$. In an analogous manner we can show the existence of $\mathrm{e}^s_+(t_0)$. □

Theorem 7.10 ([4, Section 1.4.2, Theorem 5]) *For any* $a(\tau, \eta) \in S^{\mu,l}_V(\mathbb{R}^{1+q}; H, \tilde{H})$ *the operator* $\mathrm{Op}(a)$ *induces continuous operators*

$$\mathrm{Op}(a) : \mathcal{W}^{s,l}_0([t_0, \infty) \times \mathbb{R}^q, H) \to \mathcal{W}^{s-\mu,l}_0([t_0, \infty) \times \mathbb{R}^q, \tilde{H})$$

for all $s \in \mathbb{R}$, *and* $t_0 \in \mathbb{R}$.

Theorem 7.11 *Consider* $a(\tau, \eta) \in S^{\mu,l}_V(\mathbb{R}^{1+q}; H, \tilde{H})$; *then* $\mathrm{Op}(a)$ *induces continuous operators*

$$\mathrm{Op}_+(a) := \mathrm{r}_-(t_1)\mathrm{Op}(a)\mathrm{e}^s_+(t_1) : \mathcal{W}^{s,l}((-\infty, t_1) \times \mathbb{R}^q, H) \to \mathcal{W}^{s-\mu,l}((-\infty, t_1) \times \mathbb{R}^q, \tilde{H}) \tag{7.11}$$

for every $t_1 \in \mathbb{R}, s \in \mathbb{R}$, *which are independent of the choice of the extension operator* $\mathrm{e}^s_+(t_1)$.

Proof The continuity is an immediate consequence of the mapping properties of $\mathrm{e}^s_+(t_1)$ of Proposition 7.9 and of (7.8), and of continuity of

$$\mathrm{Op}(a) : \mathcal{W}^{s,l}(\mathbb{R}^{1+q}, H) \to \mathcal{W}^{s-\mu,l}(\mathbb{R}^{1+q}, \tilde{H}).$$

In order to see the independence on the specific choice of the extension operator $\mathrm{e}^s_+(t_1)$ we choose another one $\tilde{\mathrm{e}}^s_+(t_1)$. Then we have $u_+ := \tilde{\mathrm{e}}^s_+(t_1)u - \mathrm{e}^s_+(t_1)u \in \mathcal{W}^{s,l}_0([t_1, \infty) \times \mathbb{R}^q, H)$ and because of Theorem 7.10 also $\mathrm{Op}(a)u_+ \in \mathcal{W}^{s-\mu,l}_0([t_1, \infty) \times \mathbb{R}^q, H)$. This entails $\mathrm{r}_-(t_1)\mathrm{Op}(a)u_+ = 0$ and hence for every $u \in \mathcal{W}^{s,l}((-\infty, t_1) \times \mathbb{R}^q, H)$

$$\mathrm{r}_-(t_1)\mathrm{Op}(a)\tilde{\mathrm{e}}^s_+(t_1)u = \mathrm{r}_-(t_1)\mathrm{Op}(a)\mathrm{e}^s_+(t_1)u$$

which shows the claimed independence. □

Let us set

$$\mathcal{W}_0^{s,l}([t_0,t_1)\times\mathbb{R}^q,H):=\mathrm{r}_-(t_1)\mathcal{W}^{s,l}([t_0,\infty)\times\mathbb{R}^q,H), \tag{7.12}$$

for $H=\mathcal{K}^{s,\gamma}(X^\wedge)$ or $H=\mathcal{K}^{\infty,\gamma-\mu}(X^\wedge)$. Then for $t_0<t_1$ we obtain the following mapping property which is essential for the parabolic theory, namely

Theorem 7.12 *For every $a(\tau,\eta)\in R^{\mu,l}_{\mathrm{M+G},V}(\mathbb{R}^{1+q},\boldsymbol{g})$ for $g=(\gamma,\gamma-\mu,\Theta)$ and any $t_0<t_1$, the operator* (7.11) *induces continuous operators*

$$\mathrm{Op}_+(a):\mathcal{W}_0^{s,l}([t_0,t_1)\times\mathbb{R}^q,\mathcal{K}^{s,\gamma}(X^\wedge))\to\mathcal{W}_0^{s-\mu,l}([t_0,t_1)\times\mathbb{R}^q,\mathcal{K}^{\infty,\gamma-\mu}(X^\wedge))$$

for all $s\in\mathbb{R}$.

Proof The proof is a consequence of Theorems 7.10 and 7.11. □

Corollary 7.13 *Since Volterra symbols are specific symbols in the sense of* (4.6) *from the resulting continuity of associated operators in anisotropic edge spaces and the continuity of*

$$S^{\mu,l}(\mathbb{R}^{1+q}_{\tau,\eta};H,\tilde{H})\to\mathcal{L}(\mathcal{W}^{s,l}(\mathbb{R}^{1+q},H),\mathcal{W}^{s-\mu,l}(\mathbb{R}^{1+q},\tilde{H}))$$

we conclude that the map Op_+ *induces continuous operators*

$$\begin{aligned}\mathrm{Op}_+:&R^{\mu,l}_{\mathrm{M+G},V}(\mathbb{R}^{1+q},\boldsymbol{g})\\ &\to\mathcal{L}(\mathcal{W}_0^{s,l}([t_0,t_1)\times\mathbb{R}^q,\mathcal{K}^{s,\gamma}(X^\wedge)),\mathcal{W}_0^{s-\mu,l}([t_0,t_1)\times\mathbb{R}^q,\mathcal{K}^{\infty,\gamma-\mu}(X^\wedge)))\end{aligned} \tag{7.13}$$

for every $s\in\mathbb{R}$. The preceding discussion concerned the case of Volterra symbols with constant coefficients. By virtue of relation

$$R^{\mu,l}_{\mathrm{M+G},V}(\mathbb{R}^{1+q}_{t,y}\times\mathbb{R}^{1+q}_{\tau,\eta},\boldsymbol{g})=C^\infty(\mathbb{R}^{1+q}_{t,y})\hat{\otimes}_\pi R^{\mu,l}_{\mathrm{M+G},V}(\mathbb{R}^{1+q}_{\tau,\eta},\boldsymbol{g}),$$

the assertions of Theorems 7.10, 7.11 *and* 7.12 *extend to Volterra symbols with variable coefficients, by tensor product arguments. In an analogous manner we can argue for double symbols*

$$a(t,y,t',y',\tau,\eta)\in R^{\mu,l}_{\mathrm{M+G},V}(\mathbb{R}^{1+q}_{t,y}\times\mathbb{R}^{1+q}_{t',y'}\times\mathbb{R}^{1+q}_{\tau,\eta},\boldsymbol{g}). \tag{7.14}$$

Thus we have altogether

Theorem 7.14 *For every symbol* (7.14) *and $t_0<t_1$, the operator*

$$\mathrm{Op}_+(a)=\mathrm{r}_-(t_1)\mathrm{Op}(a)\mathrm{e}^s_+(t_1)$$

induces continuous maps

$$\mathrm{Op}_+ : \mathcal{W}_0^{s,l}([t_0, t_1) \times \mathbb{R}^q, \mathcal{K}^{s,\gamma}(X^\wedge)) \to \mathcal{W}_0^{s-\mu,l}([t_0, t_1) \times \mathbb{R}^q, \mathcal{K}^{\infty,\gamma-\mu}(X^\wedge))$$

for all $s \in \mathbb{R}$. *Those are independent of the specific choice of* $\mathrm{e}_+^s(t_1)$.

The straightforward proof is left to the reader.

8 Global Operators on $\mathbb{R} \times$ Y and $\mathbb{R} \times$ B

There are also variants of the assertions of the preceding Sect. 7, first with respect to symbols which have not necessarily constant coefficients and then globalized with respect to operators between spaces

$$\mathcal{W}^{s,l}(\mathbb{R} \times Y, \mathcal{K}^{s,\gamma}(X^\wedge))$$

obtained by gluing together $\mathcal{W}^{s,l}(\mathbb{R} \times \mathbb{R}^q, \mathcal{K}^{s,\gamma}(X^\wedge))$ by means of a partition of unity on Y and applying push forwards of $\mathcal{W}^{s,l}(\mathbb{R} \times \mathbb{R}^q, \mathcal{K}^{s,\gamma}(X^\wedge))$ to Y via corresponding charts. This also gives us derived spaces

$$\mathcal{W}_0^{s,l}([t_0, t_1) \times Y, \mathcal{K}^{s,\gamma}(X^\wedge))$$

and other spaces of similar kind as in (7.9), (7.10) for Y rather than $\mathbb{R}^q$, etc. Those in turn generate spaces

$$H^{s,\gamma,l}([t_0, t_1) \times B) \tag{8.1}$$

which are contained in $H^{s,l}_{\mathrm{loc}}((\tilde{t}_0, \tilde{t}_1) \times (B \setminus Y))$ for any $\tilde{t}_0 < t_0, \tilde{t}_1 > t_1$. Anisotropic pseudo-differential operators and also operators of Volterra type exist over the open manifold $(\tilde{t}_0, \tilde{t}_1) \times (B \setminus Y)$, namely, $L^{\mu,l}((\tilde{t}_0, \tilde{t}_1) \times (B \setminus Y))$ for any $\mu \in \mathbb{R}$, and also Volterra subspaces

$$L_V^{\mu,l}((\tilde{t}_0, \tilde{t}_1) \times (B \setminus Y)),$$

as well as classical operators

$$L_{\mathrm{cl},V}^{\mu,l}((\tilde{t}_0, \tilde{t}_1) \times (B \setminus Y)).$$

This material partly belongs to the calculus of non-smoothing Volterra and parabolic operators on $\mathbb{R} \times B$ which is not yet studied here in detail. The corresponding general classes of Volterra operators over $\mathbb{R} \times B$ themselves are denoted by

$$L_V^{\mu,l}(\mathbb{R} \times B, \boldsymbol{g})$$

or with subscript "cl, V", for weight data $\boldsymbol{g} = (\gamma, \gamma - \mu, \Theta)$. Subspaces of smoothing Mellin plus Green operators, indicated by subscripts "$M + G$" over $\mathbb{R} \times B$ will be formulated below, and then we come back to the corresponding anisotropic Sobolev spaces over $\mathbb{R} \times B$, cf. formula (8.1). The anisotropic version of edge calculus gives us, in particular, global operators on $\mathbb{R} \times Y$ of smoothing Mellin plus Green type, namely, global analogues of operators in Definition 6.2 (i), denoted by

$$L^{\mu,l}_{\mathrm{M+G}}(\mathbb{R} \times Y, \boldsymbol{g}) \tag{8.2}$$

and

$$L^{\nu,l}_{\mathrm{G}}(\mathbb{R} \times Y, \boldsymbol{g}), \tag{8.3}$$

respectively. In particular, we have spaces

$$L^{-\infty}(\mathbb{R} \times Y, \boldsymbol{g}) \tag{8.4}$$

of smoothing operators over $\mathbb{R} \times Y$. Recall that the spaces (8.2) belong to

$$L^{\mu,l}(\mathbb{R} \times Y; H, \tilde{H})$$

for

$$H = \mathcal{K}^{s,\gamma}(X^\wedge), \tilde{H} = \mathcal{K}^{s-\mu,\gamma-\mu}(X^\wedge).$$

The latter consist of operators of the form

$$A = \sum_{j=1}^{N} \varphi_j((1 \times \chi_j)_*^{-1}\mathrm{Op}_{t,y}(a_j))\varphi_j' + C \tag{8.5}$$

for arbitrary $a_j(t, y, \tau, \eta) \in S^{\mu,l}(\mathbb{R} \times \mathbb{R}^q \times \mathbb{R}^{1+q}; H, \tilde{H})$ and $C \in L^{-\infty}(\mathbb{R} \times Y; H, \tilde{H})$. In order to prepare constructions for parabolicity globally with respect to Y we study a Volterra variant of $L^{\mu,l}_{\mathrm{M+G}}(\mathbb{R} \times Y, \boldsymbol{g})$. Without loss of generality we assume that the local symbols a_j are invariant under symbol push forwards with respect to χ_j.

Definition 8.1 Let Y be a closed smooth Riemannian manifold, $q = \dim Y$. Then

$$L^{-\infty}_{V}(\mathbb{R} \times Y, \boldsymbol{g})$$

for $\boldsymbol{g} = (\gamma, \gamma - \mu, \Theta)$ is defined as the set of all integral operators with kernels

$$c(t, y, t', y') \in C^\infty((\mathbb{R} \times Y) \times (\mathbb{R} \times Y), \mathcal{L}(\mathcal{K}^{s,\gamma;e}(X^\wedge), \mathcal{K}_{\mathcal{P}}^{\infty,\gamma-\mu;\infty}(X^\wedge)))$$

for all $s, e \in \mathbb{R}$ and some asymptotic type $\mathcal{P}$, such that

$$c(t, y, t', y')|_{\{(t,y,t',y') \in (\mathbb{R}\times Y)^2 : t<t'\}} = 0$$

and its formal adjoint $c^*(t', y', t, y)$ belongs to

$$C^\infty((\mathbb{R} \times Y) \times (\mathbb{R} \times Y), \mathcal{L}(\mathcal{K}^{s,-\gamma+\mu;e}(X^\wedge), \mathcal{K}_{\mathcal{Q}}^{\infty,-\gamma;\infty}(X^\wedge))),$$

for all $s, e \in \mathbb{R}$ and an asymptotic type $\mathcal{Q}$. The Volterra subspace

$$L^{\mu,l}_{\text{M+G},V}(\mathbb{R} \times Y, \boldsymbol{g}) \tag{8.6}$$

of $L^{\mu,l}_{\text{M+G}}(\mathbb{R} \times Y, \boldsymbol{g})$ is defined as the set of all operators (8.5) for arbitrary $a_j(t, y, \tau, \eta) \in R^{\mu,l}_{\text{M+G},V}(\Omega \times \mathbb{R}^{1+q}, \boldsymbol{g})$, cf. formula (6.11), and

$$C \in L_V^{-\infty}(\mathbb{R} \times Y, \boldsymbol{g}).$$

Moreover, $L^{\nu,l}_{\text{G},V}(\mathbb{R} \times Y, \boldsymbol{g})$ is the space of those (8.5) for arbitrary $a_j(t, y, \tau, \eta) \in R^{\nu,l}_{\text{G},V}(\Omega \times \mathbb{R}^{1+q}, \boldsymbol{g})$ and $C \in L_V^{-\infty}(\mathbb{R} \times Y, \boldsymbol{g})$, cf. formula (6.13).

Theorem 8.2 *Let*

$$A \in L^{\mu,l}_{\text{M+G},V}(\mathbb{R} \times Y, \boldsymbol{g_0}), \quad B \in L^{\mu,l}_{\text{M+G},V}(\mathbb{R} \times Y, \boldsymbol{g_1})$$

for $\boldsymbol{g_1} := (\gamma, \gamma - \nu, \Theta)$, $\boldsymbol{g_0} := (\gamma - \nu, \gamma - (\mu + \nu), \Theta)$ *and let* A *or* B *be properly supported. Then we have* $AB \in L^{\mu+\nu,l}_{\text{M+G},V}(\mathbb{R} \times Y, \boldsymbol{g})$ *for* $\boldsymbol{g} = (\gamma, \gamma - (\mu + \nu), \Theta)$. *In addition if* A *and* B *are represented analogously as* (8.5) *where the involved charts are the same for* A, B *and with the same set of functions* $\varphi_j \prec \varphi'_j$, *then* AB *can be written in a similar manner with symbols* $a_j \sharp b_j$, $j = 1, \ldots, N$.

The proof is a special case of corresponding relations in [4, Subsections 1.2.3, 1.2.4].

Remark 8.3 From [4, Theorem 15, page 56 and Definition 1, page 47] we see that for every

$$A \in L^{\mu,l}_{\text{M+G},V}(\mathbb{R} \times Y, \boldsymbol{g}) \text{ for } \boldsymbol{g} = (\gamma, \gamma - \mu, \Theta)$$

it follows that $Au(t, y)|_{t<t_0} \equiv 0$ for every $u \in C_0^\infty(\mathbb{R} \times Y, \mathcal{K}^{s,\gamma}(X^\wedge))$ satisfying $u(t', y')|_{t'<t_0} \equiv 0$ for any $t_0 \in \mathbb{R}$.

We now consider global Sobolev spaces over $\mathbb{R} \times Y$ and observe that Volterra pseudo-differential operators act in the expected way. We may refer to Kegel spaces with group action, and arguments on coordinate invariance under different charts on Y are similar to the isotropic edge calculus. The case of Fréchet subspaces is analogous. Remember that on $\tilde{H} := \mathcal{K}^{0,0}(X^\wedge)$ the group κ_δ^0 is unitary and for every fixed $s, \gamma \in \mathbb{R}$ there is an isomorphism $a : \mathcal{K}^{s,\gamma}(X^\wedge) \to \mathcal{K}^{0,0}(X^\wedge)$ such that $\delta \to \kappa_\delta^0 a \kappa_\delta^{-1}$ belongs to $C^\infty(\mathbb{R}_+, \mathcal{L}(\mathcal{K}^{s,\gamma}(X^\wedge), \mathcal{K}^{0,0}(X^\wedge)))$, cf. [28].

Because of

$$L^{\mu,l}_{\mathrm{M+G},V}(\mathbb{R} \times Y; \boldsymbol{g}) \subset L^{\mu,l}(\mathbb{R} \times Y; H, \tilde{H})$$

where the space of operators $L^{\mu,l}(\mathbb{R} \times Y; H, \tilde{H})$ is known from [5] an operator $A \in L^{\mu,l}_{\mathrm{M+G},V}(\mathbb{R} \times Y; \boldsymbol{g})$ induces continuous operators

$$A : \mathcal{W}^{s,l}_{\mathrm{comp}}(\mathbb{R} \times Y, \mathcal{K}^{s,\gamma}(X^\wedge)) \to \mathcal{W}^{s-\mu,l}_{\mathrm{loc}}(\mathbb{R} \times Y, \tilde{H}),$$

both for $\tilde{H} = \mathcal{K}^{\infty,\gamma-\mu}(X^\wedge)$ and $\tilde{H} = \mathcal{K}^{\infty,\gamma-\mu}_{\mathcal{P}}(X^\wedge)$ for asymptotic types $\mathcal{P}$. For a $t_0 \in \mathbb{R}$ we denote by $\mathcal{W}^{s,l}_0([t_0, \infty) \times Y, H)$ the subspace of all H-valued distributions supported by $[0, \infty) \times Y$. Analogously as the local versions of those spaces we set

$$\mathcal{W}^{s,l}((-\infty, t_0) \times Y, \mathcal{K}^{s,\gamma}(X^\wedge)) := \mathcal{W}^{s,l}(\mathbb{R} \times Y, \mathcal{K}^{s,\gamma}(X^\wedge))/\mathcal{W}^{s,l}_0([t_0, \infty) \times Y, \mathcal{K}^{s,\gamma}(X^\wedge))$$

and for any t_0, t_1 with $t_0 < t_1$

$$\mathcal{W}^{s,l}_0([t_0, t_1) \times Y, \mathcal{K}^{s,\gamma}(X^\wedge)) := \mathrm{r}_-(t_1)\mathcal{W}^{s,l}_0([t_0, \infty) \times Y, \mathcal{K}^{s,\gamma}(X^\wedge)),$$

with $\mathrm{r}_-(t_1)$ being the restriction operator to $(-\infty, t_1) \times Y$. There is then again an extension operator

$$\mathrm{e}^s_+(t_1) : \mathcal{W}^{s,l}_0([t_0, t_1) \times Y, \mathcal{K}^{s,\gamma}(X^\wedge)) \to \mathcal{W}^{s,l}_0([t_0, \infty) \times Y, \mathcal{K}^{s,\gamma}(X^\wedge))$$

which is a right inverse of $\mathrm{r}_-(t_1)$. The corresponding local mapping behaviour gives us the following property.

Theorem 8.4 *Every* $\tilde{A} \in L^{\mu,l}_{\mathrm{M+G},V}(\mathbb{R} \times Y, \boldsymbol{g})$ *induces through* $\mathrm{r}_-(t_1)\tilde{A}\mathrm{e}^s_+(t_1)$ *continuous operators*

$$A : \mathcal{W}^{s,l}_0([t_0, t_1) \times Y, H) \to \mathcal{W}^{s-\mu,l}_0([t_0, t_1) \times Y, \tilde{H})$$

for every $s \in \mathbb{R}$ *which are independent of the specific choice of extension operators* e^s_+*, where* $H = \mathcal{K}^{s,\gamma}(X^\wedge)$, $\tilde{H} = \mathcal{K}^{\infty,\gamma-\mu}_{\mathcal{P}}(X^\wedge)$.

In the proof we could ignore comp/loc aspects in the Sobolev space since Y is a closed manifold.

Theorem 8.5 *Let $A \in L^{\mu,l}_{\mathrm{M+G},V}(\mathbb{R} \times Y, \boldsymbol{g}_0)$ and $B \in L^{\nu,l}_{\mathrm{M+G},V}(\mathbb{R} \times Y, \boldsymbol{g}_1)$ for $\boldsymbol{g}_0, \boldsymbol{g}_1$ as in Theorem 8.2, and let A or B be properly supported. Then for every $s \in \mathbb{R}$ and $t_0 < t_1$ we have*

$$\mathrm{r}_-(t_1)A\mathrm{e}^{s-\nu}_+(t_1)\mathrm{r}_-(t_1)B\mathrm{e}^s_+(t_1) = \mathrm{r}_-(t_1)AB\mathrm{e}^s_+(t_1),$$

for $AB \in L^{\mu+\nu,l}_{\mathrm{M+G},V}(\mathbb{R} \times Y, \boldsymbol{g})$, $\boldsymbol{g} = (\gamma, \gamma - (\mu + \nu), \Theta)$.

Proof For $u(t, y) \in C^\infty_0((t_0, t_1) \times Y, \mathcal{K}^{s,\gamma}(X^\wedge))$ and $\mathrm{r} := \mathrm{r}_-(t_1)$, $\mathrm{e}^s := \mathrm{e}^s_+(t_1)$ in local coordinates on Y the difference

$$\mathrm{r}AB\mathrm{e}^s u(t, y) - \mathrm{r}A\mathrm{e}^{s-\nu}\mathrm{r}B\mathrm{e}^s u(t, y) = (\mathrm{r}A(I - \mathrm{e}^{s-\nu}\mathrm{r})B\mathrm{e}^s)u(t, y)$$

vanishes for $t < t_0$ because of the Volterra property of the involved operators and for $t \in [t_0, t_1]$ since $\mathrm{e}^{s-\nu}\mathrm{r} = \mathrm{id}$ in $[t_0, t_1] \times Y$ and for $t > t_1$ because of the factor r on the right-hand side. □

Let us now extend the operator classes over $\mathbb{R} \times Y$ to the case $\mathbb{R} \times B$ for a manifold B with edge, cf. notation in Sect. 2. By ω_{glob} we denote global cut-off functions on $\mathbb{B}$, i.e., elements $\omega_{\mathrm{glob}} \equiv 1$ in the collar neighbourhood, $\omega_{\mathrm{glob}} \equiv 0$ off some other collar neighbourhood of $\partial\mathbb{B}$. Let $\omega'_{\mathrm{glob}} \succ \omega_{\mathrm{glob}}$ be another such global cut-off function. Then we define the spaces of Mellin plus Green operators

$$L^{\mu,l}_{\mathrm{M+G}}(\mathbb{R} \times B, \boldsymbol{g}) \tag{8.7}$$

consisting of all

$$\omega_{\mathrm{glob}} A \omega'_{\mathrm{glob}} + C$$

for any $A \in L^{\mu,l}_{\mathrm{M+G}}(\mathbb{R} \times Y, \boldsymbol{g})$, $C \in L^{-\infty}(\mathbb{R} \times (\mathbb{R} \times B, \boldsymbol{g}))$. Because of $\varphi L^{\mu,l}_{\mathrm{M+G}}(\mathbb{R} \times Y, \boldsymbol{g})\varphi' \subseteq L^{-\infty}(\mathbb{R} \times (B \setminus Y))$ for every φ, φ' in $C^\infty_0(\mathbb{R} \times Y)$ the space (8.7) is a subspace of $L^{-\infty}(\mathbb{R} \times (B \setminus Y))$ and (8.7) is independent of the specific choice of $\omega_{\mathrm{glob}}, \omega'_{\mathrm{glob}}$. The space (8.7) contains subspaces

$$L^{\mu,l}_{\mathrm{G}}(\mathbb{R} \times B, \boldsymbol{g}) \tag{8.8}$$

of Green operators and also

$$L^{-\infty}(\mathbb{R} \times B, \boldsymbol{g})$$

consisting of the intersection of spaces (8.7) over $\mu \in \mathbb{R}$ where as in (8.4) the dependence on l disappears. In a similar manner we can define Volterra subspaces $L^{\mu,l}_{\mathrm{M+G},V}(\mathbb{R} \times B, \boldsymbol{g})$, $L^{\mu,l}_{\mathrm{G},V}(\mathbb{R} \times B, \boldsymbol{g})$ and $L^{-\infty}_V(\mathbb{R} \times B, \boldsymbol{g})$, respectively, with

$L_V^{-\infty}(\mathbb{R} \times B, \boldsymbol{g})$ being the set of all elements in $L^{-\infty}(\mathbb{R} \times B, \boldsymbol{g})$ which are integral operators with kernels

$$c(t, t') \in \bigcap_{s \in \mathbb{R}} C^\infty\big(\mathbb{R} \times \mathbb{R}, \mathcal{L}(H^{s,\gamma,l}(B), H_{\mathcal{P}}^{\infty,\gamma-\mu,l}(B))\big)$$

for some asymptotic type $\mathcal{P}$, such that

$$c(t, t')|_{\{(t,t') \in \mathbb{R}^2 : t < t'\}} = 0.$$

The above material on

$$L_V^{-\infty}(\mathbb{R} \times Y, \boldsymbol{g}) \text{ and } L_{\mathrm{M+G},V}^{\mu,l}(\mathbb{R} \times Y, \boldsymbol{g})$$

in spaces $\mathcal{W}_0^{0,l}([t_0, t_1) \times Y, \mathcal{K}^{s,\gamma}(X^\wedge))$ extend in a straightforward manner to

$$L_V^{-\infty}(\mathbb{R} \times B, \boldsymbol{g}) \text{ and } L_{\mathrm{M+G},V}^{\mu,l}(\mathbb{R} \times B, \boldsymbol{g}),$$

respectively, in spaces $H^{s,\gamma,l}([t_0, t_1) \times B)$. In particular, there are analogues of Theorem 8.2, Remark 8.3 and Theorems 8.4, 8.5 in spaces globally over $\mathbb{R} \times B$.

9 Inversion of 1+M+G in Anisotropic Edge Spaces

In this section we specify the consideration to anisotropic smoothing Mellin plus Green operators of order zero and weight data $\boldsymbol{g} = (\gamma, \gamma, \Theta)$, and we focus on operators of the form

$$1 + M + G. \tag{9.1}$$

Those have an interior symbol which is $\equiv 1$, and in the parabolic theory for more general non-smoothing terms rather than 1 the subclass of operators (9.1) is reached in the process of constructing Volterra parametrices and parabolic inverses in the general case. This step will be carried out in a forthcoming chapter which is postponed for the moment, but the idea is quite simple. If $A \in L_V^{\mu,l}(B, \boldsymbol{g})$ is parabolic on a manifold B with edge Y then we have, in particular, a Volterra parametrix $A_0^{(-1)}$, and $A_0^{(-1)}A$ as well as $AA_0^{(-1)}$ is just of the form (9.1). Then the inverse $(1 + M + G)^{-1}$ of (9.1) can be composed to A_0^{-1} and this composition gives rise the parabolic inverse A^{-1} of A. In this step we will employ Theorem 9.10 below.

Definition 9.1 A Volterra symbol

$$a(t, y, \tau, \eta) \in 1 + R^{0,l}_{\text{M+G},V}(\mathbb{R}^{1+q} \times \mathbb{R}^{1+q}, \boldsymbol{g})$$

for $\boldsymbol{g} = (\gamma, \gamma, \Theta)$ is called parabolic if for every compact subset $K \subset \mathbb{R}^{1+q}$ there is a constant $R = R(K) > 0$ such that the analytic extension $a(t, y, v, \eta) \in 1 + R^{0,l}_{\text{M+G},V}(\mathbb{R}^{1+q} \times \overline{\mathbb{C}^-} \times \mathbb{R}^q, \boldsymbol{g})$ has the following properties:

(i) The operators $a(t, y, v, \eta) : H \to H$ for $H = \mathcal{K}^{0,\gamma}(X^\wedge)$ are isomorphisms for all $(t, y) \in K$, $(v, \eta) \in \overline{\mathbb{C}^-} \times \mathbb{R}^q$ where $[v, \eta]_l \geq R(K)$,

(ii) There are constants $c = c(K) > 0$ such that

$$\|\kappa^{-1}_{[v,\eta]_l} a(t, y, v, \eta)\kappa_{[v,\eta]_l}\|_{\mathcal{L}(\mathcal{K}^{0,\gamma}(X^\wedge))} \geq c$$

for all $(t, y) \in K$, $(v, \eta) \in \overline{\mathbb{C}^-} \times \mathbb{R}^q$, for $[v, \eta]_l \geq R(K)$.

An $A = \text{Op}(a) + C$ for $a(t, y, v, \eta) \in 1 + R^{0,l}_{\text{M+G},V}(\mathbb{R}^{1+q} \times \overline{\mathbb{C}^-} \times \mathbb{R}^q, \boldsymbol{g})$, $C \in L^{-\infty}_V(\mathbb{R}^{1+q}, \boldsymbol{g})$ is called parabolic if $a(t, y, \tau, \eta)$ is parabolic.

Lemma 9.2 *Let*

$$a(t, y, v, \eta) \in 1 + R^{0,l}_{\text{M+G},V}(\mathbb{R}^{1+q} \times \overline{\mathbb{C}^-} \times \mathbb{R}^q, \boldsymbol{g}).$$

Then for every $R \geq 0$ *we have*

$$(a - T_{-iR}a)|_{\mathbb{R}^{1+q}\times\mathbb{R}^{1+q}} \in R^{-l,l}_{\text{M+G},V}(\mathbb{R}^{1+q} \times \mathbb{R}^{1+q}, \boldsymbol{g})$$

where $(T_{-iR}a)(t, y, v, \eta) = a(t, y, v - iR, \eta)$.

Proposition 9.3 *Let*

$$a(t, y, \tau, \eta) \in 1 + R^{0,l}_{\text{M+G},V}(\mathbb{R}^{1+q} \times \mathbb{R}^{1+q}, \boldsymbol{g}) \tag{9.2}$$

for $\boldsymbol{g} = (\gamma, \gamma, \Theta)$ *be parabolic, cf. Definition* 9.1. *Then there exists a symbol*

$$b(t, y, \tau, \eta) \in 1 + R^{0,l}_{\text{M+G},V}(\mathbb{R}^{1+q} \times \mathbb{R}^{1+q}, \boldsymbol{g}) \tag{9.3}$$

such that

$$b(t, y, \tau, \eta)a(t, y, \tau, \eta) - 1 \in R^{-1,l}_{\text{M+G},V}(\mathbb{R}^{1+q} \times \mathbb{R}^{1+q}, \boldsymbol{g}) \tag{9.4}$$

and

$$a(t, y, \tau, \eta)b(t, y, \tau, \eta) - 1 \in R^{-1,l}_{\text{M+G},V}(\mathbb{R}^{1+q} \times \mathbb{R}^{1+q}, \boldsymbol{g}) \tag{9.5}$$

with 1 *being the symbol representing the identity operator in* $\mathcal{K}^{s,\gamma}(X^\wedge)$ *for any fixed* s.

Proof For the proof we employ symbols which follow from translations in the complex v-variable and take into account Lemma 9.2. We set $\tilde{a}(t, y, v, \eta) := a(t, y, v - iR, \eta)$ for the constant R from Definition 9.1. By virtue of Lemma 9.2 we have $\tilde{a}(t, y, v, \eta) \in 1 + R^{0,l}_{\text{M+G},V}(\mathbb{R}^{1+q} \times \mathbb{R}^{1+q}, \boldsymbol{g})$ and $a - \tilde{a} \in R^{-l,l}_{\text{M+G},V}(\mathbb{R}^{1+q} \times \mathbb{R}^{1+q}, \boldsymbol{g})$. Because of condition (i) in Definition 9.1 the symbol $\tilde{a}(t, y, v, \eta) : \mathcal{K}^{s,\gamma}(X^\wedge) \to \mathcal{K}^{s,\gamma}(X^\wedge)$ is (t, y, v, η)-wise invertible over compact subsets with respect to $y \in \mathbb{R}^q$ when R is sufficiently large. This allows us to form $b(t, y, v, \eta) := (\tilde{a}(t, y, v, \eta))^{-1}$. Holomorphy of $\tilde{a}$ in v yields holomorphy of b in $v \in \mathbb{C}^-$ and a standard conclusion shows that b satisfies relation (9.3). The symbolic estimates for b follow from those for $\tilde{a}$. For a itself we obtain (9.4) for

$$ba = b(\tilde{a} + (a - \tilde{a})) = b\tilde{a} + b(a - \tilde{a}).$$

Concerning multiplication ab we can argue in a similar manner. □

Theorem 9.4 *Assume that for a symbol a satisfying* (9.2) *there is a symbol b satisfying* (9.3) *such that relations* (9.4), (9.5) *hold. Then* $a(t, y, \tau, \eta)$ *is parabolic.*

Proof The arguments are completely analogous to the proof of [4, Theorem 4]. □

Corollary 9.5 *Let a symbol a satisfying* (9.2) *be decomposed as* $a = a_0 + a_{-1}$, *where* $a_0 \in 1 + R^{0,l}_{\text{M+G},V}(\mathbb{R}^{1+q} \times \mathbb{R}^{1+q}, \boldsymbol{g})$ *is parabolic and* $a_{-1} \in R^{-1,l}_{\text{M+G},V}(\mathbb{R}^{1+q} \times \mathbb{R}^{1+q}, \boldsymbol{g})$. *Then a is parabolic if and only if* a_0 *is parabolic.*

Because of

$$1 + R^{0,l}_{\text{M+G},V}(\mathbb{R}^{1+q} \times \mathbb{R}^{1+q}, \boldsymbol{g}) \subset S^{0,l}_{\text{cl},V}(\mathbb{R}^{1+q} \times \mathbb{R}^{1+q}; H, \tilde{H})$$

for $H = \tilde{H} = \mathcal{K}^{s,\gamma}(X^\wedge)$ the observations on general classical parabolic symbols from [4] are valid also for symbols of Definition 9.1. In particular, we have the following remark.

Remark 9.6 An element $a(t, y, \tau, \eta) \in 1 + R^{0,l}_{\text{M+G},V}(\mathbb{R}^{1+q} \times \mathbb{R}^{1+q}, \boldsymbol{g})$ is parabolic if and only if its anisotropic homogeneous principal symbol

$$a_{(0)}(t, y, v, \eta) : \mathcal{K}^{s,\gamma}(X^\wedge) \to \mathcal{K}^{s,\gamma}(X^\wedge)$$

is invertible for all (t, y) and all $(v, \eta) \in (\overline{\mathbb{C}^-} \times \mathbb{R}^q) \setminus \{0\}$.

Theorem 9.7 *For every parabolic operator* $A \in 1 + L^{0,l}_{\mathrm{M+G},V}(\mathbb{R} \times \mathbb{R}^q, \boldsymbol{g})$ *there exists a parametrix* $B \in 1 + L^{0,l}_{\mathrm{M+G},V}(\mathbb{R} \times \mathbb{R}^q, \boldsymbol{g})$ *such that* $BA - 1, AB - 1 \in L^{-\infty}_V(\mathbb{R} \times \mathbb{R}^q, \boldsymbol{g})$.

Proof The assertion is an immediate consequence of Proposition 9.3, using the Leibniz rule expressing symbols of compositions of anisotropic pseudo-differential operators which preserves the Volterra property, cf. Theorem 8.2. □

Let us now pass to global parabolicity. Similarly as the preliminary assumptions before Definition 8.1, now in terms of Volterra symbols, we assume compatibility of local symbols under coordinate changes.

Definition 9.8 An $A = 1 + M + G \in 1 + L^{0,l}_{\mathrm{M+G},V}(\mathbb{R} \times Y, \boldsymbol{g})$, $\boldsymbol{g} = (\gamma, \gamma, \Theta)$, for

$$M + G = \sum_{j=1}^{N} \varphi_j((1 \times \chi_j)^{-1}_{*}\mathrm{Op}(a_j))\varphi'_j + C \in L^{0,l}_{\mathrm{M+G},V}(\mathbb{R} \times Y, \boldsymbol{g}) \tag{9.6}$$

for some $C \in L^{-\infty}_V(\mathbb{R} \times Y, \boldsymbol{g})$ is called parabolic if the symbols a_j, $j = 1, \dots, N$, are parabolic in the sense of Definition 9.1.

We then have

Theorem 9.9 *Every parabolic* $A \in 1 + L^{0,l}_{\mathrm{M+G},V}(\mathbb{R} \times Y, \boldsymbol{g})$ *for* $\boldsymbol{g} = (\gamma, \gamma, \Theta)$, *has a parametrix* $B \in 1 + L^{0,l}_{\mathrm{M+G},V}(\mathbb{R} \times Y, \boldsymbol{g})$ *such that* $BA - 1, AB - 1 \in L^{-\infty}_V(\mathbb{R} \times Y, \boldsymbol{g})$.

In Definition 9.1 we established the space of parabolic operators

$$A \in 1 + L^{0,l}_V(\mathbb{R} \times Y, \boldsymbol{g}). \tag{9.7}$$

Because of the Volterra property they induce continuous operators

$$A : \mathcal{W}^{s,l}_0([0, T) \times Y, \mathcal{K}^{s,\gamma}(X^\wedge)) \to \mathcal{W}^{s,l}_0([0, T) \times Y, \mathcal{K}^{s,\gamma}(X^\wedge)) \tag{9.8}$$

for

$$A := \mathrm{r}_-(T)\tilde{A}\mathrm{e}^s_+(T) \tag{9.9}$$

with the operators $\mathrm{r}_-(T), \mathrm{e}^s_+(T)$ for any fixed $T > 0$, cf. Theorem 8.4. As it has been explained the operator (9.8) does not depend on the specific choice of $\mathrm{r}_-(T), \mathrm{e}^s_+(T)$. An operator (9.9) is called parabolic if (9.7) is parabolic.

Theorem 9.10 *A parabolic operator* (9.9) *induces isomorphisms* (9.8) *between the respective anisotropic Sobolev spaces for every* $s \in \mathbb{R}$ *and any finite* $T > 0$, *and the inverse is of analogous structure for a Volterra parametrix* $\tilde{B}$ *of* $\tilde{A}$.

The proof will refer to

Lemma 9.11 *Consider a cylinder* $Z := [T_0, T_1] \times Y, T_1 > T_0$, *for a compact Riemannian manifold* Y *of dimension* q. *Let*

$$k(t, y, t', y') \in C^N(Z^2, \mathcal{L}(\mathcal{K}^{0,\gamma}(X^\wedge)),$$

for $N \in \mathbb{N}$ *a kernel satisfying for every* $\alpha \in \mathbb{N}^{2(1+q)}$, $|\alpha| \le N$, *the relations*

$$k(t, y, t', y') = 0 \text{ for } T_0 \le t < t' \le T_1, \tag{9.10}$$

and

$$\|\partial^\alpha_{t,y,t',y'} k(t, y, t', y')\|_{\mathcal{L}(\mathcal{K}^{0,\gamma}(X^\wedge))} \le C \text{ for } T_0 \le t < t' \le T_1. \tag{9.11}$$

Then for

$$k_1(t, y, t', y') := k(t, y, t', y'),$$

$$k_{j+1}(t, y, t', y') := \int_Y \int_{T_0}^{T_1} k_j(t, y, s, \tilde{y}) k_1(s, \tilde{y}, t', y') ds dy$$

the series $K(t, y, t', y') = \sum_{j=1}^{\infty} (-1)^j k_j(t, y, t', y')$ *of iterated kernels absolutely converges in* $C^N(Z^2, \mathcal{L}(\mathcal{K}^{0,\gamma}(X^\wedge)))$, *and we have*

$$K(t, y, t', y') = 0 \text{ for } T_0 \le t < t' \le T_1.$$

Proof By induction we show that for every $j \ge 1$ and all summands of the series

$$k_j(t, y, t', y') = 0 \text{ for } T_0 \le t < t' \le T_1,$$

$$\|k_j(t, y, t', y')\|_{\mathcal{L}(H,\tilde{H})} \le C \frac{(MC)^{j-1}}{(j-1)!} (t - t')^{j-1} \text{ for } T_0 \le t < t' \le T_1,$$

and $M := \int_Y 1 dy$. For $j = 1$ these are just the properties (9.10), (9.11). For $j > 1$ and $T_0 \le t < t' \le T_1$ we have

$$k_{j+1}(t, y, t', y') := \int_Y \int_{T_0}^{T_1} k_j(t, y, s, \tilde{y}) k(s, \tilde{y}, t', y') ds dy$$

$$= \int_Y \{ \int_{T_0}^{t'} k_j(t, y, s, \tilde{y}) k(s, \tilde{y}, t', y') ds + \int_{t'}^{T_1} k_j(t, y, s, \tilde{y}) k(s, \tilde{y}, t', y') ds \} dy = 0 \tag{9.12}$$

which entails the first identity. In fact, relation (9.10) for $k(s, \tilde{y}, t', y')$ shows that the first integral vanishes, since by induction assumption on k the first integral vanishes.

The second integral vanishes as well, because $s < t'$ over the integration interval, where $k(s, \tilde{y}, t', y')$ vanishes. This also gives us

$$K(t, y, t', y') = 0 \ \text{for} \ T_0 \leq t < t' \leq T_1.$$

For $T_0 \leq t < t' \leq T_1$ we obtain for $\| \cdot \| := \| \cdot \|_{\mathcal{L}(H,\tilde{H})}$, for $H = \tilde{H} = \mathcal{K}^{0,\gamma}(X^\wedge)$

$$\begin{aligned}
\|k_j(t, y, t', y')\| &= \| \int_Y \int_{T_0}^{T_1} k_j(t, y, s, \tilde{y}) k(s, \tilde{y}, t', y') ds dy \| \\
&= \| \int_Y \{ \int_{T_0}^{t'} k_j(t, y, s, \tilde{y}) k(s, \tilde{y}, t', y') ds + \int_{t'}^{t} k_j(t, y, s, \tilde{y}) k(s, \tilde{y}, t', y') ds \\
&+ \int_t^{T_1} k_j(t, y, s, \tilde{y}) k(s, \tilde{y}, t', y') ds \} dy \| = \| \int_Y \int_{t'}^{t} k_j(t, y, s, \tilde{y}) k(s, \tilde{y}, t', y') ds dy \| \\
&\leq \int_Y \int_{t'}^{t} \|k_j(t, y, s, \tilde{y})\| \|k(s, \tilde{y}, t', y')\| ds dy \leq \int_Y \int_{t'}^{t} C \frac{(MC)^{j-1}}{(j-1)!} (t-s)^{j-1} C ds dy \\
&= C \frac{C(MC)^{j-1}}{(j-1)!} \int_{t'}^{t} (t-s)^{j-1} ds \int_Y 1 dy = C \frac{(MC)^j}{j!} (t-t')^j,
\end{aligned} \tag{9.13}$$

such that also the second condition is satisfied. This implies for $T_0 \leq t < t' \leq T_1$ the relation

$$\sum_{j=1}^{\infty} \|(-1)^j k_j(t, y, t', y')\| \leq C \sum_{j=1}^{\infty} \frac{(MC)^{j-1}}{(j-1)!} (t-t')^{j-1} = Ce^{MC(t-t')} \leq Ce^{MC(T_1-T_0)},$$

which implies together with $K(t, y, t', y') = 0$ for $T_0 \leq t < t' \leq T_1$ the absolute convergence of the series in question in the space $C(Z^2, \mathcal{L}(\mathcal{K}^{0,\gamma}(X^\wedge)))$.

For $N > 0$, according to the assumptions on the corresponding derivatives we also have absolute convergence of the series $\sum_{j=1}^{\infty} (-1)^j \partial^\alpha k_j(t, y, t', y')$ for any $|\alpha| \leq N$, which gives us the assertion also for $N > 0$. □

Proposition 9.12 *For any fixed* $0 < T \in \mathbb{R}$ *and every* $C \in L_V^{-\infty}(\mathbb{R} \times Y, \boldsymbol{g})$ *there exists a* $G \in L_V^{-\infty}(\mathbb{R} \times Y, \boldsymbol{g})$ *such that*

$$\mathrm{r}_-(T)(1 + C)(1 - G)\mathrm{e}_+^s(T)u = u$$

for every $u \in \mathcal{W}_0^{0,l}([0, T) \times Y, \tilde{H})$.

Proof We construct an element $G \in L_V^{-\infty}(\mathbb{R} \times Y, \boldsymbol{g})$ such that

$$(1 + C)(1 - G)u(t, y) = u(t, y)$$

for $t \le T$. Taking into account the fact that

$$C_0^\infty((t_0,\infty)\times\mathbb{R}^q,\mathcal{K}^{0,\gamma}(X^\wedge)) \subset \mathcal{W}_0^{s,l}([t_0,\infty)\times\mathbb{R}^q,\mathcal{K}^{0,\gamma}(X^\wedge))$$

is a dense embedding, cf. Lemma 7.7, the assertion will follow, since in view of the Volterra property the map does not depend on the values of the function for $t > T$.

We first observe that the kernel $k(t,y,t',y')$ of $C \in L_V^{-\infty}(\mathbb{R}\times Y,\boldsymbol{g})$ for $T_0 = -T, T_1 = 2T$ satisfies the assumptions of Lemma 9.11 for any $N \in \mathbb{N}$. We then form an integral operator $\tilde{K}$ with the resulting kernel $\tilde{k}(t,y,t',y') \in C^\infty(([-T,2T]\times Y)^2,\mathcal{L}(\mathcal{K}^{0,\gamma}(X^\wedge)))$, and applying a formal Neumann series argument for $-T \le T_0 < T_1 \le 2T$ we obtain the equation

$$(1+C)(1-\tilde{K})|_{[T_0,T_1]\times Y} = \mathrm{Id}|_{[T_0,T_1]\times Y}.$$

Next we fix a cut-off function $\omega(r) \in C_0^\infty(\mathbb{R})$ with $\omega(r) \equiv 1$ for $|r| \le 1$, $\omega(r) \equiv 0$ for $|r| \ge 2$, and form

$$g(t,y,t',y') := \omega(\frac{2t}{T}-1)\tilde{k}(t,y,t',y')\omega(\frac{2t'}{T}-1).$$

Then the associated integral operator G belongs to $L^{-\infty}(\mathbb{R}\times Y,\boldsymbol{g})$ and has the Volterra property. Moreover, we have

$$g|_{([0,T]\times Y)^2} = \tilde{k}|_{([0,T]\times Y)^2}$$

such that $(1+C)(1-G)$ has the desired mapping property. □

Proof of Theorem 9.10 By definition, the operator $A = \mathrm{r}_-(T)\tilde{A}\mathrm{e}_+^s(T)$ is parabolic if $\tilde{A} \in 1 + L_{\mathrm{M+G},V}^{0,l}(\mathbb{R}\times Y,\boldsymbol{g})$ is parabolic. Then, according to Theorem 9.9 there is a global parametrix $\tilde{B} \in 1 + L_{\mathrm{M+G},V}^{0,l}(\mathbb{R}\times Y,\boldsymbol{g})$. In particular, it follows that

$$C := \tilde{A}\tilde{B} - 1 \in L_V^{-\infty}(\mathbb{R}\times Y,\boldsymbol{g}).$$

Let for abbreviation $\mathrm{r} := \mathrm{r}_-(T), \mathrm{e} := \mathrm{e}_+^s(T)$. Then, by virtue of Proposition 9.12 there is a $G \in L_V^{-\infty}(\mathbb{R}\times Y,\boldsymbol{g})$ such that

$$\begin{aligned}1 &= \mathrm{r}(1+C)(1-G)\mathrm{e} = \mathrm{r}(1+C)(\mathrm{e}\,\mathrm{r})(1-G)\mathrm{e}\\ &= \mathrm{r}\tilde{A}\tilde{B}\mathrm{e}\,\mathrm{r}(1-G)\mathrm{e} = \mathrm{r}\tilde{A}\mathrm{e}\,\mathrm{r}\tilde{B}(1-G)\mathrm{e}.\end{aligned} \tag{9.14}$$

This gives us a right inverse of A which belongs to our class of pseudo-differential operators. In an analogous manner we can construct a left inverse of A, i.e., we altogether obtain, under the assumption of parabolicity, the operator A is invertible. □

References

1. M.S. Agranovich, M.I. Vishik, Elliptic problems with parameter and parabolic problems of general type. Uspekhi Mat. Nauk **19**(3), 53–161 (1964)
2. H. Amann, *Uniformly Regular and Singular Riemannian Manifolds*. Springer Proceedings in Mathematics and Statistics (PROMS), vol. 119 (2015). https://doi.org/10.1007/978-3-319-12547-3-1
3. L. Boutet de Monvel, Boundary problems for pseudo-differential operators. Acta Math. **126**, 11–51 (1971)
4. T. Buchholz, *Parabolische Pseudodifferentialoperatoren mit operatorwertigen Symbolen*, PhD thesis, Universität Potsdam, 1996
5. T. Buchholz, B.-W. Schulze, Anisotropic edge pseudo-differential operators with discrete asymptotics. Mat. Nachr. **184**, 73–125 (1997)
6. Z. Cho Čan, G.I. Eskin, Boundary problems for parabolic systems of pseudo-differential equations. Dokl. Akad. Nauk SSSR **198**, 50–53 (1971) (in Russian)
7. Ch. Dorschfeldt, *Algebras of Pseudo-Differential Operators Near Edge and Corner Singularities*. Mathematical Research, vol. 102 (Akademie Verlag, Berlin, 1998)
8. V. Egorov, B.-W. Schulze, *Pseudo-Differential Operators, Singularities, Applications*. Operator Theory Advances and Applications, vol. 93 (Birkhäuser Verlag, Basel, 1997)
9. J.B. Gil, B.-W. Schulze, J. Seiler, in *Differential Equations, Asymptotic Analysis and Mathematical Physics*, ed. by M. Demuth et al. Mathematical Research, vol. 100 (Akademic Verlag, Berlin, 1997), pp. 113–137
10. J.B. Gil, B.-W. Schulze, J. Seiler, Cone pseudodifferential operators in the edge symbolic calculus. Osaka J. Math. **37**, 221–260 (2000)
11. G. Grubb, *Functional Calculus of Pseudo-Differential Boundary Problems*, 2nd edn. (Birkhäuser Verlag, Boston, 1996)
12. M. Hedayat Mahmoudi, B.-W. Schulze, Corner boundary value problems. Asian Eur. J. Math. (2016). https://doi.org/10.1142/S1793557117500541
13. T. Hirschmann, Functional analysis in cone and edge Sobolev spaces. Ann. Glob. Anal. Geom. **8**(2), 167–192 (1990)
14. L. Hörmander, *The Analysis of Linear Partial Differential Operators*, vols. 1 and 2 (Springer, New York, 1983)
15. D. Kapanadze, B.-W. Schulze, I. Witt, Coordinate invariance of the cone algebra with asymptotics, in *Advances in Partial Differential Equations "Parabolicity, Volterra Calculus, and Conical Singularities"*, ed. by S. Albeverio, M. Demuth, E. Schrohe, B.-W. Schulze. Operator Theory: Advances and Applications, vol. 138 (Birkhäuser Verlag, Basel, 2002), pp. 307–358
16. T. Krainer, *Parabolic Pseudodifferential Operators and Long-Time Asymptotics of Solutions*, PhD thesis, University of Potsdam, 2000
17. T. Krainer, On the inverse of parabolic systems of partial differential equations of general form in an infinite space-time cylinder, in *Advances in Partial Differential Equations "Parabolicity, Volterra Calculus, and Conical Singularities"*, ed. by S. Albeverio, M. Demuth, E. Schrohe, B.-W. Schulze. Operator Theory: Advances and Applications, vol. 138 (Birkhäuser Verlag, Basel, 2002), pp. 93–278
18. H. Kumano-go, *Pseudo-Differential Operators* (The MIT Press, Cambridge, 1981)
19. X. Lyu, B.-W. Schulze, Mellin operators in the edge calculus, Compl. Anal. Oper. Theory **11**, (2015). https://doi.org/10.1007/s11785-015-0511-6
20. H. Mikayelyan, *Parabolic Boundary and Transmission Problems*, PhD thesis, University of Potsdam, 2002
21. A. Piriou, Une classe d'opérateurs pseudo-différentiels du type de Volterra. Ann. Inst. Fourier Grenoble **20**(1), 77–94 (1970)
22. A. Piriou, Problèmes aux limites généraux pour des opérateurs différentiels paraboliques dans un domaine borné. Ann. Inst. Fourier Grenoble **21**(1), 59–78 (1971)

23. S. Rempel, B.-W. Schulze, *Index Theory of Elliptic Boundary Problems* (Akademie-Verlag, Berlin, 1982)
24. S. Rempel, B.-W. Schulze, *Asymptotics for Elliptic Mixed Boundary Problems (Pseudo-Differential and Mellin Operators in Spaces with Conormal Singularity)*. Mathematical Research, vol. 50 (Akademie-Verlag, Berlin, 1989)
25. B.-W. Schulze, A symbolic calculus for a class of parabolic boundary value problems. Mathematical structure-Computational mathematics-Mathematical modelling, 2. Papers dedicated to professor L. Iliev's 70th anniversary, Sofia (1984), pp. 279–286
26. B.-W. Schulze, Pseudo-differential operators on manifolds with edges, in *Symposium of "Partial Differential Equations, Holzhau 1988"*. Teubner-Texte zur Mathematik, vol. 112 (BSB Teubner, Leipzig, 1989), pp. 259–287
27. B.-W. Schulze, *Pseudo-Differential Operators on Manifolds with Singularities* (North-Holland, Amsterdam, 1991)
28. B.-W. Schulze, *Boundary Value Problems and Singular Pseudo-Differential Operators* (Wiley, Chichester, 1998)
29. B.-W. Schulze, J. Seiler, The edge algebra structure of boundary value problems. Ann. Glob. Anal. Geom. **22**, 197–265 (2002)
30. J. Seiler, *Pseudodifferential Calculus on Manifolds with Non-compact Edges*, PhD thesis, University of Potsdam, 1997
31. J. Seiler, Continuity of edge and corner pseudo-differential operators. Math. Nachr. **205**, 163–182 (1999)
32. M.A. Shubin, *Pseudodifferential Operators and Spectral Theory* (Springer, Berlin, 1987)
33. B.J. Sternin, Elliptische und parabolische Gleichungen auf berandeten Mannigfaltigkeiten, die aus Komponenten verschiedener Dimension bestehen (russisch). Trudy Mosk. Ob. **15**, 346–382 (1966)

Bismut's Way of the Malliavin Calculus for Non-Markovian Semi-groups: An Introduction

Rémi Léandre

Abstract We give a review of our recent works related to the Malliavin calculus of Bismut type for non-Markovian generators. Part IV is new and relates the Malliavin calculus and the general theory of elliptic pseudo-differential operators.

Keywords Malliavin calculus · Large deviations estimates · Higher-order parabolic equation · Pseudo-differential operators

Mathematics Subject Classification (2000) Primary 60H07; Secondary 35K05, 35K25

1 Introduction

Let M be a compact Riemannian manifold endowed with its natural Riemannian measure dx (x is the generic element of M). In local coordinates, we can think at the linear space $\mathbb{R}^d$ endowed with the metric $g_{i,j}(x)dx^i \otimes dx^j$ where x are the local coordinates and $x \to (g_{.,.}(x))$ is a smooth function from $\mathbb{R}^d$ into the space of symmetric strictly positive matrix. The Riemannian measure associated is

$$dx = det(g_{.,.})^{-1/2}dx^1 .. \otimes dx^d \tag{1.1}$$

We consider a **linear** symmetric positive operator densely defined on $L^2(dx)$ acting on a space which separates the point on M. This means if f and g belong to this space,

$$\int_M g(x)Lh(x)ds = \int_M h(x)Lg(x)dx \tag{1.2}$$

R. Léandre (✉)
Laboratoire de Mathématiques, Université de Bourgogne-Franche-Comté, Besançon, France
e-mail: remi.leandre@univ-fcomte.fr

S. Molahajloo, M. W. Wong (eds.), *Analysis of Pseudo-Differential Operators*, Trends in Mathematics, https://doi.org/10.1007/978-3-030-05168-6_7

$$\int_M h(x)Lh(x)dx \geq 0 \tag{1.3}$$

It has by abstract theory a self-adjoint extension on $L^2(dx)$, which generates a contraction semi-group P_t on $L^2(dx)$ which solves the heat equation for $t > 0$

$$\frac{\partial}{\partial t}P_t h = -LP_t h \tag{1.4}$$

with initial condition

$$P_0 h = h \tag{1.5}$$

It is a natural question to know if there is a heat kernel:

$$P_t h(x) = \int_M p_t(x, y)h(y)dy \tag{1.6}$$

There are several ways to solve this problem:

- The microlocal analysis [12, 18, 19], which uses as basic tool the Fourier transform and some regularity on the coefficients of L. In the case of a partial differential operator on $\mathbb{R}^d$, this means that $L = \sum a_{(\alpha)}(x)\frac{\partial^{(\alpha)}}{\partial x^{(\alpha)}}$ where (α) is a multiindex and $x \to a_{(\alpha)}(x)$ is smooth.
- The harmonic analysis, which uses as basic tools functional inequalities and does not need any regularity on the coefficients of L [3, 13, 51].
- The Malliavin calculus [20, 44, 49], which works for Markov semi-groups: $P_t f \geq 0$ if $f \geq 0$. The Malliavin calculus requires moreover that the semi-group is represented by a stochastic differential equation.

More precisely, the Malliavin calculus needs a probabilistic representation of the semi-group P_t by using the theory of stochastic differential equations where a flat Brownian motion or a Poisson process plays a fundamental role.

Let us recall the main idea of the Malliavin calculus in the case of the flat Brownian motion. Let us consider the Hilbert space $\mathbb{H}$ of finite energy maps starting from 0 from $[0, 1]$ into $\mathbb{R}^m$ $t \to r_t = (r_t^i)$ endowed with the Hilbert norm

$$\|r\|^2 = \sum_{i=1}^m \int_0^1 |d/dt r_t^i|^2 dt \tag{1.7}$$

We consider the formal Gaussian measure on $\mathbb{H}$ (written in the heuristic way of Feynman path integral)

$$d\mu(r) = 1/Z \exp| - \|r\|^2/2]dD(r) \tag{1.8}$$

where $dD(r)$ is the formal Lebesgue measure on $\mathbb{H}$. Haar measure satisfying all the axioms of measure theory on a group exists if and only if the group is locally compact. (We refer to [2] and [30] to define Haar measure in infinite dimension in a generalized way). This explains that we need to construct this measure on a bigger space, the space of continuous function $C([0,1],\mathbb{R}^m)t \to B_t$ issued from 0 from $[0,1]$ into $\mathbb{R}^m$. There are a lot of Gaussian measures on $C([0,1],\mathbb{R}^m)$ [48] but the law of the Brownian motion is related to the heat equation on $\mathbb{R}^m$

$$\frac{\partial}{\partial t} P_t f(x) = 1/2 \sum_{i=1}^{m} \frac{\partial^2}{\partial x_i^2} P_t f(x) \tag{1.9}$$

We have, namely,

$$P_t h(x) = E[h(B_t + x)] \tag{1.10}$$

if f is a bounded continuous function on $\mathbb{R}^m$. In such a case we have a semi-group operating on continuous function on $\mathbb{R}^m$.

We consider m smooth vector fields on $\mathbb{R}^d$ with bounded derivatives at each order. Vector fields here are considered as first order partial differential operators. We consider the operator

$$L = 1/2 \sum_{i=1}^{m} X_i^2 \tag{1.11}$$

We introduce the Stratonovich differential equation [20, 49] starting from x (vector fields here are considered as vectors which depend smoothly on x).:

$$dx_t(x) = \sum_{i=1}^{m} X_i(x_t(x))dB_t^i \tag{1.12}$$

This is (and not the Itô equation) the correct equation associated to

$$dx_t(r)(x) = \sum_{i=1}^{m} X_i(x_t(h)(x))dr_t^i \tag{1.13}$$

for $r \in \mathbb{H}$ endowed with the formal Gaussian measure $d\mu(r)$.

By Itô Calculus [20, 49], we can show that the semi-group P_t generated by $L = 1/2 \sum_{i=1}^{m} X_i^2$ is related to the diffusion $x_t(x)$ by the formula

$$P_t(h)(x) = E[h(x_t(x))] \tag{1.14}$$

if h is a continuous function on $\mathbb{R}^d$ (in such a case, the semi-group acts on continuous bounded functions on $\mathbb{R}^d$).

Malliavin idea is the following [44]: he differentiates in a generalized sense the Itô map $B_. \to x_t(x)$. If this Itô map is a submersion in a generalized sense (the inverse of the Malliavin matrix belongs to all the L^p), the law of $x_t(x)$ has a smooth density and therefore the semi-group has a heat kernel. Malliavin for that uses a heavy apparatus of differential operations on the Wiener space. Let us recall that there are several pioneering works of the Malliavin calculus [1, 6, 16] motivated by mathematical physics, but only Malliavin calculus is adapted to the study of stochastic differential equations and fits very well to the study of all measures of stochastic analysis.

Bismut [7] don't use this heavy apparatus of differential operations on the Wiener space, by using a suitable Girsanov transformation and a system of convenient stochastic differential equations in cascade associated to the original stochastic differential equation. This allows Bismut's way to get in a simpler way the Malliavin integration by parts for diffusions: if (α) is a multiindex, if $t > 0$,

$$E[h^{(\alpha)} x_t(x))] = E[h(x_t(x)) Q_t^{(\alpha)}] \tag{1.15}$$

where $Q_t^{(\alpha)}$ is a polynomial in the extra components of the system of stochastic differential equations in cascade and in the inverse of the Malliavin matrix.

The fact that only stochastic differential equations in cascade (therefore a system of semi-groups in cascade) appear in Bismut's approach of the Malliavin calculus allows us to interpret Bismut's way of the Malliavin calculus in the theory of semi-group by expulsating the probabilistic language in [31]. We refer to [32, 33] for reviews with some applications.

Léandre [31] uses an elementary integration by parts, which has to be optimized. The main remark is that we can adapt this elementary integration by parts for non-Markovian semi-groups. It is possible to adapt Bismut's way of the Malliavin calculus for non-Markovian semi-groups.

It is divided into two steps:

- An algebra on the semi-group. Only existences on the semi-group are required.
- Estimates on the enlarged semi-group, which are necessary because polynomial function appears in the Malliavin integration by parts which are not bounded, but are performed in the non-Markovian case by the Davies gauge transform (in the Markovian case, they were done by an adaptation in semi-group on the classic Burkholder-Davies-Gundy inequalities of stochastic analysis).

Moreover, Bismut in his seminal work [9] has done an intrinsic integration by part formula for the Brownian motion on a manifold, which overcame the problem that in the standard Malliavin calculus there are a lot of stochastic differential equations which represent the **same** semi-group. In Part IV we perform an intrinsic Malliavin calculus associated to a wide class of pseudo-differential elliptic operator, by performing a variation of the original pseudo-differential operator by a fractional power of it **intrinsically** associated to the original operator. We exhibit the relation between the Malliavin calculus of Bismut type and the general theory of elliptic pseudo-differential operators.

Bismut in his seminal work [9] pointed out the relation between the Malliavin calculus and the large deviation theory for the study of short time asymptotics of the heat-kernel associated to diffusion semi-groups. We refer to the reviews [26, 29, 53], the book [5], and the seminal work [47] for probabilistic methods in short time asymptotics of semi-groups.

Let us recall quickly the main goal of large deviation theory, here of Wentzel-Freidlin type [4, 52] and [54]. We introduce a small parameter and consider the stochastic differential equation with a small parameter starting from x:

$$dx_t^\epsilon(x) = \epsilon \sum_{i=1}^{m} X_i(x_t^\epsilon)(x) dB_t^i \tag{1.16}$$

Wentzel-Freidlin theory allows to get estimates of the type, when $\epsilon \to 0$

$$\lim 2\epsilon^2 Log[P[x_.^\epsilon(x) \in 0] = - \inf_{x_.(h)(x) \in O} \|r\|^2 \tag{1.17}$$

if O is an open subset of $C([0,1], \mathbb{R}^d)$ equipped with the uniform norm. We don't give details of the lot of technicalities in this estimate.

It is possible to adapt [35, 37–40] Wentzel-Freidlin estimates to the case of non-Markovian semi-groups with the normalization of W.K.B. analysis of Maslov school [45] (see [17, 27] for seminal works on W.K.B. analysis). The main remark is that we can get only upper-bounds, because the semi-group does not preserve the positivity in this case. The second remark is that these estimates are valid only for the semi-group, because in this case path space functional integrals are not defined (see [36] for a review and the work [11, 25, 46]). The normalizations are standard in semi-classical analysis but the type of estimates is different. They work for the heat equation and not for the Schrodinger equation.

This allows to fulfill in this non-Markovian context the beautiful request of Bismut's book [5] and to do the marriage between the Malliavin calculus and Wentzel-Freidlin estimates. The main difference is that we have to consider the absolute value of the heat-kernel because in such a case the semi-group does not preserve the positivity such that we get only upper-bound in the studied Varadhan type estimates (Wentzel-Freidlin estimates are still valid for the heat-kernel).

This work is a review paper of several of our works. The main novelty is part IV, which is new.

2 The Case of a Formal Stochastic Differential Equation

Let us consider an elliptic differential operator of order l on a compact manifold M of dimension d. If we perturb it by a strictly lower order operator L_p, it results by the theory of pseudo-differential operator (which is given by the role of the principal symbol of an elliptic operator) that the qualitative behavior (hypoellipticity..) is the same than the qualitative behavior of $L + L_p$. See [12, 18, 19] for various textbooks in analysis about this problem.

Recently, we have introduced an elliptic operator of order $2k$ $L_0 = \sum f_i^{2k}$ where f_i is an orthonormal basis of the Lie algebra of a compact Lie group G of dimension m with generic element g. f_i are considered as right invariant vector fields. We have established the Malliavin calculus of Bismut type for L_0. We consider a polynomial Q of degree strictly smaller than $2k$ in the vector fields f_i with constant components. We consider the total operator

$$L = L_0 + Q \tag{2.1}$$

The goal of this part by using a small interpretation of [41] and [42] is to adapt in this present situation the strategy of [41] for diffusions. (Léandre [41, 42] used the machinery of the Malliavin calculus [7] translated in semi-group theory for diffusions in [31].) Malliavin matrix plays here a fundamental role in the optimization of the integration by parts in order to arrive to full Malliavin integration by parts. All formulas are **formally the same** if we add or do not add the perturbation of the main operator.

We consider the elliptic operator on $G \times \mathbb{R}$

$$Q + \sum_i f_i^{2k} + \sum r_{i,t} f_i \frac{\partial}{\partial u} + \frac{\partial^{2k}}{\partial u^{2k}} = \tilde{L}_t^r \tag{2.2}$$

It generates by elliptic theory a semi-group on $C_b(G \times \mathbb{R})$, the space of bounded continuous function on $G \times \mathbb{R}$ endowed with the uniform norm.

Theorem 2.1 (Elementary Integration by Parts Formula) *We have if h is smooth with compact support*

$$\int_0^t P_{t-s} \sum h_{s,i} f_i P_s[h] ds = \tilde{P}_t^h[uh](.,0) \tag{2.3}$$

Proof It is the same proof than the proof of Theorem 3 of [42]. □

Let $V = G \times M_d$. M_d is the space of symmetric matrices on $LieG$. $(x, v) \in V$. v is called the Malliavin matrix. We consider

$$\hat{X}_0 = (0, \sum < g^{-1} f_i, . >^2) \tag{2.4}$$

We consider the Malliavin generator (we skip the problems of signs)

$$\hat{L} = \sum f_i^{2k} - \hat{X}_0 \tag{2.5}$$

Theorem 2.2 *$\hat{L}$ spans a semi-group. $\hat{P}_t$ called the Malliavin semi-group on $C_b(M)$.*

Proof It is the same proof of theorem 4 of [42] since Q is a polynomial with constant components in the f_i and L generates a $C_b(G)$ semi-group. The proof leads to some difficulties because the Malliavin operator is not the perturbation of an elliptic operator and uses the Volterra expansion. □

The Malliavin semi-group will allow us to get suitable integration by parts formulas 2. We have the main theorem of this paper:

Theorem 2.3 (Malliavin) *If the Malliavin condition holds*

$$|\hat{P}_t][v^{-p}](g, 0) < \infty \tag{2.6}$$

for all integer positive integer p, P_t has a heat-kernel.

Proof It is the same proof as in the beginning of the proof of theorem 6 of [42]. Under Malliavin assumption, we can optimize the elementary integration by part of Theorem 2, in order to get, according to the framework of the Malliavin calculus, the inequality for any smooth function h on G

$$|P_t[< dh, f_i >]| \leq C\|f\|_\infty \tag{2.7}$$

□

Remark Let us explain quickly the philosophy of this theorem, when there is no perturbation term. We consider a set of path in $\mathbb{R}^m$ denoted r_t^i which represent the semi-group associated to $\sum_i \frac{\partial^{2k}}{\partial u_i^{2k}}$. We don't enter into the problem of signs. We consider the formal stochastic differential equation

$$dx_t(r)(e) = \sum_i f_i dr_t^i \tag{2.8}$$

issued from e. Formally, this represents the semi-group P_t without the perturbation term

$$P_t[h](e) = \text{“}E\text{”}[f(x_t(e)] \tag{2.9}$$

Malliavin assumption expresses in some sense that the “Itô” map $r_. \to x_t(e)$ is a submersion.

By this inequality, we deduce according to the framework of the Malliavin calculus that

$$P_t[h](e) = \int_G h(g) p_t(e, g) dg \tag{2.10}$$

for a nonstrictly positive heat-kernel p_t (dg) denotes the normalized Haar measure on G), if the Malliavin assumption is satisfied.

Theorem 2.4 *Under the previous elliptic assumptions,*

$$|\hat{P}_t|[|v^{-p}|]](g_0, 0) < \infty \tag{2.11}$$

if $t > 0$

Proof It is the same proof than the proof of theorem 8 of [42]. It is based upon the initial strategy to invert the Malliavin matrix in stochastic analysis by slicing the time interval in small time intervals. Only the main part of the generator plays the main role in this strategy because we are in an elliptic case. □

We can iterate the integration by parts formulas, by introducing a system of semi-groups in cascade. We deduce the theorem:

Theorem 2.5 *If $t > 0$, the semi-group P_t has a smooth heat kernel*

$$P_t([h](g) = \int_G p_t(g, g')h(g')dg' \tag{2.12}$$

We remark that the heat kernel can change of sign. This theorem is classical in analysis [51] but it enters in our general strategy to implement stochastic tools in the general theory of linear semi-groups.

In order to simplify the computation, we have used the symmetry of the group. In the next part, we will use fully the symmetry of the group to simplify the computations.

3 The Full Use of the Symmetry of the Group

Let us recall what is a pseudo-differential operator on $\mathbb{R}^d$ [12, 17, 18]. Let be a smooth function $a(x, \xi)$ from $\mathbb{R}^d \times \mathbb{R}^d$ with values in $\mathbb{C}$. We suppose that

$$\sup_{x \in \mathbb{R}^d} |D_x^r D_\xi^l a(x, \xi)| \leq C|\xi|^{m-l} + C \tag{3.1}$$

We suppose that

$$\inf_{x \in \mathbb{R}^d} |a(x, \xi)| \geq C|\xi|^{m'} \tag{3.2}$$

for $|\xi| > C$ for a suitable $m' > 0$. Let $\hat{h}$ be the Fourier transform of the continuous function h. We consider the operator L defines on smooth function h by :

$$\hat{Lh}(x) = \int_{\mathbb{R}^d} a(x, \xi)\hat{h}(\xi)d\xi \tag{3.3}$$

L is said to be a pesudodifferential operator elliptic of order larger than m' with symbol a. This property is invariant if we do a diffeomorphism on $\mathbb{R}^d$ with bounded derivatives at each order. This remark allows to define by using charts a pseudo-differential operator elliptic of order larger than m' on a compact manifold M.

Let f^i be a basis of T_eG. We can consider rightinvariant vector fields. This means that if we consider the action $R_{g_0}\ h \to (g \to h(gg_0))$ on smooth function h on G, we have

$$R_{g_0}(f^i h) = f^i(R_{g_0}h). \tag{3.4}$$

We consider a rightinvariant elliptic pseudo-differential positive operator L of order larger than $2k$ on G . It generates by elliptic theory a semi-group P_t on $L^2(dg)$ and even on $C_b(G)$ the space of continuous functions on G endowed with the uniform norm.

Theorem 3.1 *If $t > 0$,*

$$P_t h(g_0) = \int_G p_t(g_0, g)h(g)dg \tag{3.5}$$

where $g \to p_t(g_0, g)$ is smooth if h is continuous.

This theorem is classical in analysis , but it enters in our general program to implement stochastic analysis tool in the theory of non-Markovian semi-group. See the review [36] for that. See [41, 42] for another presentation where the Malliavin matrix plays a key role. Here we don't use the Malliavin matrix. See [43] for the case of rightinvariant differential operators. The proof is divided into two steps.

3.1 Algebraic Scheme of the Proof: Malliavin Integration by Parts

We consider the family of operators on $C^\infty(G \times \mathbb{R}^n)$:

$$\tilde{L}_t^n = L + \sum_{i=1}^n f^{j_i} \frac{\partial}{\partial u_i} \alpha_t^i + \sum_{i=1}^n \frac{\partial^{2k}}{\partial u_i^{2k}} \tag{3.6}$$

α_t^i are smooth function from $\mathbb{R}^+$ into $\mathbb{R}$. By elliptic theory, $\tilde{L}_t^n$ generates a semi-group $\tilde{P}_t^n$ on $C_b(G \times \mathbb{R}^n)$. This semi-group is time inhomogeneous.

$$\tilde{P}_t^{n+1}[h(g)h^n(u)v](.,.,0) = \int_0^t \tilde{P}_{t,s}^n[f^{j+1}\alpha_s^{n+1}\tilde{P}_s^n[h(g)h^n(u)](.,.) \tag{3.7}$$

Moreover

$$\tilde{P}_t^{n+1}[uh(.)h^n(.)](.,.,u_{n+1}) = \tilde{P}_t^{n+1}[uh(.)h^n(.)](.,.,0) + \tilde{P}_t^n[h(.)h^n(.)](.,.)u_{n+1} \tag{3.8}$$

h is a function of g, h^n a function of $u_1, \ldots, u_n$. This comes from the fact that $\frac{\partial}{\partial u_{n+1}}$ commute with the considered operator.

Therefore the two sides of (3.8) satisfy the same parabolic equation with second member. We deduce that

$$\tilde{P}_t^{n+1}[u_{n+1}\prod_{j=1}^n u_j h(.)](.,.,0) = \int_0^t ds\tilde{P}_{t,s}^n[f^{j_{n+1}}\alpha_s^{n+1}\tilde{P}_s^n[h\prod_{j=1}^n u_j]](.,.) \tag{3.9}$$

This is an integration by parts formula. We would like to present this formula in a more appropriate way for our object.

We consider the operator

$$\overline{L}^n = L + \sum_{j=1}^n \frac{\partial^{2k}}{\partial u_j^{2k}} \tag{3.10}$$

It generates a semi-group $\overline{P}_t^n$. In the sequel we will skip the problem of sign coming if k is even or not.

We introduce a suitable generator

$$\tilde{R}_t^{n+1} = \overline{L}^n + F_s \tag{3.11}$$

by taking care of the relation $[f^i, f^j] = \sum_k \lambda_k^{i,j} f^k$. It is an operator of the type studied. It generates therefore a time inhomogeneous semi-group $\tilde{Q}_t^n$. Therefore the integration by parts formula (3.9) can be written in a more suitable way

$$\tilde{P}_t^{n+1}[u_{n+1}\prod_{j=1}^n u_j h(.)](.,.,0) = \int_0^t \alpha_s^{n+1} ds\tilde{P}_t^n[f^{j_{n+1}}h\prod_{i=1}^n u_i](.,.)+$$

$$\int_0^t \alpha_s^{n+1} ds\tilde{P}_{t,s}^n\tilde{Q}_s^n[h\prod_{i=1}^n u_i](.,.) \tag{3.12}$$

We do the following recursion hypothesis on l:

Hypothesis (l) There exists a positive real r_l such that if (α) is a multiindex of length smaller than l

$$|\tilde{P}_t^n[f^{(\alpha)}h\prod_{i=n}^n u_i](g,v.)| \leq Ct^{-r_l}\|h\|_\infty(1+\prod_{i=n}^n |v_i|) \tag{3.13}$$

where $\|.\|_\infty$ is the uniform norm of h.

It is true for $l = 1$ by (3.9) and the estimates which follow.

If it is true for l, it is still true for $l + 1$, by using (3.12) for $f^{(\alpha)}h$ and taking $\alpha_s^{n+1} = s^{r_l}$.

By choosing suitable α_t^j, we have according to the framework of the Malliavin calculus for any multiindex (α)

$$|P_t[f^{(\alpha)}h](g_0)| \le C_{(\alpha)}\|h\|_\infty \tag{3.14}$$

in order to conclude.

3.2 Estimates: The Davies Gauge Transform

We do as in [43] (26). The problem is that in $\tilde{P}_t^n[h\prod_{j=1}^n u_j](.,.)$ the test function u_j are not bounded and that $\tilde{P}_t^n$ acts only on $C_b(G\times\mathbb{R}^n)$. We do as in [3] the Davies gauge transform $\prod_l^n g(u_i)$ where

$$g(u) = (|u|) \tag{3.15}$$

if u is big and g is smooth strictly positive.

This gauge transform acts on the original operator by the simple formula $(\prod_{i=1}^n g(u_i))^{-1}\tilde{L}_1^n((\prod_{i=1}^n g(u_i).)$. On the semi-group it acts as

$$(\prod_{i=1}^n g(.))^{-1}\tilde{P}_t^n[(\prod_{i=1}^n g(u_i)h(.)h^n(.)](.,.) \tag{3.16}$$

But

$$(g(u_i))^{-1}\frac{\partial}{\partial u_i}(g(u_i).) = \frac{\partial}{\partial u_i} + C(u_i) \tag{3.17}$$

where the potential $C(u_i)$ is smooth with bounded derivatives at each order. Therefore the transformed semi-group acts on $C_b(G \times R^n)$.

Remark We can consider a particular case [43] Let G be a compact connected Lie group, with generic element g endowed with its bi-invariant Riemannian structure and with its normalized Haar measure dg. e is the unit element of G.

Let f^i be a basis of T_eG. We can consider rightinvariant vector fields. This means that if we consider the action $R_{g_0}\ h \to (g \to h(gg_0))$ on smooth function h on G, we have

$$R_{g_0}(f^i h) = f^i(R_{g_0}h). \tag{3.18}$$

Let be $\xi^{(\alpha)} = \xi^{\alpha_1} \ldots .\xi^{\alpha_{|\alpha|}}$ and let be $f^{(\alpha)} = f^{\alpha_1}..f^{\alpha_{|\alpha|}}$. (α) is a multi-index of length $|\alpha|$.

We consider a matrix $a_{\alpha,\beta}$ for multiindices of length k, which is supposedly symmetric strictly positive.

We consider the operator

$$L = \sum_{(\alpha),(\beta)} f^{(\alpha)} a_{(\alpha),(\beta)} f^{(\beta)} \tag{3.19}$$

According to [51], $(-1)^k L$ is a positive symmetric densely elliptic defined operator on $L^2(G)$, which generates by elliptic theory a semi-group acting on $C_b(G)$, the space of continuous function on G. In such a case, we have a heat-kernel associated to the semi-group (See [43]). The case of a rightinvariant differential operator has exactly the same proof than the case of theorem 6, where the details will be presented elsewhere. See [14] for the general case.

4 The Case of an Intrinsic Variation

Let L be a strictly positive self-adjoint operator on a compact manifold M. We suppose that L is a pseudo-differential elliptic operator of order $l \geq 2k$ for an integer $k \geq 1$. It generates a contraction semi-group on $L^2(M)$ and by ellipticity a semi-group on $C_b(M)$. See [8] and [23, 24] in the Markovian case.

Theorem 4.1 *There is a heat-kernel $p_t(x, y)$ associated to P_t. If $t > 0$*

$$P_t(h)(x) = \int_M p_t(x, y)h(y)dy \tag{4.1}$$

where $y \to p_t(x, y)$ is smooth.

The proof is divided into two steps:

4.1 Algebraic Scheme of the Proof: Malliavin Integration by Parts

Let α belong to $]0, 1[$. The fractional power [50] L^α is still a strictly positive pseudo-differential operator elliptic of order αl, which commutes with L. We skip up later the problem if k is even or not. We consider the operator on $C^\infty(M \times \mathbb{R}^n)$

$$\tilde{L}_s^n = L + s^r L^\alpha \frac{\partial}{\partial u_n} + \sum_{i=1}^n \frac{\partial^{2k}}{\partial u_i^{2k}} \tag{4.2}$$

It is an elliptic operator of order $2k$ on $M \times \mathbb{R}^n$. The main part

$$\overline{L}^n = L + \sum_{i=1}^{n} \frac{\partial^{2k}}{\partial u_i^{2k}} \tag{4.3}$$

is positive and is essentially self-adjoint. Therefore the main part generates a semi-group on $C_b(M \times \mathbb{R}^n)$. This remains true for $\tilde{L}^n$ because $\tilde{L}^n$ is a perturbation of $\overline{L}^n$ by a strictly lower operator. We call this semi-group $\tilde{P}_t^n$.

The main remark is that L^α commutes with $\tilde{L}^n$ such that

$$L^\alpha \tilde{P}_t^n = \tilde{P}_t^n L^\alpha \tag{4.4}$$

According to the beginning of the previous part, we get the elementary integration by part

$$\tilde{P}_t^{n+1}[h \prod_{i=1}^{n} u_i u](x, v_i, 0) = \int_0^t P_{t-s}^n[s^r L^\alpha \tilde{P}_s^n[h \prod_{i=1}^{n} u_i]](x, v_i) =$$
$$\tilde{P}_t^n[L^\alpha h \prod_{i=1}^{n} u_i](x, u_i) \int_0^t s^r ds \tag{4.5}$$

Suppose by induction on l that

$$|\tilde{P}_t^n[(L^\alpha)^l h \prod_{i=1}^{n} u_i](x, v_i)| \le C t^{-r(l)} \|h\|_\infty (1 + \prod_{i=1}^{n} |v_i|) \tag{4.6}$$

By applying the elementary integration by parts (4.5) to $(L^\alpha)^l)f$, and choosing $r = r(l)$, we deduce our result. Therefore we have the inequality

$$|P_t[(L^\alpha)^l h](x)| \le C t^{-r(l)} \|h\|_\infty \tag{4.7}$$

The result follows from the fact that L^α is an elliptic operator.

4.2 Estimates: The Davies Gauge Transform

We do as in [43] (26). The problem is that in $\tilde{P}_t^n[h \prod_{j=1}^n u_j](.,.)$ the test function u_j are not bounded and that $\tilde{P}_t^n$ acts only on $C_b(G \times \mathbb{R}^n)$. We do as in [35] the Davies gauge transform $\prod_I^n g(u_i)$ where

$$g(u) = (|u|) \tag{4.8}$$

if u is big and g is smooth strictly positive.

This gauge transform acts on the original operator by the simple formula $(\prod_{i=1}^n g(u_i))^{-1}\tilde{L}_1^n((\prod_{i=1}^n g(u_i).)$. On the semi-group it acts as

$$(\prod_{i=1}^{n} g(.))^{-1}\tilde{P}_t^n[(\prod_{i=1}^{n} g(u_i)h(.)h^n(.)](.,.) \tag{4.9}$$

But

$$(g(u_i))^{-1}\frac{\partial}{\partial u_i}(g(u_i).) = \frac{\partial}{\partial u_i} + C(u_i) \tag{4.10}$$

where the potential $C(u_i)$ is smooth with bounded derivatives at each order. Therefore the transformed semi-group acts on $C_b(G \times R^n)$. It remains to choose

$$h^n(u_.) = \prod_{j=1}^{n} \frac{u_j}{g(u_j)} \tag{4.11}$$

in order to conclude. We deduce the bound:

$$|\tilde{P}_t^n|[h\prod_{j=1}^{n}|u_j|](.;v_.) \leq C(\|h\|_\infty(1+\prod_{i=n}^{n}|v_i|) \tag{4.12}$$

where $|\tilde{P}_t^n|$ is the absolute value of the semi-group $\tilde{P}_t^n$.

Remark We could show that $(x, y) \to p_t(x, y)$ is smooth if $t > 0$ by the same argument.

Remark We can replace the hypothesis L strictly positive by the hypothesis L positive by replacing L^α by $(L + CI_d)^\alpha$ where $C > 0$.

5 Wentzel-Freidlin Estimates for the Semi-Group Only

We consider a differential operator of order $2k$ on the compact manifold M which is supposedly elliptic of order $2k$ and strictly positive. We suppose we can write it as

$$L = \sum_{j=0}^{2k}\sum_{i=0}^{r(j)}(X_{i,j})^j \tag{5.1}$$

where $X_{i,j}$ are smooth vector fields on M. The ellipticity assumption states that

$$\sum_{i=0}^{r(2k)} < X_{i,2k}, \xi >^{2k} = H(x,\xi) \geq C|\xi|^{2k} \tag{5.2}$$

To the Hamiltonian H, we introduce the Lagrangian

$$L(x, p) = \sup_{\xi}(< p, \xi > -H(x, \xi)) \tag{5.3}$$

We get the estimate

$$-C + C|p|^{\frac{2k}{2k-1}} \leq L(x, p) \leq C + |p|^{\frac{2k}{2k-1}} \tag{5.4}$$

for some strictly positive constants C.

If ϕ is a continuous piecewise differentiable path on M, we put:

$$S(\phi) = \int_0^1 L(\phi(t), d/dt\phi(t))dt \tag{5.5}$$

and we put

$$l(x, y) = \inf_{\phi(0)=x, \phi(1)=y} S(\phi) \tag{5.6}$$

By Ascoli theorem, $(x, y) \to l(x, y)$ is a continuous function on $M \times M$.

Theorem 5.1 (Wentzel-Freidlin) *If O is an open ball of M, we have when $t \to 0$*

$$\overline{\lim} t^{\frac{1}{2k-1}} \log |P_t|(1_O](x) \leq - \inf_{y \in O} l(x, y) \tag{5.7}$$

Proof We put $\epsilon = t^{\frac{1}{2k-1}}$. According to the normalization of Maslov school [37], we consider the semi-group P_s^ϵ associated to $L_\epsilon = \epsilon^{2k-1}L$. Moreover

$$P_t = P_1^t \tag{5.8}$$

where P_s^t is associated to tL ([10]). The result will arise if we show when $\epsilon \to 0$

$$\overline{\lim} \epsilon \log |P_1^\epsilon|(1_O](x) \leq - \inf_{y \in O} l(x, y) \tag{5.9}$$

The main ingredient is: □

Lemma 5.2 *For all $\delta > 0$, all C, there exists s_δ such that if $s < s_\delta$*

$$|P_s^\epsilon|[1_{B(x,\delta)^c}](x) \leq \exp[-C/\epsilon] \tag{5.10}$$

where $B(x, \delta)$ is the ball of radius δ and center x.

Proof We imbed M in a linear space. We consider the semi-group

$$Q_s^\epsilon(h)(x) = \exp[- < x, \xi > /\epsilon] P_s^\epsilon[\exp[< x', \xi > /\epsilon] h(x')](x) \tag{5.11}$$

Its generator is

$$\overline{L}_\epsilon + H(x, \xi)/\epsilon \tag{5.12}$$

$$\overline{L}_\epsilon = L_\epsilon + R_\epsilon \tag{5.13}$$

In the perturbation term R_ϵ, there are only differential operators of order l, $l \in]0, 2k[$. When a differential operator of degree l appears, there is a power of at least $l - 1$ of ϵ which appears and a power of ξ at most $2k$ which appears.

Let us consider in a small neighborhood of x the diffeomorphism

$$\Psi_\epsilon : y \to x + \frac{y - x}{\epsilon^{\frac{2k-1}{2k}}} \tag{5.14}$$

Outside a big neighborhood of x, Ψ_ϵ is the identity.

We consider the measure μ_ϵ

$$f \to P_1^\epsilon[F(\Psi_\epsilon(x))](x) \tag{5.15}$$

Under the transformation Ψ_ϵ, the vector fields $\epsilon^{\frac{2k-1}{2k}} X_{i,j}$ are transformed in the vector field $X_{i,j}(x + \epsilon^{\frac{2k-1}{2k}}(y - x))$. Therefore we can apply the machinery of the previous part in order to show that the measure μ_ϵ has a bounded density $q_\epsilon(x, .)$ when $\epsilon \to 0$.

Let R be a differential operator of order l. We have

$$\int_M g(x) R P_1^\epsilon[h](x) dx = \int_{M \times M} g(x) h(y) R_x p_1^\epsilon(x, y) dx dy \tag{5.16}$$

By symmetry

$$p_1^\epsilon(x, y) = p_1^\epsilon(y, x) \tag{5.17}$$

Then

$$\int_M g(x) R P_1^\epsilon[h](x) dx = \int_M h(y) P_1^\epsilon[Rg](y) dy \tag{5.18}$$

By the previous remark

$$|P_1^\epsilon[Rh](y)| \leq \frac{C}{\epsilon^{l\frac{2k-1}{2k}}} \|h\|_\infty \tag{5.19}$$

Therefore

$$|\int_M g(x)RP_1^\epsilon[h](x)dx| \leq \frac{C}{\epsilon^{l\frac{2k-1}{2k}}}\|g\|_\infty\|h\|_\infty \tag{5.20}$$

We deduce that

$$|RP_1^\epsilon[h](x)| \leq \frac{C}{\epsilon^{l\frac{2k-1}{2k}}}\|h\|_\infty \tag{5.21}$$

We deduce a bound of $R_\epsilon P_s^\epsilon$

$$|R_\epsilon P_s^\epsilon h(x)| \leq \frac{|\xi|^{2k-1}}{s^{\frac{l}{2k}}}\epsilon^{-1+1/k}\|h\|_\infty \tag{5.22}$$

We apply Volterra expansion to Q_s^ϵ. We get

$$|Q_s^\epsilon h| \leq |P_s^\epsilon h| + \sum_{i=1}^\infty |\int_{\Delta_l(s)} I_{s_1,..,s_l}ds_1\ldots ds_l| \tag{5.23}$$

where $\Delta_l(s)$ is the simplex $0 < s_1 < .. < s_l < s$ and

$$I_{s_1,..,s_l} = P_{s_1}^\epsilon(R_\epsilon + H/\epsilon)\ldots P_{s_l-s_{l-1}}^\epsilon(R_\epsilon + H/\epsilon)P_{s-s_{l-1}}^\epsilon h \tag{5.24}$$

We deduce a bound of $|\int_{\Delta_l(s)} I_{s_1,..,s_l}ds_1\ldots ds_l|$ by

$$\frac{|\xi|^{2lk}}{\epsilon^l}\int_{\Delta_l(s)}\prod(s_{i+1}-s_i)^{-\frac{2k-1}{2k}}ds_1..ds_l = \frac{|\xi|^{2lk}}{\epsilon^l}I_l(s) \tag{5.25}$$

We suppose by induction that

$$I_l(s) = \alpha_l s^{l(1+\beta_k)} \tag{5.26}$$

where $\beta_k \in]-1, 0[$. It is still true by the recursion formula

$$I_{l+1}(s) = \int_0^s I_l(u)(s-u)^{-\frac{2k-1}{2k}}du \tag{5.27}$$

We deduce the bound

$$\alpha_l \leq \frac{C^l}{l!} \tag{5.28}$$

Therefore

$$|Q_s^\epsilon h(x)| \leq \exp[Cs|\xi|^{2k}/\epsilon]\|h\|_\infty \tag{5.29}$$

It remains to remark that we have the bound

$$|P_s^\xi|[1_{B(x,\delta)^c}](x) \leq \exp[-\frac{C\delta|\xi|}{\epsilon} + Cs|\xi|^{2k}/\epsilon] \tag{5.30}$$

and to extremize in $|\xi|$ to conclude. □

End of the Proof of Theorem 5.1 We operate as in Freidlin-Wentzel book [54] and as in [35, 38] and [39]. We slice the time interval [0, , 1] in a finite number of time intervals $[s_i, s_{i+1}]$ where we can apply the previous lemma. We deduce a positive measure on the set of polygonal paths, where we can repeat exactly the considerations of [35].◇

Remark This estimate is a semi-classical estimate with different type of estimates of W.K.B. estimates a la Maslov and with a different method. We consider in W.K.B. estimate a symbol of an operator $a(x, \xi)$ and we consider the generator L_ϵ associated with the normalized symbol (a la Maslov) $1/\epsilon a(x, \epsilon\xi)$. Let us suppose that L_ϵ generates a semi-group P_t^ϵ. The object of WKB method is to get **precise** estimates of the semi-group P_1^ϵ when $\epsilon \to 0$. For that people look at a formal asymptotic expansion (we omit to write the initial conditions) of P_1^ϵ of the type

$$\epsilon^{-r} \exp[-l(y)/\epsilon] \sum \epsilon^i C_i(y) \tag{5.31}$$

The function l satisfy a highly non-linear equation (the Hamilton-Jacobi-Belman equation) and $c_i(y)$ satisfy formally a system of linear partial differential equation in cascade. The cost function in theorem $l(x, y)$ is the solution of the highly non-linear Hamilton-Jacobi-Belman equation, which is difficult to solve. We don't have precise asymptotics, we are interested by logarithmic estimates which are totally different with a method totally different. On the other hand, generally semi-classical asymptotics considers the case of the Schrodinger equation.

On $\mathbb{R}^d$ we can speak without any difficulty of the symbol of an operator. Poisson processes, Lévy processes, and jump processes are more or less generated by pseudo-differential operators whose generator satisfy the maximum principle (See [10, 13, 21, 22, 24, 28]). We will present pseudo-differential operators with a type of compensation of stochastic analysis which do not satisfy the maximum principle. The end of this part is extracted from [35] and [40]. Let us consider the generator on $C_\infty(\mathbb{R}^d)$

$$Lf(x) = (-)^{l+1} \int_{\mathbb{R}^d} (f(x+y) - f(x)) - \sum_{i=1}^{2l} < y^{\otimes i}, h^{\otimes i}(x) >) \frac{h(x, y)}{|y|^{2l+d+\alpha}} dy \tag{5.32}$$

$\alpha \in]-1, 0[$ $h(x, y) = 0$ if $|y| > C$ and $h \geq 0$. The measure $\frac{h(x,y)}{|y|^{2l+d+\alpha}} dy$ is called the Lévy measure.

Theorem 5.2 *If $h(x, 0) = 1$, L is an elliptic pseudo-differential generator.*

Definition 5.3 If $h(x, y) = h(y)$, we will say that L is a generalized Lévy generator.

Theorem 5.4 *Suppose that L is of Lévy type and that $h(y) = h(-y)$. L is positive symmetric, and therefore admits by ellipticity a self-adjoint extension on $L^2(\mathbb{R}^d)$, which generates a contraction semi-group on $L^2(\mathbb{R}^d)$ which is still a semi-group on $C_b(\mathbb{R}^d)$.*

Remark The symbol $a(x, \xi)$ of the generator is given by

$$(-)^{l+1} \int_{\mathbb{R}^d} (\exp[\sqrt{-1} < y, \xi > - \sum_{i=1}^{2l} (\frac{(\sqrt{-1} < \xi, y >^i)}{i!} \frac{h(x, y)}{|y|^{2l+d+\alpha}} dy \tag{5.33}$$

The Hamiltonian associated is the symbol in real phase. Let us consider a generator of Lévy type of the previous theorem: it is

$$(-)^{l+1} \int_{\mathbb{R}^d} (\exp[< y, \xi > - \sum_{i=1}^{l} (\frac{< \xi, y >^{2i}}{2i!}) \frac{h(x, y)}{|y|^{2l+d+\alpha}} dy \tag{5.34}$$

The Hamiltonian is smooth, convex, equal to 1 in 0. Associated to it, we consider the Lagrangian:

$$L(p) = \sup_{\xi} (< \xi, p > -H(\xi)) \tag{5.35}$$

If $t \to \phi_t$ is a piecewise differentiable continuous curve in $\mathbb{R}^d$, we consider its action $\int_0^1 dt L(\phi_t, d/dt\phi_t) = S(\phi_.)$ We introduce the control function

$$l(x, y) = \inf_{\phi_0 = x; \phi_1 = y} S(\phi) \tag{5.36}$$

Let us recall that $(x, y) \to l(x, y)$ is positive finite continuous.

We consider the generator associated to $1/\epsilon a(\epsilon\xi)$. This corresponds in the classical case of jump process where the compensation is only of one term to the case of a jump process with more and more jumps which are more and more small [54]. We consider the generator L^ϵ associated to $1/\epsilon a(\epsilon\xi)$. It generates a semi-group P_t^ϵ. We get:

Theorem 5.5 *[Wentzel-Freidlin [35, 40]] When $\epsilon \to 0$, we get if O is an open ball of $\mathbb{R}^d$ if $l + 1$ is even:*

$$\overline{\lim} \epsilon \log |P_1^\epsilon| [1_O](x) \leq - \inf_{y \in O} l(x, y) \tag{5.37}$$

Remark For this type of operator, Wentzel-Freidlin estimates are not related to short time asymptotics.

6 Application: Some Varadhan Estimates

This part follows closely [43]. Only the mechanism of the integration by part is different from [39]. For large deviation estimates with respect to W.K.B normalization in the manner of Maslov [45] for non-Markovian operators, we refer to [38] for instance.

Let us consider the Hamiltonian function from $T^*(G)$ into $\mathbb{R}^+$

$$H(g,\xi)= \sum_{|\alpha|=k,|\beta|=k} < f^{(\alpha)1},\xi > \cdots < f^{(\alpha)_k},\xi > a_{(\alpha),(\beta)} < f^{(\beta)1},\xi > \cdots < f^{(\beta)_k},\xi > \tag{6.1}$$

$H(g,p)$ is positive convex in p. According to the theory of large deviation, we consider the associated Lagrangian

$$L(g,\xi) = \sup_p (< \xi, p > -H(g,\xi)) \tag{6.2}$$

If $t \to \phi_t$ is a curve in the group, we consider its action $\int_0^1 dtL(\phi_t, d/dt\phi_t) = S(\phi_.)$ We introduce the control function

$$l(g_0,g_1) = \inf_{\phi_0=g_0;\phi_1=g_1} S(\phi) \tag{6.3}$$

Let us recall that $(g_0,g_1) \to l(g_0,g_1)$ is positive finite continuous.

We have shown in the previous part that if we consider a small parameter ϵ and if we consider the generator $\epsilon^{2k-1}L$ and the semi-group P_t^ϵ associated and if g_0 and g_1 are not closed , we get for any small ball centered in g_1 uniformly:

$$\overline{Lim}_{\epsilon\to 0}\epsilon Log|P_1^\epsilon|[1_O](g_0) \leq -\inf_{g_1\in O} l(g_0,g_1) \tag{6.4}$$

where$|P_1^\epsilon|$ is the absolute value of the semi-group (See [38]). See for that the previous part.

But $P_t = P_1^t$ where P_s^t is the semi-group associated to tL (See [15]). We put $\epsilon = t^{1/2k-1}$ such that

$$\overline{Lim}_{t\to 0}t^{1/2k-1}Log|P_t|[1_O](g_0) \leq -\inf_{g_1\in O} l(g_0,g_1) \tag{6.5}$$

We consider a smooth positive function χ equal to 0 outside O and equal to 1 on a small open ball centered in g_1 smaller than 1.

We would like to apply the mechanism of Malliavin integration by parts to the measure

$$h \to P_t[h\chi](g_0) \tag{6.6}$$

such that

$$|P_t[\chi f^{(\alpha)}h](g_0)| \leq Ct^{(-r_{(\alpha)})} \exp[\frac{-l(g_0, g_1) + \delta}{t^{1/2k-1}}]\|h\|_\infty \tag{6.7}$$

for a small δ. Since (6.7) is true, we have:

Theorem 6.1 *When $t \to 0$*

$$\overline{Lim}_{t \to 0} t^{1/2k-1} Log|p_t(g_0, g_1)| \leq -l(g_0, g_1) \tag{6.8}$$

References

1. S. Albeverio, R. Hoegh-Krohn, Dirichlet forms and diffusion processes on rigged Hilbert spaces. Z.W. **40**, 1–57 (1977)
2. A. Asada, Regularized calculus: an application of zeta regularization to infinite dimensional geometry. Int. J. Geom. Methods Mod. Phys. **1**, 107–157 (2004)
3. P. Auscher, P. Tchamitchian, Square root problem for divergence operators and related topics. Asterisque **249** (1998) pp. 1–172
4. R. Azencott, Grandes déviations et applications, in *Ecole de probabilités de Saint-Flour*. Lecture Notes in Mathematics, vol. 778 (Springer, Berlin, 1978), pp. 2–176
5. F. Baudoin, *An Introduction to the Geometry of Stochastic Flows* (Imperial College Press, London, 2000)
6. Y. Berezanskii, The self-adjointes of elliptic operators with infinite number of variables. Ukr. Math. J. **27**, 729–742 (1975)
7. J.M. Bismut, Martingales, the Malliavin Calculus and hypoellipticity under general Hoermander's condition. Z.W. **56**, 469–505 (1981)
8. J.M. Bismut, Calcul des variations stochastiques et processus de sauts. Z. Wahr. Verw. Gebiete **63**, 147–235 (1983)
9. J.M. Bismut, *Large Deviations and the Malliavin Calculus* (Birkhauser, Boston, 1984)
10. N. Bouleau, L. Denis, *Dirichlet Forms Methods for Poisson Measures and Lévy Processes with Emphasis on the Creation-Annihilation Techniques* (Springer, London, 2015)
11. K. Burdzy, Some path properties of iterated Brownian motion, in *Seminar on Stochastic Processes 1992*. Progress in Probability, vol. 33 (Birkhauser, Boston, 1993), pp. 67–87
12. J. Chazarain, A. Piriou, *Introduction a la théorie des équations au dérivées partielles linéaires* (Gauthier-Villars, Paris, 1981)
13. E.B. Davies, *Heat Kernels and Spectral Theory* (Cambridge University Press, Cambridge, 1989)
14. E.B. Davies, Uniformly elliptic operators with measurable coefficients. J. Funct. Anal. **132**, 141–169 (1995)
15. P. Greiner, An asymptotic expansion for the heat equation. Arch. Ration. Mech. Anal. **41**, 163–218 (1971)
16. T. Hida, *Analysis of Brownian Functionals*. Carleton Mathematical Lecture Notes, vol. 13 (Carleton University Press, Ottawa, 1975)
17. L. Hörmander, Fourier integral operators. I. Acta. Math. **127**, 78–183 (1971)
18. L. Hörmander, *The Analysis of Linear Partial Operators III* (Springer, Berlin, 1984)
19. L. Hörmander, *The Analysis of Linear Partial Operators IV* (Springer, Berlin, 1984)
20. N. Ikeda, S. Watanabe, *Stochastic Differential Equations and Diffusion Processes* (North-Holland, Amsterdam, 1989)
21. Y. Ishikawa, *Stochastic Calculus of Variations for Jump Processes* (de Gruyter, Berlin, 2012)

22. N. Jacob, *Pseudo-Differential Operators and Markov Processes. I. Fourier Analysis and Semigroups* (Imperial College Press, London, 2001)
23. N. Jacob, *Pseudo-Differential Operators and Markov Processes. II. Generators and Their Potential Theory* (Imperial College Press, London, 2003)
24. N. Jacob, *Pseudo-Differential Operators and Markov Processes. III. Markov Processes and Their Applications* (Imperial College Press, London, 2005)
25. N. Kumano-Go, Phase space Feynman path integrals of higher order parabolic type with general functional as integrand. Bull. Sci. Math. **139**, 495–537 (2015)
26. S. Kusuoka, More recent theory of Malliavin Calculus. Sugaku Expositions **5**, 155–173 (1992)
27. P. Lax, Asymptotic solutions of oscillatory initial value problems. Duke Math. J. **25**, 627–646 (1957)
28. R. Léandre, Extension du théoreme de Hoemander a divers processus de sauts, PHD Thesis, Université de Besançon, 1984
29. R. Léandre, Applications quantitatives et qualitatives du Calcul de Malliavin, in *Séminaire Franco-Japonais*. Lectures Notes in Maths, vol. 1322 (Springer, Berlin, 1988), pp. 109–134. English translation: *Geometry of random motion*. Contemporary Maths, vol. 73. (AMS, 1988), pp. 173–196
30. R. Léandre, Path integrals in noncommutative geometry. *Encyclopedia of Mathematical Physics* (Elsevier, Amsterdam, 2006), pp. 8–12
31. R. Léandre, Malliavin Calculus of Bismut type without probability. Festchrift in honour of K. Sinha. Proc. Indian Acad. Sci. Math. Sci. **11**, 507–518 (2006)
32. R. Léandre, Applications of the Malliavin Calculus without probability. WSEAS Trans. Math. **5**, 1205–1206 (2006)
33. R. Léandre, Malliavin Calculus of Bismut type in semi-group theory. Far East J. Math. Sci. **30**, 1–26 (2008)
34. R. Léandre, Wentzel-Freidlin estimates for an operator of order four, in *2014 International Conference on Computational Science and Computational Intelligence* (IEEE Computer Society, Piscataway, 2014), pp. 360–364. (IEEE X-plore)
35. R. Léandre, Large deviation estimates for a non-Markovian generator of big order, in *4th International Conference on Mathematical Models in Physical Sciences*. Journal of Physics: Conference Series, **633** (2015) paper 012085
36. R. Léandre, Stochastic analysis for a non-Markovian generator: an introduction. Russ. J. Math. Phys. **22**, 39–52 (2015)
37. R. Léandre, *The Itô-Stratonovitch for an operator of order four. Festchrift in honour of A. Mukharjea, Ph. Feinsilver and S. Mohammed*. Contemporary Mathematics, vol. 668 (American Mathematical Society, Providence, 2016), pp. 165–169
38. R. Léandre, Large deviation estimates for an operator of order four with a potential, in *XII International Work Differential Geometry and Application*. Revue Roumaine de Math. Pures et Appliquées, **61** (2016), pp. 233–240
39. R. Léandre, Varadhan estimates for an operator of order four on a Lie group, in *Control, Decision and Information Technologies*. IEEE-Xplore.IEEE. (2016)
40. R. Léandre, A class of non-Markovian generator of Lévy type, in *Pseudo-Differential Operators: Groups, Geometry and Application* (Birkhauser, Basel, 2017), pp. 149–159
41. R. Léandre, Perturbation of the Malliavin Calculus of Bismut type of large order, in *Physical and Mathematical, Aspects of Symmetries* (Springer, Berlin, 2017), pp. 221–225
42. R. Léandre, Malliavin Calculus of Bismut type for an operator of order four on a Lie group. J. Pseudo-Differ. Oper. Appl. **9** (2018), pp. 419–430
43. R. Léandre, Bismut's way of the Malliavin Calculus of large order generators on a Lie group, in *6th International Eurasian Conference on Mathematical Sciences and Applications*. A.I.P. Proceedings, vol. 1926 (American Institute of Physics, Melville, 2018) 020026
44. P. Malliavin, Stochastic calculus of variations and hypoelliptic operators, in *Proceedings of the Symposium on Stochastic Differential Equations* (Kinokuyina, 1978), pp. 195–263
45. V.P. Maslov, M.V. Fedoriuk, *Semiclassical Approximation in Quantum Mechanics* (Reidel, Dordrecht, 1981)

46. S. Mazzucchi, Infinite dimensional oscillatory integrals with polynomial phase and application to higher-order heat equation. Potential Anal. **49** (2018), pp. 209–233
47. S. Molchanov, Diffusion processes and Riemannian geometry. Russ. Math. Surv. **30** 1–63 (1975)
48. J. Neveu, Processus aléatoires Gaussiens, in *Séminaire Maths Supérieures*, vol. 34 (The University Press of Montreal, Montreal, 1968)
49. D. Nualart, *The Malliavin Calculus and Related Topics* (Springer, Berlin, 1995)
50. R.T. Seeley, Complex powers of an elliptic operator, in *Singular Integrals*. Proceedings of Symposia in Pure Mathematics (American Mathematical Society, Providence, 1970), pp. 288–307
51. A. Ter Elst, D.W. Robinson, Subcoercive and subelliptic operators on Lie groups: variable coefficients. Publ. Res. Inst. Math. Sci. **29**, 745–801 (1993)
52. S.R.S. Varadhan, *Large Deviations and Applications* (Society for Industrial and Applied Mathematics, Philadelphia, 1984)
53. S. Watanabe, Stochastic analysis and its applications. Sugaku Expositions **5**, 51–69 (1992)
54. A.D. Wentzel, M.J. Freidlin, *Random Perturbations of Dynamical Systems* (Springer, Berlin, 1984)

Operator Transformation of Probability Densities

Leon Cohen

Abstract We describe an operator relation that relates two arbitrary probability densities. The relation may be thought of as a generalization of the Edgeworth and Gram–Charlier series of probability theory. We apply the relation to a number of issues. We generalize to relate a probability distribution with itself but at different times and show that it can be used to obtain approximate solutions. We apply the scale operator to the case of the product of two independent random variables, and generalize the concept of cumulants for that case. An operator relation between the energy density of a signal and the energy density of the spectrum is obtained. In addition, we show that the spectral moments of a signal may be expressed in terms of the Bell polynomials.

Keywords Operator transformation · Probability densities · Gram–Charlier series · Bell polynomials

Mathematics Subject Classification (2000) 47G30, 60A05

1 Introduction

This paper is based on the following result: *Any* two probability densities, $P_1(x)$ and $P_2(x)$, may be related by

$$P_2(x) = \Omega(\mathbf{A}) P_1(x) \tag{1}$$

L. Cohen (✉)
Department of Physics, Hunter College and Graduate Center of CUNY, New York, NY, USA
e-mail: leon.cohen@hunter.cuny.edu

S. Molahajloo, M. W. Wong (eds.), *Analysis of Pseudo-Differential Operators*, Trends in Mathematics, https://doi.org/10.1007/978-3-030-05168-6_8

where **A** is *any* self-adjoint (Hermitian) operator and where Ω is a function that will be explicitly given. We call Eq. (1) an operator transformation equation and we will show that it generalizes the classical Edgeworth and Gram–Charlier methods of probability theory [18]. It is the aim of this paper to review the issues associated with Eq. (1) and give a number of new consequences and applications [5–7].

We also generalize the above to the same probability distribution but at different times,

$$P(x,t) = \Omega(\mathbf{A}(t))P(x,0) \tag{2}$$

where $P(x,t)$ is a time dependent probability distribution that is usually obtained by solving a partial differential equation that governs it. We will show that Eq. (2) can sometimes be used to circumvent solving the governing equation and to obtain an approximation.

Notation Operators will be designated by boldface, an exception being the differentiation operator

$$D = \frac{d}{dx} \tag{3}$$

as that is the standard notation. Also, to indicate differentiation with a different variable we use, for example,

$$D_y = \frac{d}{dy} \tag{4}$$

All integrals go from $-\infty$ to ∞ unless otherwise indicated.

A number of our results will involve the so-called complete Bell polynomials, B_n, which may be defined by the expansion

$$\exp\left(\sum_{i=1}^{\infty} a_i \frac{x^i}{i!}\right) = \sum_{n=0}^{\infty} B_n(a_1,\ldots,a_n)\frac{x^n}{n!} \tag{5}$$

We review the Bell polynomials in the Appendix. The nth Bell polynomial is a function of n numbers $a_1,\ldots,a_n$. We will often abbreviate the notation and write

$$B_n(a) = B_n(a_1,\ldots,a_n) \tag{6}$$

No confusion arises because the subscript in $B_n(a)$ indicates the number of a's, namely, n of them.

2 Operator Transformation

The eigenvalue problem for a Hermitian operator, **A**,

$$\mathbf{A}\, u_\theta(x) \;=\; \theta\, u_\theta(x) \tag{7}$$

results in real eigenvalues, θ, and eigenfunctions, $u_\theta(x)$, that are complete and orthogonal

$$\int u^*_{\theta'}(x)\, u_\theta(x)\, dx \;=\; \delta(\theta - \theta') \tag{8}$$

$$\int u^*_\theta(x')\, u_\theta(x)\, d\theta \;=\; \delta(x - x') \tag{9}$$

Any function, $f(x)$, can be expanded as

$$f(x) = \int F(\theta)\; u_\theta(x)\, d\theta \tag{10}$$

where

$$F(\theta) = \int f(x)\, u^*_\theta(x)\, dx \tag{11}$$

The function $F(\theta)$ is called the transform of $f(x)$ in the domain of the operator **A**.

Now consider any two densities P_1 and P_2 and expand them as per Eqs. (10) and (11)

$$P_1(x) = \int N_1(\theta) u_\theta(x)\, d\theta \tag{12}$$

$$P_2(x) = \int N_2(\theta)\, u_\theta(x)\, d\theta \tag{13}$$

where N_1 and N_2 are given by

$$N_1(\theta) = \int P_1(x) u^*_\theta(x)\, dx \tag{14}$$

$$N_2(\theta) = \int P_2(x) u^*_\theta(x)\, dx \tag{15}$$

Starting with $P_2(x)$ as given by Eq. (13), we have

$$P_2(x) = \int N_2(\theta)u_\theta(x)d\theta \tag{16}$$

$$= \int \frac{N_2(\theta)}{N_1(\theta)} N_1(\theta)u_\theta(x)d\theta \tag{17}$$

$$= \int \Omega(\theta)N_1(\theta)u_\theta(x)d\theta \tag{18}$$

where we have defined

$$\Omega(\theta) = \frac{N_2(\theta)}{N_1(\theta)} \tag{19}$$

In general for any function $\Omega(\theta)$, we have that

$$\Omega(\theta)u_\theta(x) = \Omega(\mathbf{A})u_\theta(x) \tag{20}$$

and therefore

$$P_2(x) = \int \Omega(\mathbf{A})N_1(\theta)u_\theta(x)d\theta \tag{21}$$

$$= \Omega(\mathbf{A}) \int N_1(\theta)u_\theta(x)d\theta \tag{22}$$

$$= \Omega(\mathbf{A})P_1(x) \tag{23}$$

which is Eq. (1). We also have that

$$P_1(x) = \Omega^{-1}(\mathbf{A})P_2(x) \tag{24}$$

where

$$\Omega^{-1}(\theta) = \frac{N_1(\theta)}{N_2(\theta)} \tag{25}$$

2.1 Expectation Values

In certain fields, like time-frequency analysis [2, 4] and the phase-space formulation of quantum mechanics [11], we also want functions to transform in a manner that keeps expectation values the same. In particular, suppose we have a real function

$g_1(x)$ whose expectation value with $P_1(x)$ is given by

$$\langle g_1(x)\rangle = \int g_1(x)P_1(x)dx \tag{26}$$

then we seek a function $g_2(x)$ so that its expectation value with $P_2(x)$ also gives $\langle g_1(x)\rangle$. That is, we want

$$\int g_2(x)P_2(x)dx = \int g_1(x)P_1(x)dx \tag{27}$$

We emphasize though that this is not usually required for our considerations, but it is the case in the field of time-frequency distributions such as when we want to transform from the Wigner distribution to the Margenau–Hill distribution and keep the expectation values the same. The result we now derive is the one-dimensional analogue but where the operator is general.

In the left-hand side of Eq. (27) substitute Eq. (1) to obtain

$$\int g_2(x)P_2(x)dx = \int g_2(x)\Omega(\mathbf{A})P_1(x)dx \tag{28}$$

$$= \int \left[\Omega^{\dagger}(\mathbf{A})g_2(x)\right]^* P_1(x)dx \tag{29}$$

where $\Omega^{\dagger}$ is the adjoint of $\Omega(\mathbf{A})$. Hence the relation between g_1 and g_2 is

$$g_1(x) = \left[\Omega^{\dagger}(\mathbf{A})g_2(x)\right]^* \tag{30}$$

Also

$$g_2(x) = \left[\Omega^{-1\dagger}(\mathbf{A})g_1(x)\right]^* \tag{31}$$

Since $\mathbf{A}$ is Hermitian Eq. (30) simplifies to

$$g_1(x) = \Omega(\mathbf{A})g_2(x) = \frac{F_2(\mathbf{A})}{F_1(\mathbf{A})}g_2(x) \tag{32}$$

This is the case even if $\Omega(\theta)$ is a complex function.

3 Edgeworth and Gram–Charlier Cases

In standard probability theory, the Edgeworth and Gram–Charlier series are methods for "correcting" probability distributions [18]. We now show that they are special cases of our formulation. For the operator **A** take

$$\mathbf{A} = iD \tag{33}$$

We note that iD is Hermitian and the eigenvalue problem

$$i\frac{d}{dx}u(\theta,x) = \theta u(\theta,x) \tag{34}$$

results in the Fourier basis

$$u(\theta,x) = \frac{1}{\sqrt{2\pi}}e^{-i\theta x} \tag{35}$$

Expanding the probability in terms if the eigenfunctions

$$P(x) = \frac{1}{2\pi}\int M(\theta)e^{-i\theta x}\,d\theta \tag{36}$$

where

$$M(\theta) = \int P(x)e^{i\theta x}\,dx \tag{37}$$

which is the standard characteristic function [13] and where we have used $M(\theta)$ instead of $N(\theta)$ to stay with the conventional notation for characteristic functions.

Applying Eq. (1) and (19), we have

$$P_2(x) = \Omega(iD)P_1(x) \tag{38}$$

with

$$\Omega(\theta) = \frac{M_2(\theta)}{M_1(\theta)} \tag{39}$$

If we write the characteristic functions in the cumulant form [18]

$$M_1(\theta) = \exp\left[\sum_{n=1}^{\infty}\kappa_n^{(1)}\frac{i^n}{n!}\theta^n\right] \tag{40}$$

$$M_2(\theta) = \exp\left[\sum_{n=1}^{\infty}\kappa_n^{(2)}\frac{i^n}{n!}\theta^n\right] \tag{41}$$

where $\kappa_n^{(1)}$ and $\kappa_n^{(2)}$ are the respective cumulants of $P_1(x)$ and $P_2(x)$, then

$$\Omega(\theta) = \frac{M_2(\theta)}{M_1(\theta)} = \frac{\exp\left[\sum_{n=1}^{\infty} \kappa_n^{(2)} \frac{i^n}{n!} \theta^n\right]}{\exp\left[\sum_{n=1}^{\infty} \kappa_n^{(1)} \frac{i^n}{n!} \theta^n\right]} \tag{42}$$

$$= \exp\left[\sum_{n=1}^{\infty} \left(\kappa_n^{(2)} - \kappa_n^{(1)}\right) \frac{i^n}{n!} \theta^n\right] \tag{43}$$

Therefore, Eq. (1) gives

$$P_2(x) = \exp\left[\sum_{n=1}^{\infty} (\kappa_n^{(2)} - \kappa_n^{(1)}) \frac{i^n}{n!} (iD)^n\right] P_1(x) \tag{44}$$

and

$$P_2(x) = \exp\left[\sum_{n=1}^{\infty} (\kappa_n^{(2)} - \kappa_n^{(1)}) \frac{(-1)^n}{n!} D^n\right] P_1(x) \tag{45}$$

Equation (45) holds for any two densities, while historically $P_1(x)$ is taken to be Gaussian distribution,

$$N(m, \sigma^2) = \sqrt{\frac{1}{2\pi\sigma^2}} \exp\left[-\frac{(x-m)^2}{2\sigma^2}\right] \tag{46}$$

In that case

$$P_2(x) = \exp\left[-(\kappa_1^{(2)} - m)D + \frac{1}{2}(\kappa_2^{(2)} - \sigma^2)D^2 + \sum_{n=3}^{\infty} \kappa_n^{(2)} \frac{(-i)^n}{n!} D^n\right] N(m, \sigma^2) \tag{47}$$

since for the Gaussian distribution there are only two cumulants. Equation (47) is the standard formulation of the Edgeworth series [18] while our formulation involves any two distributions.

3.1 Approximation

One of the standard uses of the Edgeworth series is the approximation of probability distributions [10]. We now derive an approximation which is somewhat more general.

Applying Eq. (5) to Eq. (45) we have that

$$\exp\left[\sum_{n=1}^{\infty}(\kappa_n^{(2)}-\kappa_n^{(1)})\frac{(-1)^n}{n!}D^n\right]=\sum_{n=0}^{\infty}B_n(\eta_1,\ldots,\eta_n)\frac{(-1)^n}{n!}D^n \tag{48}$$

where

$$\eta_n=\kappa_n^{(2)}-\kappa_n^{(1)} \tag{49}$$

Truncating the series gives an approximation. For example, if one keeps terms up to D^4, one obtains that

$$P_2(x)\sim\left[1-B_1D+\frac{1}{2}B_2D^2-\frac{1}{6}B_3D^3+\frac{1}{24}B_4D^4+\cdots\right]P_1(x) \tag{50}$$

The first few Bell polynomials are

$$B_1(\eta_1)=\eta_1 \tag{51}$$

$$B_2(\eta_1,\eta_2)=\eta_2+\eta_1^2 \tag{52}$$

$$B_3(\eta_1,\ldots,\eta_3)=\eta_3+3\eta_2\eta_1+\eta_1^3 \tag{53}$$

$$B_4(\eta_1,\ldots,\eta_4)=\eta_4+4\eta_3\eta_1+3\eta_2^2+6\eta_2\eta_1^2+\eta_1^4 \tag{54}$$

Also, since $\eta_n=\kappa_n^{(2)}-\kappa_n^{(1)}$ we have, using Eq. (269) of the Appendix

$$B_n(\kappa_1^{(2)}-\kappa_1^{(1)},\ldots,\kappa_n^{(2)}-\kappa_n^{(1)})=\sum_{i=0}^{n}\binom{n}{i}B_{n-i}(\kappa_1^{(2)},\ldots,\kappa_{n-i}^{(2)})B_i(-\kappa_1^{(1)},\ldots,-\kappa_i^{(2)}) \tag{55}$$

3.2 Gram–Charlier Series

Suppose we expand $\Omega(\theta)$ in a power series

$$\Omega(\theta)=\sum_{n=0}^{\infty}a_n\theta^n \tag{56}$$

in which case

$$a_n=\frac{1}{n!}\,D^n\Omega(\theta)\big|_{\theta=0}=\frac{1}{n!}\,D^n\frac{M_2(\theta)}{M_1(\theta)}\bigg|_{\theta=0} \tag{57}$$

For Eq. (1) with $\mathbf{A} = i D$ we have

$$P_2(x) = \sum_{n=0}^{\infty} a_n i^n D^n P_1(x) \tag{58}$$

One can show that for the case where $P_1(x)$ is Gaussian, one obtains the standard Gram–Charlier series [18].

4 Cumulants

We make some well-known remarks regarding cumulants and characteristic functions as we will generalize the concept for the case of the scale operator [13, 18]. For a random variable, Z, which is the sum of two independent random variables, X and Y,

$$Z = X + Y \tag{59}$$

the characteristic function of Z, $M_Z(\theta)$, is

$$M_Z(\theta) = \iint P_X(x) P_Y(y) e^{i\theta(x+y)}\, dx dy = \left\langle e^{i\theta Z} \right\rangle = \left\langle e^{i\theta(X+Y)} \right\rangle = \left\langle e^{i\theta X} \right\rangle \left\langle e^{i\theta Y} \right\rangle \tag{60}$$

where $P_X(x)$ is the probability of x and similarly for y. Hence, the characteristic function of Z is the product of the characteristic functions of X and Y

$$M_Z(\theta) = M_X(\theta) M_Y(\theta) \tag{61}$$

Furthermore

$$\ln M_Z(\theta) = \ln M_X(\theta) + \ln M_Y(\theta) \tag{62}$$

Now, the characteristic function is expanded in terms of cumulants

$$M_Z(\theta) = \exp\left[\sum_{n=1}^{\infty} \kappa_n^{(Z)} \frac{i^n}{n!} \theta^n \right] \tag{63}$$

where $\kappa_n^{(Z)}$ are the cumulants of Z. Taking the logarithm

$$\ln M_Z(\theta) = \sum_{n=1}^{\infty} \kappa_n^{(Z)} \frac{i^n}{n!} \theta^n \tag{64}$$

and using Eq. (62) we have that

$$\ln M_Z(\theta) = \sum_{n=1}^{\infty} \kappa_n^{(X)} \frac{i^n}{n!}\theta^n + \sum_{n=1}^{\infty} \kappa_n^{(Y)} \frac{i^n}{n!}\theta^n \tag{65}$$

and therefore

$$\kappa_n^{(Z)} = \kappa_n^{(X)} + \kappa_n^{(Y)} \tag{66}$$

Hence cumulants add while the moments do not, which is one of the fundamental reasons for the use of cumulants.

The other important issue with cumulants is that they characterize the Gaussian distribution

$$P(x) = \frac{1}{\sqrt{2\pi\sigma^2}} \exp\left[-\frac{(x-m)^2}{2\sigma^2}\right] \tag{67}$$

in a simple way. In particular, while the Gaussian has an infinite number of moments it only has two cumulants. That is the case since the characteristic function is calculated to be

$$M(\theta) = e^{im\theta - \sigma^2\theta^2/2} \tag{68}$$

and hence

$$\ln M(\theta) = im\theta - \sigma^2\theta^2/2 \tag{69}$$

giving

$$\kappa_1 = m; \qquad \kappa_2 = \sigma^2 \tag{70}$$

and all other cumulants are zero. We have reviewed these standard points about cumulants because we will show how one can have the same type of formulation for the case where we have a product of independent random variables.

5 Scale Operator and the Product of Random Variables

We now develop the above formulation for the scale operator and show that it can be used in the consideration of the product of two independent random variables. The scale operator, **C**, is [3]

$$\mathbf{C} = \frac{1}{2i}(\mathbf{x}D + D\mathbf{x}) = \frac{1}{i}\left(\mathbf{x}D + \tfrac{1}{2}\right) = \frac{1}{i}\left(D\mathbf{x} - \tfrac{1}{2}\right) \tag{71}$$

A basic property of the scale operator is the operation of $e^{i\theta \mathbf{C}}$ on an arbitrary function, $f(x)$,

$$e^{i\theta \mathbf{C}} f(x) = e^{\theta/2} f(e^{\theta} x) \tag{72}$$

That is, $e^{i\theta \mathbf{C}}$ scales the argument of the function by e^{θ}. This is analogous to

$$e^{i\theta \mathbf{K}} f(x) = f(x+\theta) \tag{73}$$

where $\mathbf{K}$ is the frequency operator

$$\mathbf{K} = \frac{1}{i}\frac{d}{dx} \tag{74}$$

That is, $e^{i\theta \mathbf{K}}$ translates while $e^{i\theta \mathbf{C}}$ scales arguments of a function.

The eigenvalue problem for the scale operator

$$\mathbf{C}\,\gamma(\theta, x) = \theta\,\gamma(\theta, x) \tag{75}$$

produces the eigenfunctions

$$\gamma(\theta, x) = \frac{1}{\sqrt{2\pi}} \frac{e^{i\theta \ln x}}{\sqrt{x}}, \qquad x \geq 0 \tag{76}$$

which are complete and orthogonal,

$$\int_0^\infty \gamma^*(\theta', x)\,\gamma(\theta, x) dx = \delta(\theta - \theta') \tag{77}$$

$$\int \gamma^*(\theta, x')\,\gamma(\theta, x) d\theta = \delta(x - x') \qquad x, x' \geq 0 \tag{78}$$

For a one-sided probability density we define the generalized scale characteristic function, $N(\theta)$, by

$$N(\theta) = \int_0^\infty \frac{e^{i\theta \ln x}}{\sqrt{x}} P(x) dx = \int_0^\infty P(x) x^{-i\theta - 1/2} dx \tag{79}$$

with

$$P(x) = \frac{1}{2\pi} \int \frac{e^{-i\theta \ln x}}{\sqrt{x}} N(\theta) d\theta \tag{80}$$

We note that $N(\theta)$ is a simply invertible Mellin transform with the complex argument $-i\theta + 1/2$.

5.1 Product of Random Variables

Consider a random variable which is the product of two independent random variables

$$Z = XY \tag{81}$$

Then

$$N_Z(\theta) = \frac{1}{\sqrt{2\pi}} \int_0^\infty P_X(x) P_Y(y) \frac{e^{-i\theta \ln z}}{\sqrt{z}} dx dy \tag{82}$$

$$= \frac{1}{\sqrt{2\pi}} \int_0^\infty P_X(x) P_Y(y) \frac{e^{-i\theta \ln xy}}{\sqrt{xy}} dx dy \tag{83}$$

from which it follows that

$$N_Z(\theta) = N_X(\theta) N_Y(\theta) \tag{84}$$

and

$$\ln N_Z(\theta) = \ln N_X(\theta) + \ln N_Y(\theta) \tag{85}$$

This is analogous to Eq. (62) but note that this would *not* be the case if we considered the standard characteristic function instead of $N(\theta)$. Expanding $N_X(\theta)$ and $N_Y(\theta)$ as

$$N_X(\theta) = \exp\left[\sum_{n=0}^{\infty} \lambda_n^{(X)} \frac{i^n}{n!} \theta^n\right] \tag{86}$$

$$N_Y(\theta) = \exp\left[\sum_{n=0}^{\infty} \lambda_n^{(Y)} \frac{i^n}{n!} \theta^n\right] \tag{87}$$

where $\lambda_n^{(1)}$ and $\lambda_n^{(2)}$ are the analogues to cumulants, we have

$$\ln N_Z(\theta) = \sum_{n=1}^{\infty} \lambda_n^{(X)} \frac{i^n}{n!} \theta^n + \sum_{n=1}^{\infty} \lambda_n^{(Y)} \frac{i^n}{n!} \theta^n \tag{88}$$

Hence

$$\lambda_n^{(Z)} = \lambda_n^{(X)} + \lambda_n^{(Y)} \tag{89}$$

In addition we have

$$\Omega(\theta) = \frac{N_2(\theta)}{N_1(\theta)} = \frac{\exp\left[\sum_{n=1}^{\infty} \lambda_n^{(Y)} \frac{i^n}{n!}\theta^n\right]}{\exp\left[\sum_{n=1}^{\infty} \lambda_n^{(X)} \frac{i^n}{n!}\theta^n\right]} \tag{90}$$

$$= \exp\left[\sum_{n=1}^{\infty} \left(\lambda_n^{(Y)} - \lambda_n^{(X)}\right) \frac{i^n}{n!}\theta^n\right] \tag{91}$$

giving

$$\Omega(\mathbf{C}) = \exp\left[\sum_{n=1}^{\infty} \left(\lambda_n^{(Y)} - \lambda_n^{(X)}\right) \frac{i^n}{n!}\mathbf{C}^n\right] \tag{92}$$

Therefore any two one-sided densities may be related by

$$P_2(x) = \exp\left[\sum_{n=1}^{\infty} \left(\lambda_n^{(Y)} - \lambda_n^{(X)}\right) \frac{1}{2^n n!} (\mathbf{x}D + D\mathbf{x})^n\right] P_1(x) \tag{93}$$

5.2 Log-Normal Distribution

Suppose we take the log-normal distribution for $P_1(x)$,

$$P_1(x) = \frac{1}{x\sqrt{2\pi b^2}} \exp\left[-\frac{(\ln x - a)^2}{2b^2}\right] \tag{94}$$

The moments are given by

$$\left\langle x^n \right\rangle = e^{na+n^2b^2/2} \tag{95}$$

and we note that there are an finite number of them. The first and second moments and standard deviation are

$$\langle x \rangle = e^{a+b^2/2} \tag{96}$$

$$\left\langle x^2 \right\rangle = e^{2a+2b^2} \tag{97}$$

$$\sigma^2 = e^{2a+b^2}(e^{b^2} - 1) \tag{98}$$

Straightforward calculation gives that

$$N(\theta) = \exp\left[\frac{b^2/4 - a + i\theta(b^2 - 2\alpha) - b^2\theta^2}{2}\right] \tag{99}$$

and hence

$$\ln N(\theta) = \frac{b^2/4 - a + i(b^2 - 2\alpha)\theta - b^2\theta^2}{2} \tag{100}$$

We see that $\ln N(\theta)$ has only three terms, that is there are only three λ_n's while it has an infinite number of moments and cumulants. Using Eqs. (96)–(98), one can express these λ_n in terms of the moments.

6 Time Dependent Operator Transformation

We now consider applying Eq. (1) to the *same* distribution but at different times. That is, we take

$$P(x,t) = \Omega(\mathbf{A},t)P(x,0) \tag{101}$$

where now the generalized characteristic functions are time dependent given by

$$N(\theta,0) = \int P(x,0)u_\theta^*(x)\,dx \tag{102}$$

$$N(\theta,t) = \int P(x,t)u_\theta^*(x)\,dx \tag{103}$$

and where

$$\Omega(\theta,t) = \frac{N(\theta,t)}{N(\theta,0)} \tag{104}$$

The transformation operator is therefore time dependent

$$\Omega(\mathbf{A},t) = \frac{N(\mathbf{A},t)}{N(\mathbf{A},0)} \tag{105}$$

We explore the possibility of using Eq. (101) to approximate $P(x,t)$ given $P(x,0)$. We consider here the case where $\mathbf{A} = iD$. We rewrite Eqs. (1) - (2) taking into account that we are dealing with the same distribution but at different times.

For Eq. (45) we have

$$P(x,t) = \exp\left[\sum_{n=1}^{\infty}(\kappa_n(t) - \kappa_n(0))\frac{(-1)^n}{n!}D^n\right] P(x,0) \tag{106}$$

where now κ_n are the time dependent cumulants. For Eq. (48) we have

$$\exp\left[\sum_{n=1}^{\infty}(\kappa_n(t) - \kappa_n(0))\frac{(-1)^n}{n!}D^n\right] = \sum_{n=0}^{\infty} B_n(\eta_1,\ldots,\eta_n)\frac{(-1)^n}{n!}D^n \tag{107}$$

where

$$\eta_n(t) = \kappa_n(t) - \kappa_n(0) \tag{108}$$

and the Bell polynomials are time dependent by virtue of the time dependence of $\eta_n(t)$. In particular, for example,

$$B_1(\eta_1) = \eta_1 = \kappa_1(t) - \kappa_1(0) \tag{109}$$

$$B_2(\eta_1,\eta_2) = \eta_2 + \eta_1^2 = \kappa_2(t) - \kappa_2(0) + (\kappa_1(t) - \kappa_1(0))^2 \tag{110}$$

and so forth. We note that since the first cumulant is the mean of a distribution and the second is the square of the standard deviation we can write

$$\eta_1(t) = \langle x\rangle_t - \langle x\rangle_0 \tag{111}$$

$$\eta_2(t) = \sigma_t^2 - \sigma_0^2 \tag{112}$$

For the scheme to work, one must be able to obtain the moments without solving the governing evolution equation for $P(x,t)$. To illustrate the method we give two examples.

6.1 Example 1

Consider the standard diffusion equation for the probability density, $P(x,t)$,

$$\frac{\partial P}{\partial t} = \alpha\frac{\partial^2 P}{\partial x^2} \tag{113}$$

where α is the diffusion constant [1, 19]. Multiply Eq. (113) by x and integrate

$$\int x \frac{\partial P}{\partial t} dx = \alpha \int x \frac{\partial^2 P}{\partial x^2} dx \tag{114}$$

The left-hand side gives

$$\int x \frac{\partial P}{\partial t} dx = \frac{d}{dt} \int x P dx = \frac{d}{dt} \langle x \rangle_t \tag{115}$$

and for the right-hand side we have, by integration by parts,

$$\alpha \int x \frac{\partial^2 P}{\partial x^2} dx = \alpha \int P \frac{\partial^2}{\partial x^2} x dx = 0 \tag{116}$$

Therefore

$$\frac{d}{dt} \langle x \rangle_t = 0 \tag{117}$$

and

$$\langle x \rangle_t = \langle x \rangle_0 \tag{118}$$

Now, multiply Eq. (113) by x^2 and integrate

$$\int x^2 \frac{\partial P}{\partial t} dx = \alpha \int x^2 \frac{\partial^2 P}{\partial x^2} dx \tag{119}$$

Following the procedure above, we have

$$\int x^2 \frac{\partial P}{\partial t} dx = \frac{d}{dt} \int x^2 P dx = \frac{d}{dt} \left\langle x^2 \right\rangle_t \tag{120}$$

and

$$\alpha \int x^2 \frac{\partial^2 P}{\partial x^2} dx = \alpha \int P \frac{\partial^2}{\partial x^2} x^2 dx = 2\alpha \int P dx = 2\alpha \tag{121}$$

Therefore

$$\frac{d}{dt} \left\langle x^2 \right\rangle_t = 2\alpha \tag{122}$$

giving

$$\left\langle x^2\right\rangle_t = 2\alpha t + \left\langle x^2\right\rangle_0 \tag{123}$$

The standard deviation is given by

$$\sigma_t^2 = \left\langle x^2\right\rangle_t - \langle x\rangle_t^2 = 2\alpha t + \sigma_0^2 \tag{124}$$

Now

$$B_1 = \eta_1 = \kappa_1^{(2)} - \kappa_1^{(1)} = \langle x\rangle_t - \langle x\rangle_0 = 0 \tag{125}$$

$$B_2 = \eta_2 + \eta_1^2 = \sigma_t^2 - \sigma_0^2 + \langle x\rangle_t - \langle x\rangle_0^2 = 2\alpha t \tag{126}$$

Suppose in Eq. (45) we truncate after two terms, then

$$P(x,t) = \exp\left[\alpha t D^2\right] P(x,0) \tag{127}$$

Further if we approximate by way of Eq. (50) then

$$P(x,t) \sim \left[1 - B_1 D + \frac{1}{2} B_2 D^2\right] P(x,t) \tag{128}$$

$$= \left[1 + a t D^2\right] P(x,t) \tag{129}$$

which is seen to be the first two terms of the expansion in Eq. (127).

We now show that in fact Eq. (127) is the exact solution. Let

$$P(x,0) = \frac{1}{2\pi}\int M(\theta) e^{-ix\theta} d\theta \tag{130}$$

and substitute into the right-hand side of Eq. (127)

$$P(x,t) = \exp\left[\alpha t D^2\right] \frac{1}{2\pi}\int M(\theta,0) e^{-ix\theta} d\theta \tag{131}$$

$$= \frac{1}{2\pi}\int M(\theta,0) e^{-ix\theta - \alpha t\theta^2} d\theta \tag{132}$$

$$= \frac{1}{2\pi}\iint P(x',0) e^{ix'\theta} e^{-ix\theta - \alpha t\theta^2} d\theta dx' \tag{133}$$

$$= \sqrt{\frac{1}{4\pi\alpha t}}\int e^{-\frac{(x-x')^2}{4\alpha t}} P(x',0) dx' \tag{134}$$

which we write as

$$P(x,t) = \sqrt{\frac{1}{4\pi\alpha t}} \int K(x-x',t)P(x',0)dx' \tag{135}$$

where

$$K(x-x',t) = \sqrt{\frac{1}{4\pi\alpha t}} \int e^{-\frac{(x-x')^2}{4\alpha t}} \tag{136}$$

Now, it is well known that $K(x,t)$ is indeed the exact propagator for diffusion equation (113).

6.2 Example 2

Suppose a probability distribution, $P(x,t)$, satisfies the following differential equation

$$\frac{\partial^2 P}{\partial t^2} = (v+at)^2 \frac{\partial^2 P}{\partial x^2} - a\frac{\partial P}{\partial x} \tag{137}$$

where v and a are constants. The basic idea is to solve this differential equation for $P(x,t)$ given the density, $P(x,0)$at time zero. This equation can be solved exactly but our aim here is to show how the methods developed in the previous section can be used to approximate the solution. From the differential equations we can obtain $\langle x \rangle_t$. Multiply Eq. (137) by x and integrate with respect to x

$$\int x\frac{\partial^2 P}{\partial t^2}dx = (v+at)^2 \int x\frac{\partial^2 P}{\partial x^2}dx - a\int x\frac{\partial P}{\partial x}dx \tag{138}$$

The first term gives

$$\int x\frac{\partial^2 P}{\partial t^2}dx = \frac{d^2}{dt^2}\int xP(x,t) = \frac{d^2}{dt^2}\langle x \rangle_t \tag{139}$$

Integration by parts gives zero for the first term on the right-hand side and for the second term we have

$$\int x\frac{\partial P}{\partial x}dx = -\int P\frac{\partial x}{\partial x}dx = -\int Pdx = -1 \tag{140}$$

since P is a probability density. Therefore we have

$$\frac{d^2}{dt^2}\langle x\rangle_t = a \tag{141}$$

whose solution is

$$\langle x\rangle_t = \langle x\rangle_0 + vt + at^2/2 \tag{142}$$

Substituting these values into Eq. (50), and keeping only the first term we have

$$P(x,t) \sim [1 - B_1 D_x]\, P(x,0) \tag{143}$$

$$= \left(1 - \eta_1(t)\frac{d}{dx}\right) P(x,0) \tag{144}$$

and therefore

$$P(x,t) \sim \left(1 - (vt + at^2/2)\frac{d}{dx}\right) P(x,0) \tag{145}$$

Now the exact solution of Eq. (137) is

$$P(x,t) = P(x - vt - at^2/2, 0) \tag{146}$$

and we can see that the approximate solution is a Taylor expansion in power of $(vt + at^2/2)$ but of course, that would not be known if the exact solution is not known.

6.3 Example 3

We consider Schrödinger type equations of evolution [14]

$$i\frac{\partial u(x,t)}{\partial t} = \mathbf{H}u(x,t) \tag{147}$$

where

$$P(x,t) = |u(x,t)|^2 \tag{148}$$

In Eq. (147), $\mathbf{H}$ is a Hermitian operator that is a function of $\mathbf{x}$ and D

$$\mathbf{H} = \mathbf{H}(\mathbf{x}, D) \tag{149}$$

If we solve the differential equation, Eq. (147), then the expected value of an operator **A** is given by

$$\langle \mathbf{A}^n \rangle = \int u^*(x,t)\mathbf{A}^n u(x,t)dx \tag{150}$$

but of course, the purpose of the method is to circumvent solving Eq. (147). It is possible to get the moments without solving the equation for $u(x,t)$. That is done by solving Heisenberg's equation of motion [14]

$$i\frac{d\mathbf{A}}{dt} = [\mathbf{A},\mathbf{H}] = \mathbf{AH} - \mathbf{HA} \tag{151}$$

The formal solution of this equation is

$$\mathbf{A}(t) = e^{it\mathbf{H}}\mathbf{A}_0 e^{-it\mathbf{H}} \tag{152}$$

which can be expanded in a power series

$$\mathbf{A}(t) = \sum_{n=0}^{\infty} \frac{1}{n!}[\mathbf{A}, H]_n \left(\frac{t}{i\hbar}\right)^n \tag{153}$$

$$= \mathbf{A(0)} + [\mathbf{A},\mathbf{H}]\frac{t}{i\hbar} + \frac{1}{2!}[[\mathbf{A},\mathbf{H}],\mathbf{H}]\left(\frac{t}{i\hbar}\right)^2 + \cdots \tag{154}$$

where $[\mathbf{A},\mathbf{H}]_n$ is the repeated commutator, evaluated at time zero. Therefore, in general,

$$\langle \mathbf{A}(t) \rangle = \sum_{n=0}^{\infty} \frac{1}{n!} \langle [\mathbf{A}, H]_n \rangle \left(\frac{t}{i\hbar}\right)^n \tag{155}$$

Specializing to the case of position and momentum we have

$$\mathbf{x}(t) = e^{it\mathbf{H}}\mathbf{x}_0 e^{-it\mathbf{H}} \tag{156}$$

$$= \sum_{n=0}^{\infty} \frac{1}{n!}[\mathbf{x},\mathbf{H}]_n \left(\frac{t}{i\hbar}\right)^n \tag{157}$$

$$= \mathbf{x}_0 + [\mathbf{x},\mathbf{H}]\frac{t}{i\hbar} + \frac{1}{2!}[[\mathbf{x},\mathbf{H}],\mathbf{H}]\left(\frac{t}{i\hbar}\right)^2 + \cdots \tag{158}$$

and

$$\mathbf{p}(t) = e^{it\mathbf{H}}\mathbf{p}_0 e^{-it\mathbf{H}} \tag{159}$$

$$= \sum_{n=0}^{\infty} \frac{1}{n!}[\mathbf{p}, \mathbf{H}]_n \left(\frac{t}{i\hbar}\right)^n \tag{160}$$

$$= \mathbf{p}_0 + [\mathbf{p}, \mathbf{H}]\frac{t}{i\hbar} + \frac{1}{2!}[[\mathbf{p}, \mathbf{H}], \mathbf{H}]\left(\frac{t}{i\hbar}\right)^2 + \cdots \tag{161}$$

where $\mathbf{x}_0$ and $\mathbf{p}_0$ are the operators at time zero.

Consider the example where we take

$$\mathbf{H} = -\frac{D^2}{2m} - F\mathbf{x} \tag{162}$$

where F is a real constant and m is the mass. The exact solution to the Heisenberg equation of motion are

$$\mathbf{p}(t) = \mathbf{p}_0 + Ft \tag{163}$$

$$\mathbf{x}(t) = \mathbf{x}_0 + \frac{\mathbf{p}_0}{m}t + \frac{F}{2m}t^2 \tag{164}$$

Therefore, the position and momentum time-dependent moments are

$$\left\langle \mathbf{p}^n(t) \right\rangle = \left\langle (\mathbf{p}_0 + Ft)^n \right\rangle \tag{165}$$

$$\left\langle \mathbf{x}^n(t) \right\rangle = \left\langle \left(\mathbf{x}_0 + \frac{\mathbf{p}_0}{m}t + \frac{F}{2m}t^2 \right)^n \right\rangle \tag{166}$$

Before continuing we note that the exact solution to Eq. (147) in momentum space is

$$\widehat{u}(p,t) = \widehat{u}(p - Ft, 0) \exp\left[i \frac{(p - Ft)^3 - p^3}{6mF} \right] \tag{167}$$

$$= \widehat{u}(p - Ft, 0) \exp\left[-i \left(\frac{p^2 t}{2m} - \frac{pFt^2}{2m} + \frac{F^2 t^3}{6m} \right) \right] \tag{168}$$

where

$$\widehat{u}(p,t) = \frac{1}{\sqrt{2\pi}} \int u(x,t) e^{-ixp} dx \tag{169}$$

The solution in position space, $u(x,t)$, is given by

$$u(x,t) = \frac{1}{\sqrt{2\pi}} \int \widehat{u}(p,t) e^{ixp} dp \tag{170}$$

and can be expressed in terms of Airy functions but we do not do so here.

The momentum probability density at time t, $P(p,t)$, is

$$P(p,t) = |\widehat{u}(p,t)|^2 = P(p - Ft, 0) \tag{171}$$

and the position probability density is

$$P(x,t) = |u(x,t)|^2 = \left| \frac{1}{\sqrt{2\pi}} \int \widehat{u}(p,t) e^{ixp} dp \right|^2 \tag{172}$$

We now examine the application of our method. We first consider the momentum distribution. For the Bell polynomials we find that

$$B_1 = \eta_1 = \kappa_1^{(2)} - \kappa_1^{(1)} = \mu_1(t) - \mu_1(0) = \langle \mathbf{p}_0 \rangle + Ft - \langle \mathbf{p}_0 \rangle = Ft \tag{173}$$

$$B_2 = \eta_2 + \eta_1^2 = \kappa_2(t) - \kappa_2(0) + (Ft)^2 = (Ft)^2 \tag{174}$$

$$B_n = (Ft)^n \tag{175}$$

Substituting these values into Eq. (50), we have

$$P(p,t) \sim \left[1 - B_1 D_p + \frac{1}{2} B_2 D_p^2 + \cdots \right] P(p,0) \tag{176}$$

$$= \left[1 - Ft D_p + \frac{1}{2} (Ft)^2 D_p^2 + \cdots \right] P(p,0) \tag{177}$$

which is a Taylor series of the exact solution, Eq. (171). Note that we have avoided solving the equation of motion, Eq. (147).

We now consider the same problem in position space. We just keep the first and second moments

$$\mathbf{x}^2(t) = \mathbf{x}_0^2 + \frac{t}{m}(\mathbf{x}_0\mathbf{p}_0 + \mathbf{p}_0\mathbf{x}_0) + \left(\mathbf{x}_0 \frac{F}{m} + \frac{\mathbf{p}_0^2}{m^2} \right) t^2 + \mathbf{p}_0 \frac{F}{m^2} t^3 + \frac{F^2}{4m^2} t^4 \tag{178}$$

$$\left\langle \mathbf{x}^2(t) \right\rangle = \left\langle \mathbf{x}_0^2 \right\rangle + \frac{t}{m} \langle (\mathbf{x}_0\mathbf{p}_0 + \mathbf{p}_0\mathbf{x}_0) \rangle + \left(\langle \mathbf{x}_0 \rangle \frac{F}{m} + \frac{\langle \mathbf{p}_0^2 \rangle}{m^2} \right) t^2 + \langle \mathbf{p}_0 \rangle \frac{F}{m^2} t^3 + \frac{F^2}{4m^2} t^4 \tag{179}$$

to obtain

$$B_1 = \eta_1 = \kappa_1^{(2)} - \kappa_1^{(1)} = \mu_1(t) - \mu_1(0) \tag{180}$$

$$= \langle \mathbf{x}_0 \rangle + \frac{\langle \mathbf{p}_0 \rangle}{m} t + \frac{F}{2m} t^2 - \langle \mathbf{x}_0 \rangle = \frac{\langle \mathbf{p}_0 \rangle}{m} t + \frac{F}{2m} t^2 \tag{181}$$

$$B_2 = \eta_2 + \eta_1^2 = \kappa_2(t) - \kappa_2(0) + \left(\frac{\langle \mathbf{p}_0 \rangle}{m} t + \frac{F}{2m} t^2 \right)^2 \tag{182}$$

$$= \sigma_x^2(t) - \sigma_x^2(0) + \left(\frac{\langle \mathbf{p}_0 \rangle}{m} t + \frac{F}{2m} t^2 \right)^2 \tag{183}$$

$$= \frac{t}{m} \left\{ \langle (\mathbf{x}_0 \mathbf{p}_0 + \mathbf{p}_0 \mathbf{x}_0) \rangle - 2 \langle \mathbf{x}_0 \rangle \frac{\langle \mathbf{p}_0 \rangle}{m} \right\} + \frac{\sigma_p^2}{m^2} t^2 + \left(\frac{\langle \mathbf{p}_0 \rangle}{m} t + \frac{F}{2m} t^2 \right)^2 \tag{184}$$

Thus, the probability density of position at time t goes as

$$P(x,t) \sim \left[\begin{array}{c} 1 - \left(\frac{\langle \mathbf{p}_0 \rangle}{m} t + \frac{F}{2m} t^2 \right) D \\ +\frac{1}{2} \left(\begin{array}{c} \frac{t}{m} \left\{ \langle (\mathbf{x}_0 \mathbf{p}_0 + \mathbf{p}_0 \mathbf{x}_0) \rangle - 2 \langle \mathbf{x}_0 \rangle \frac{\langle \mathbf{p}_0 \rangle}{m} \right\} \\ + \frac{\sigma_p^2}{m^2} t^2 + \left(\frac{\langle \mathbf{p}_0 \rangle}{m} t + \frac{F}{2m} t^2 \right)^2 \end{array} \right)^2 D^2 + \cdots \end{array} \right] P(x,0)$$

where we have terms up to D^2. We have not examined the accuracy of the approximation.

7 Signal Processing Case

One of the remarkable aspects of signal analysis and quantum mechanics is that while they deal with densities and expectation values, the method of calculation is strange as they deal with wave functions and signals. Our aim in this section is to explore the application of Eq. (1) to wave functions and signals and their corresponding densities. The quantum case and the signal analysis case are similar, and for the sake of concreteness we deal with the signal analysis case.

First, we point that the steps leading to Eq. (1) did not use any particular aspects of a probability density, and therefore we can write that for any two *functions*, $f_2(x)$ and $f_1(x)$, we have

$$f_2(x) = \Omega(\mathbf{A}) f_1(x) \tag{185}$$

where

$$f_1(x) = \int N_1(\theta) u_\theta(x)\, d\theta \tag{186}$$

$$f_2(x) = \int N_2(\theta)\, u_\theta(x)\, d\theta \tag{187}$$

and where N_1 and N_2 are given by

$$N_1(\theta) = \int f_1(x) u_\theta^*(x)\, dx \tag{188}$$

$$N_2(\theta) = \int f_2(x) u_\theta^*(x)\, dx \tag{189}$$

We now take $f_1(x)$ to be a time signal and call it $s(t)$. Instead of writing $f(x)$ we write $f(k)$ since we are now dealing with physical quantities and naming with a different letter namely, k, is helpful and makes the variables clearer. Hence we write the above as

$$f(k) = \Omega(\mathbf{A}) s(t) \tag{190}$$

where

$$\Omega(\theta) = \frac{N_f(\theta)}{N_s(\theta)} \tag{191}$$

and where

$$s(t) = \int N_s(\theta) u_\theta(t)\, d\theta \tag{192}$$

$$f(k) = \int N_f(\theta)\, u_\theta(k)\, d\theta \tag{193}$$

N_1 and N_2 are given by

$$N_s(\theta) = \int s(t) u_\theta^*(t)\, dt \tag{194}$$

$$N_f(\theta) = \int f(k) u_\theta^*(k)\, dk \tag{195}$$

7.1 Operator Relation Between the Energy Density and Energy Density Spectrum

For a signal $s(t)$ whose spectrum is defined by

$$S(\omega) = \frac{1}{\sqrt{2\pi}} \int s(t)\, e^{-i\omega t}\, dt \tag{196}$$

the energy density $|s(t)|^2$ and the energy density spectrum is $|S(\omega)|^2$. Since they are both proper densities we have

$$|S(\omega)|^2 = \Omega(iD_t)\, |s(t)|^2 \tag{197}$$

where now

$$\Omega(\theta) = \frac{M_\omega(\theta)}{M_t(\theta)} \tag{198}$$

and where one has to substitute ω for t after the operation in the left-hand side of Eq. (197) is carried out. The characteristic functions are

$$M_t(\theta) = \int |s(t)|^2\, e^{i\theta t}\, dt \tag{199}$$

$$M_\omega(\theta) = \int |S(\omega)|^2 e^{i\theta\omega}\, d\omega \tag{200}$$

Notational Issue In the usual formulation as per Eq. (1), the variable x appears on both sides, and of course the mathematics does not care what symbols are used. On the other hand, Eq. (197) seemingly does not make sense because it mixes time and frequency which of course are not the same from a physical point of view. However, we can write

$$|S(\omega)|^2 = \Omega(iD_\omega)\, |s(\omega)|^2 \tag{201}$$

keeping in mind that we have substituted ω for t in Eq. (197) only for notational consistency.

Example Consider the signal

$$s(t) = (\alpha/\pi)^{1/4}\, e^{-\alpha t^2/2 + i\beta t^2/2 + i\omega_0 t} \tag{202}$$

whose spectrum is

$$S(\omega)=\frac{(\alpha/\pi)^{1/4}}{\sqrt{\alpha-i\beta}}\exp\left[-\frac{\alpha(\omega-\omega_0)^2}{2(\alpha^2+\beta^2)}-i\frac{\beta(\omega-\omega_0)^2}{2(\alpha^2+\beta^2)}\right] \tag{203}$$

The energy density and the energy density spectrum are, respectively,

$$|s(t)|^2 = (\alpha/\pi)^{1/2} e^{-\alpha t^2} \tag{204}$$

and

$$|S(\omega)|^2=\sqrt{\frac{\alpha}{\pi(\alpha^2+\beta^2)}}\, e^{-\alpha(\omega-\omega_0)^2/(\alpha^2+\beta^2)} \tag{205}$$

The corresponding characteristic functions are calculated to be

$$M_t(\theta) = (\alpha/\pi)^{1/2}\int e^{-\alpha t^2+i\theta t}dt = e^{-\theta^2/(4\alpha)} \tag{206}$$

and

$$M_\omega(\theta) = \sqrt{\frac{\alpha}{\pi(\alpha^2+\beta^2)}}\int e^{-\alpha(\omega-\omega_0)^2/(\alpha^2+\beta^2)}e^{i\theta\omega}d\omega \tag{207}$$

$$= e^{i\theta\omega_0}e^{-(\alpha^2+\beta^2)\theta^2/(4a)} \tag{208}$$

giving that

$$\frac{M_2(\theta)}{M_1(\theta)} = e^{i\theta\omega_0}e^{-(\alpha^2+\beta^2-1)\theta^2/(4a)} \tag{209}$$

Hence

$$\Omega(iD) = e^{-D\omega_0+(\alpha^2+\beta^2-1)D^2/(4a)} \tag{210}$$

We therefore have that

$$|S(\omega)|^2= e^{-D\omega_0+(\alpha^2+\beta^2-1)D^2/(4a)}\,|s(\omega)|^2 \tag{211}$$

or explicitly

$$\sqrt{\frac{\alpha}{\pi(\alpha^2+\beta^2)}}\, e^{-\alpha(\omega-\omega_0)^2/(\alpha^2+\beta^2)} = (\alpha/\pi)^{1/2}e^{-D\omega_0+(\alpha^2+\beta^2-1)D^2/(4a)}\,e^{-\alpha\omega^2} \tag{212}$$

7.2 Operator Relation Between a Signal and Spectrum

In this section we explore the possibility of connecting the spectrum with the signal by way of

$$S(\omega) = \Omega(iD_t)s(t) \tag{213}$$

where

$$S(\omega) = \frac{1}{\sqrt{2\pi}}\int s(t)\, e^{-i\omega t}\, dt \tag{214}$$

For this case

$$\Omega(\theta) = \frac{N_f(\theta)}{N_s(\theta)} \tag{215}$$

where N_1 and N_2 are given by

$$N_s(\theta) = \int s(t)e^{i\theta t}\, dt = \sqrt{2\pi}\, S(-\theta) \tag{216}$$

$$N_f(\theta) = \int S(\omega)e^{i\theta\omega}\, d\omega = \sqrt{2\pi}\, s(\theta) \tag{217}$$

Therefore we have that

$$S(\omega) = \Omega(iD)s(t) \tag{218}$$

or

$$S(\omega) = \Omega(iD)s(t) = \frac{N_f(iD_t)}{N_s(iD_t)}s(t) \tag{219}$$

Using the notation indicated in the previous section

$$S(\omega) = \frac{S(-iD_\omega)}{s(iD_\omega)}s(\omega) \tag{220}$$

Another interesting operational formula derived by Ben-Benjamin (private communication, 2018) is to start with

$$S(\omega) = \frac{1}{\sqrt{2\pi}}\int s(t)\, e^{-i\omega t}\, dt \tag{221}$$

and write

$$S(\omega) = \frac{1}{\sqrt{2\pi}} \int s(iD_\omega)\, e^{-i\omega t}\, dt \tag{222}$$

$$= \frac{1}{\sqrt{2\pi}} s(iD_\omega) \int e^{-i\omega t}\, dt \tag{223}$$

and therefore

$$S(\omega) = \sqrt{2\pi} s(iD_\omega)\delta(\omega) \tag{224}$$

where $\delta(\omega)$ is the Dirac delta function.

Suppose now

$$N_1(\theta) = \exp\left[\sum_{n=0}^{\infty} \delta_n^{(1)} \frac{i^n}{n!} \theta^n\right] \tag{225}$$

$$N_2(\theta) = \exp\left[\sum_{n=0}^{\infty} \delta_n^{(2)} \frac{i^n}{n!} \theta^n\right] \tag{226}$$

where $\delta_n^{(1)}$ and $\delta_n^{(2)}$ are physical quantities that characterize the signal and spectrum. Using Eq. (215) we have

$$\Omega(\theta) = \frac{N_2(\theta)}{N_1(\theta)} = \exp\left[\sum_{n=0}^{\infty} \left(\delta_n^{(2)} - \delta_n^{(1)}\right) \frac{i^n}{n!} \theta^n\right] \tag{227}$$

and

$$\Omega(iD) = \exp\left[\sum_{n=0}^{\infty} \left(\delta_n^{(2)} - \delta_n^{(1)}\right) \frac{(-1)^n}{n!} D_t^n\right] \tag{228}$$

Therefore

$$S(\omega) = \exp\left[\sum_{n=0}^{\infty} (\delta_n^{(2)} - \delta_n^{(1)}) \frac{(-1)^n}{n!} D_\omega^n\right] s(\omega) \tag{229}$$

This relates the spectrum with the signal by way of an operator relation. We emphasize that λ_n are not cumulants or moments and their meaning will be discussed in another paper.

8 Relation Between Spectral Moments and Phase and Amplitude

One of the strange aspects of signal analysis is that one cannot relate the spectral moments directly to the time moments. That is the same situation in quantum mechanics where one cannot simply connect the momentum moments with the spatial moments. We consider here relating the spectral moments to the phase and amplitude of the signal. In this section

$$D = \frac{d}{dt} \tag{230}$$

The spectral moments are given by

$$\langle \omega^n \rangle = \int \omega^n \, |\varphi(\omega)|^2 \, d\omega = \frac{1}{i^n} \int s^*(t) D^n s(t) dt \tag{231}$$

We want to express the spectral moments in terms of the amplitude, $a(t)$,and phase, $\varphi(t)$, of the signal

$$s(t) \; = a(t) e^{i\varphi(t)} \tag{232}$$

This problem has been considered previously [8, 9, 12, 16, 17], but here we give a formulation that involves the Bell polynomials. Starting with

$$\langle \omega^n \rangle = \frac{1}{i^n} \int a(t) e^{-i\varphi(t)} D^n a(t) e^{i\varphi(t)} dt \tag{233}$$

we have

$$\langle \omega^n \rangle = \frac{1}{i^n} \int a(t) \sum_{k=0}^{n} \binom{n}{k} \left(D^{n-k} a(t) \right) e^{-i\varphi(t)} D^k e^{i\varphi(t)} \tag{234}$$

Using Eq. (271) of the appendix we have

$$e^{-i\varphi(t)} D^k e^{i\varphi(t)} = B_k(i\varphi', i\varphi'', \ldots, i\varphi^{(k)}) \tag{235}$$

and therefore

$$\langle \omega^n \rangle = \frac{1}{i^n} \int a(t) \sum_{k=0}^{n} \binom{n}{k} \left(D^{n-k} a(t) \right) B_k(i\varphi', i\varphi'', \ldots, i\varphi^{(k)}) dt \tag{236}$$

Also,

$$\langle\omega^n\rangle = \frac{1}{i^n}\int e^{\ln a}\sum_{k=0}^{n}\binom{n}{k}\left(D^{n-k}e^{\ln a}\right) B_k(i\varphi', i\varphi'', \ldots, i\varphi^{(k)})dt \tag{237}$$

If we call

$$\mu(t) = \ln a(t) \tag{238}$$

then

$$\left(D^{n-k}e^{\ln a}\right) = e^{\ln a} B_{n-k}(\mu', \mu'', \ldots, \mu^{(n-k)}) \tag{239}$$

and we have that

$$\langle\omega^n\rangle = \frac{1}{i^n}\int a^2(t)\sum_{k=0}^{n}\binom{n}{k} B_{n-k}(\mu', \mu'', \ldots, \mu^{(n-k)}) B_k(i\varphi', i\varphi'', \ldots, i\varphi^{(k)})dt = \tag{240}$$

$$\frac{1}{i^n}\sum_{k=0}^{n}\binom{n}{k}\left\langle B_{n-k}(\mu', \mu'', \ldots, \mu^{(n-k)}) B_k(i\varphi', i\varphi'', \ldots, i\varphi^{(k)})\right\rangle \tag{241}$$

In addition, depending on whether n is odd or even, further simplification is possible. For even $n = 2N$, Eq. (236) becomes

$$\left\langle\omega^{2N}\right\rangle = (-1)^N\int a(t)\sum_{k=0}^{2N}\binom{2N}{k}\left(D^{2N-k}a(t)\right) B_k(i\varphi', i\varphi'', \ldots, i\varphi^{(k)})dt \tag{242}$$

However, since the expectation value is real we have that

$$\left\langle\omega^{2N}\right\rangle = (-1)^N\int a(t)\sum_{k=0}^{2N}\binom{2N}{k}\left(D^{2N-k}a(t)\right) \operatorname{Re} B_k(i\varphi', i\varphi'', \ldots, i\varphi^{(k)})dt \tag{243}$$

$$0 = (-1)^N\int a(t)\sum_{k=0}^{2N}\binom{2N}{k}\left(D^{2N-k}a(t)\right) \operatorname{Im} B_k(i\varphi', i\varphi'', \ldots, i\varphi^{(k)})dt \tag{244}$$

Also, for this case we can start with

$$\left\langle\omega^{2N}\right\rangle = \int \left|D^N s(t)\right|^2 dt \tag{245}$$

which leads to

$$\left\langle \omega^{2N} \right\rangle = \int \left\{ \left(\sum_{k=0}^{N} \binom{N}{k} \left(D^{N-k} a(t) \right) \operatorname{Re} B_k \right)^2 + \left(\sum_{k=0}^{N} \binom{N}{k} \left(D^{N-k} a(t) \right) \operatorname{Im} B_k \right)^2 \right\} dt \tag{246}$$

If n is odd and equal to $2N + 1$

$$\left\langle \omega^{2N+1} \right\rangle = \frac{(-1)^N}{i} \int a(t) \sum_{k=0}^{n} \binom{n}{k} \left(D^{n-k} a(t) \right) B_k(i\varphi', i\varphi'', \ldots, i\varphi^{(k)}) dt \tag{247}$$

which gives

$$\left\langle \omega^{2N+1} \right\rangle = (-1)^{N+1} \int a(t) \sum_{k=0}^{2N+1} \binom{2N+1}{k} \left(D^{2N+1-k} a(t) \right) \operatorname{Re} i B_k(i\varphi', i\varphi'', \ldots, i\varphi^{(k)}) \tag{248}$$

$$= (-1)^N \int a(t) \sum_{k=0}^{2N+1} \binom{2N+1}{k} \left(D^{2N+1-k} a(t) \right) \operatorname{Im} B_k(i\varphi', i\varphi'', \ldots, i\varphi^{(k)}) \tag{249}$$

In addition, by writing

$$\left\langle \omega^{2N+1} \right\rangle = \frac{1}{i^{2N+1}} \int s^*(t) D^{2N+1} s(t) dt \tag{250}$$

$$= \int \left\{ \left(\frac{1}{i^N} D^N s(t) \right) \left(\frac{1}{i^{N+1}} D^{N+1} s(t) \right) \right\} dt \tag{251}$$

$$= \int \left\{ \left(\frac{1}{i^N} D^N a(t) e^{-i\varphi(t)} \right) \left(\frac{1}{i^{N+1}} D^{N+1} a(t) e^{i\varphi(t)} \right) \right\} \tag{252}$$

one obtains

$$\left\langle \omega^{2N+1} \right\rangle = \frac{1}{i^{2N+1}} \int \left(\sum_{k=0}^{N} \binom{N}{k} \left(D^{N-k} a \right) B_k(-i\varphi', -i\varphi'' \ldots) \right)$$
$$\times \left(\sum_{k=0}^{N+1} \binom{N+1}{k} \left(D^{N+1-k} a \right) B_k(i\varphi', i\varphi'' \ldots) \right) dt \tag{253}$$

Polleti Formulation Another formulation is to define, following Polleti [16, 17], the dynamical signal $\beta(t)$ by

$$\beta(t) = \frac{a'}{a} + i\varphi' = \frac{d}{dt}(\ln a + i\varphi) \tag{254}$$

which gives that

$$Ds(t) = \beta(t)s(t) \tag{255}$$

and furthermore

$$D^n s(t) = \beta_n(t)s(t) \tag{256}$$

where the $\beta_n(t)$ may be calculated recursively [16, 17]

$$\beta_n(t) = \beta_n(t)\left(\beta_1(t) + D\ln\beta_n(t)\right) \tag{257}$$

This formulation was used by Poletti in his study of conditional moments.

Writing

$$s(t) \; = e^{\ln a + i\varphi(t)} \tag{258}$$

then

$$\left\langle \omega^n \right\rangle = \frac{1}{i^n}\int e^{\ln a - i\varphi(t)} D^n e^{\ln a + i\varphi(t)} dt \tag{259}$$

$$= \frac{1}{i^n}\int e^{2\ln a} e^{-\ln a - i\varphi(t)} D^n e^{\ln a + i\varphi(t)} dt \tag{260}$$

which gives

$$\left\langle \omega^n \right\rangle = \frac{1}{i^n}\int e^{\ln a - i\varphi(t)} D^n e^{\ln a + i\varphi(t)} dt \tag{261}$$

$$= \frac{1}{i^n}\int a^2(t) B_k(\beta_1(t)s(t), \beta_2(t)s(t) \ldots \beta_n(t)s(t)) dt \tag{262}$$

$$= \frac{1}{i^n}\left\langle B_k(\beta_1(t)s(t), \beta_2(t)s(t) \ldots \beta_n(t)s(t)) \right\rangle \tag{263}$$

That is,

$$\left\langle \omega^n \right\rangle = \frac{1}{i^n}\left\langle B_k(\beta_1(t)s(t), \beta_2(t)s(t) \ldots \beta_n(t)s(t)) \right\rangle \tag{264}$$

Appendix: Bell Polynomials

The complete Bell polynomials, B_n, may be defined by the expansion [15]

$$\exp\left(\sum_{k=1}^{\infty} a_i \frac{x^i}{k!}\right) = \sum_{n=0}^{\infty} B_n(a_1, \ldots, a_n)\frac{x^n}{n!} \tag{265}$$

where

$$B_n(a_1, \ldots, a_n) = \frac{d^n}{dx^n} \exp\left(\sum_{k=1}^{\infty} a_i \frac{x^i}{k!}\right)\Bigg|_{x=0}. \tag{266}$$

They satisfy the following recurrence relation

$$B_{n+1}(a_1, \ldots, a_{n+1}) = \sum_{i=0}^{n} \binom{n}{i} B_{n-i}(a_1, \ldots, a_{n-i}) a_{i+1}, \tag{267}$$

with

$$B_0 = 1 \tag{268}$$

also,

$$B_n(a_1 + b_1, \ldots, a_n + b_n) = \sum_{i=0}^{n} \binom{n}{i} B_{n-i}(a_1, \ldots, a_{n-i}) B_i(b_1, \ldots, b_i). \tag{269}$$

For a function $f(x)$, the nth derivative of $e^{f(x)}$ may be expressed in terms of the Bell polynomials

$$D^n e^{f(x)} = e^{f(x)} B_n(f', f'', \ldots, f^{(n)}) \tag{270}$$

and

$$e^{-f(x)} D^n e^{f(x)} = B_n(f', f'', \ldots, f^{(n)}) \tag{271}$$

The first few polynomials are

$$B_0 = 1, \tag{272}$$

$$B_1(x_1) = x_1 \tag{273}$$

$$B_2(x_1, x_2) = x_1^2 + x_2 \tag{274}$$

$$B_3(x_1, x_2, x_3) = x_1^3 + 3x_1x_2 + x_3 \tag{275}$$

$$B_4(x_1, x_2, x_3, x_4) = x_1^4 + 6x_1^2x_2 + 4x_1x_3 + 3x_2^2 + x_4 \tag{276}$$

References

1. S. Chandrasekhar, Stochastic problems in physics and astronomy. Rev. Mod. Phys. **15**, 1–89 (1943)
2. L. Cohen, Time-frequency distributions - a review. Proc. IEEE **77**, 941–981 (1989)
3. L. Cohen, The scale representation. IEEE Trans. Signal Process. **41**, 3275–3292 (1993)
4. L. Cohen, *Time-Frequency Analysis* (Prentice-Hall, Englewood Cliffs, 1995)
5. L. Cohen, Generalization of the Gram-Charlier/Edgeworth series and application to time-frequency analysis, Multidim. Syst. Sign. Process. **9**, 363–372 (1998)
6. L. Cohen, Construction and transformation of probability densities. Integrated Comput. Aided Eng. **12**(1), 129–134 (2005)
7. L. Cohen, Generalization of the Edgeworth and Gran-Charlier series and quasi-probability distributions, Research Institute for Mathematical Sciences, Kyoto University (2018)
8. K. Davidson, Instantaneous moments of signals, Ph.D. dissertation, University of Pittsburgh, 2001
9. K. Davidson, P. Loughlin, Instantaneous spectral moments. J. Frankl. Inst. **337**, 421–436 (2000)
10. J.S. Kolassa, *Series Approximation Methods in Statistics* (Springer, New York, 1994)
11. H.W. Lee, Theory and application of the quantum phase-space distribution functions. Phys. Rep. **259**, 147–211 (1995)
12. P. Loughlin, K. Davidson, Instantaneous kurtosis. IEEE Signal Process. Lett. **7**(6), 156–159 (2000)
13. E. Lukacs, *Characteristic Functions* (Charles Griffin and Company, London, 1970)
14. E. Merzbacher, *Quantum Mechanics* (Wiley, New York, 1961)
15. P. Natalini, P.E. Ricci, Remarks on Bell and higher order Bell polynomials and numbers. Cogent Math. **3**, 1220670 (2016)
16. M. Poletti, Instantaneous frequency and conditional moments in the time-frequency plane. IEEE Trans. Signal Process. **39**, 755–756 (1991)
17. M. Poletti, The development of instantaneous bandwidth via local signal expansion. Signal Process. **31**(3), 273–281 (1993)
18. A. Stuart, K.K. Ord, *Kendall's, Advanced Theory of Statistics* (Halsted Press, New York, 1994)
19. G.E. Uhlenbeck, L.S. Ornstein, On the theory of Brownian Motion. Phys. Rev. **36**, 823–841 (1930)

The Time-Frequency Interference Terms of the Green's Function for the Harmonic Oscillator

Lorenzo Galleani

Abstract The harmonic oscillator is a fundamental prototype for all types of resonances, and hence plays a key role in the study of physical systems governed by differential equations. The time-frequency representation of its Green's function, obtained through the Wigner distribution, reveals the time-varying frequency structure of resonances. Unfortunately, the Wigner distribution of the Green's function is affected by strong interference terms with a highly oscillatory structure. We characterize these interference terms by evaluating the ambiguity function of the Green's function. The obtained result shows that, in the ambiguity domain, the interference terms are localized and separate from the resonance component, and hence they can be reduced by a proper filtering.

Keywords Time-frequency analysis · Interference terms · Green's function · Harmonic oscillator

Mathematics Subject Classification (2000) 60H10

1 Introduction

Differential equations model a wide variety of deterministic and random physical phenomena. A common approach to study them are transformation techniques, such as frequency analysis (the Fourier transform) [1] and the Laplace transform [2]. An effective approach is also time-frequency analysis [3, 4], a body of techniques for the characterization of signals whose frequency content changes with time. Conversely from frequency analysis, where the Fourier transform connects the time

---

L. Galleani (✉)
Department of Electronics and Telecommunications, Politecnico di Torino, Torino, Italy
e-mail: galleani@polito.it

S. Molahajloo, M. W. Wong (eds.), *Analysis of Pseudo-Differential Operators*,
Trends in Mathematics, https://doi.org/10.1007/978-3-030-05168-6_9

and frequency domains, in time-frequency analysis there are infinite time-frequency representations, or distributions, such as the Wigner distribution [3, 5, 6]

$$W_x(t,\omega) = \frac{1}{2\pi}\int_{-\infty}^{+\infty} x^*(t-\tau/2)x(t+\tau/2)e^{-i\tau\omega}d\tau. \tag{1}$$

In [7] we have obtained the Wigner distribution of the Green's function for the harmonic oscillator, a fundamental model for resonant phenomena defined as

$$\frac{d^2x(t)}{dt^2} + 2\mu\frac{dx(t)}{dt} + \omega_0^2 x(t) = f(t), \tag{2}$$

where $f(t)$ is the forcing term, or input, $x(t)$ is the solution, also referred to as output or response, and we consider the case $\mu < \omega_0$, which gives rise to a resonance at the frequency

$$\omega_c = \sqrt{\omega_0^2 - \mu^2}. \tag{3}$$

The Green's function is defined as the solution $h(t)$ when the forcing term is a Dirac delta function [8]. Since the delta function is the ideal impulse, the Green's function is also referred to as the impulse response. The advantage of the Green's function is that, for any forcing term, the solution of (2) can be written through the convolution integral

$$x(t) = \int_{-\infty}^{+\infty} h(t-t')f(t')dt'. \tag{4}$$

The convolution property holds also in the time-frequency domain [3]

$$W_x(t,\omega) = \int_{-\infty}^{+\infty} W_h(t-t',\omega)W_f(t',\omega)dt'. \tag{5}$$

The Green's function is a cornerstone for the analysis and design of physical systems and devices, and it can be used for any ordinary differential equation with constant coefficients [1], as well as for partial differential equations.

The Wigner distribution of the Green's function for the harmonic oscillator is given by [7]

$$W_h(t,\omega) = \frac{1}{4\omega_c^2}W_{h_L}(t,\omega-\omega_c) + \frac{1}{4\omega_c^2}W_{h_L}(t,\omega+\omega_c) - \frac{1}{2\omega_c^2}W_{h_L}(t,\omega)\cos 2\omega_c t, \tag{6}$$

where

$$W_{h_L}(t,\omega) = u(t)e^{-2\mu t}\frac{\sin 2\omega t}{\pi\omega} \tag{7}$$

is the Wigner distribution of the Green's function $h_L(t)$ corresponding to the first-order differential equation

$$\frac{dx(t)}{dt} + \mu x(t) = f(t), \tag{8}$$

and $u(t)$ is the Heaviside step function defined as $u(t) = 1$ for $t \geq 0$, and $u(t) = 0$ for $t < 0$. When $f(t)$ is white Gaussian noise, (8) is the Langevin equation [9], a fundamental model for random phenomena. The quantity $\mu > 0$ is referred to as the damping coefficient. Unfortunately, due to its quadratic nature, the Wigner distribution is affected by interference terms, highly oscillatory components which make the understanding and interpretation of the time-frequency structure of signals a difficult problem [10–12]. A common approach to reduce the interference terms is to Fourier transform the Wigner distribution, thus obtaining the ambiguity function [3]. Because of their oscillatory behavior, in the ambiguity domain the interference terms are mostly located away from the origin, and they can be therefore reduced by a proper lowpass filtering [4].

We obtain the ambiguity function of the Green's function for the harmonic oscillator, and we show that, similarly to the Wigner distribution $W_h(t, \omega)$, it can be written with respect to the ambiguity function of the Langevin equation. The time-frequency interference terms of the Green's function have a simple structure in the ambiguity domain, which we discuss in detail. Our results can pave the way for the design of interference mitigation filters which take advantage of the structure of the differential equation defining the signal $x(t)$.

We note that an alternative approach for the time-frequency study of differential equations is to transform the differential equation in the time domain to an equivalent differential equation in the time-frequency domain, whose structure is often more complicated than the original equation, but whose solution is often easier to get and more revealing than in the time-domain [13–16].

The article is organized as follows. In Sect. 2 we define the ambiguity function and give some of its properties. In Sect. 3 we obtain the ambiguity function for the Langevin equation and for the harmonic oscillator, and we use it to discuss the structure of the interference terms of these differential equations. Finally, Sect. 4 summarizes the obtained results.

2 The Ambiguity Function

The ambiguity function of a signal $x(t)$, also referred to as the characteristic function, is defined as [3]

$$A_x(\theta, \tau) = \int_{-\infty}^{+\infty} x^*(t - \tau/2)x(t + \tau/2)e^{i\theta t}dt, \tag{9}$$

and it plays a fundamental role in radars, where θ is the Doppler frequency and τ the time delay. This definition is known as the symmetric ambiguity function, and it is connected through a Fourier transformation to the Wigner distribution,

$$A_x(\theta, \tau) = \int_{-\infty}^{+\infty}\int_{-\infty}^{+\infty} W(t, \omega) e^{i\theta t + i\tau\omega} dt d\omega. \tag{10}$$

Therefore, the magnitude $|A_x(\theta, \tau)|$ describes the oscillatory structure of the time-frequency representation of $x(t)$. Actually, (10) is an inverse Fourier transform, but since the Wigner distribution is real, then $|A_x(\theta, \tau)|$ is even with respect to θ and τ, and therefore adopting a definition for $A_x(\theta, \tau)$ which connects it to the Wigner distribution through a direct Fourier transformation would not change $|A_x(\theta, \tau)|$. The cross-ambiguity function of two signals $x(t)$ and $y(t)$ is defined as

$$A_{x,y}(\theta, \tau) = \int_{-\infty}^{+\infty} x^*(t - \tau/2) y(t + \tau/2) e^{i\theta t} dt. \tag{11}$$

We now give some properties of the ambiguity function which are useful for our analysis. These properties can be easily proved from definition (9) and from (10).

Multiplication by a Constant If

$$y(t) = cx(t), \tag{12}$$

then

$$A_y(\theta, \tau) = |c|^2 A_x(\theta, \tau). \tag{13}$$

Multiplication by Constants (Cross-Ambiguity Function) If

$$y_1(t) = c_1 x_1(t), \tag{14}$$

$$y_2(t) = c_2 x_2(t), \tag{15}$$

then

$$A_{y_1,y_2}(\theta, \tau) = c_1^* c_2 A_{x_1,x_2}(\theta, \tau). \tag{16}$$

Complex Frequency Modulation When

$$y(t) = x(t) e^{i\omega_0 t}, \tag{17}$$

it is

$$A_y(\theta, \tau) = A_x(\theta, \tau) e^{i\omega_0 \tau}. \tag{18}$$

Complex Frequency Modulation (Cross-Ambiguity Function) When

$$x_1(t) = x(t)e^{i\omega_0 t}, \tag{19}$$

$$x_2(t) = x(t)e^{-i\omega_0 t}, \tag{20}$$

it is

$$A_{x_1,x_2}(\theta, \tau) = A_x(\theta - 2\omega_0, \tau). \tag{21}$$

Sum of Two Signals If

$$y(t) = x_1(t) + x_2(t), \tag{22}$$

then

$$A_y(\theta, \tau) = A_{x_1}(\theta, \tau) + A_{x_2}(\theta, \tau) + A_{x_1,x_2}(\theta, \tau) + A_{x_2,x_1}(\theta, \tau). \tag{23}$$

Real Frequency Modulation From the previous properties, if

$$y(t) = x(t) \sin \omega_0 t, \tag{24}$$

then

$$A_y(\theta, \tau) = \frac{1}{2} A_x(\theta, \tau) \cos \omega_0 \tau - \frac{1}{4} A_x(\theta - 2\omega_0, \tau) - \frac{1}{4} A_x(\theta + 2\omega_0, \tau). \tag{25}$$

Sum of Two Wigner Distributions If

$$W_y(t, \omega) = c_1 W_{x_1}(t, \omega) + c_2 W_{x_2}(t, \omega), \tag{26}$$

then

$$A_y(\theta, \tau) = c_1 A_{x_1}(\theta, \tau) + c_2 A_{x_2}(\theta, \tau). \tag{27}$$

Frequency Translation of the Wigner Distribution If

$$W_y(t, \omega) = W_x(t, \omega - \omega_0), \tag{28}$$

then

$$A_y(\theta, \tau) = A_x(\theta, \tau) e^{i\omega_0 \tau}. \tag{29}$$

Complex Frequency Modulation of the Wigner Distribution If

$$W_y(t, \omega) = W_x(t, \omega) e^{i\omega_0 t}, \tag{30}$$

then

$$A_y(\theta, \tau) = A_x(\theta + \omega_0, \tau). \tag{31}$$

Real Frequency Modulation of the Wigner Distribution By using the previous properties, if

$$W_y(t, \omega) = W_x(t, \omega) \cos \omega_0 t, \tag{32}$$

then

$$A_y(\theta, \tau) = \frac{1}{2} A_x(\theta + \omega_0, \tau) + \frac{1}{2} A_x(\theta - \omega_0, \tau). \tag{33}$$

3 The Interference Terms of the Harmonic Oscillator

We first obtain the ambiguity function $A_{h_L}(\theta, \tau)$ of the Green's function for the Langevin equation (8), and then we use it to obtain the ambiguity function $A_h(\theta, \tau)$ of the Green's function for the harmonic oscillator (2).

3.1 *The Ambiguity Function for the Langevin Equation*

The Green's function of the Langevin equation (8) is given by [1]

$$h_L(t) = u(t) e^{-\mu t}. \tag{34}$$

The corresponding ambiguity function is given by

$$A_{h_L}(\theta, \tau) = \int_{-\infty}^{+\infty} h_L^*(t - \tau/2) h_L(t + \tau/2) e^{i\theta t} dt, \tag{35}$$

$$= \int_{-\infty}^{+\infty} u(t - \tau/2) u(t + \tau/2) e^{(-2\mu + i\theta)t} dt. \tag{36}$$

We note that

$$u(t - \tau/2) u(t + \tau/2) = u(t - \tau/2), \quad \text{for } \tau \geq 0, \tag{37}$$

$$u(t - \tau/2) u(t + \tau/2) = u(t + \tau/2), \quad \text{for } \tau < 0. \tag{38}$$

Therefore

$$u(t - \tau/2) u(t + \tau/2) = u(t - |\tau| /2). \tag{39}$$

Substituting,

$$A_{h_L}(\theta, \tau) = \int_{-\infty}^{+\infty} u(t - |\tau|/2) e^{(-2\mu+i\theta)t} dt, \tag{40}$$

$$= \int_{|\tau|/2}^{+\infty} e^{(-2\mu+i\theta)t} dt. \tag{41}$$

Finally,

$$A_{h_L}(\theta, \tau) = \frac{e^{(-\mu+i\theta/2)|\tau|}}{2\mu - i\theta}. \tag{42}$$

In the appendix we confirm this result by (inverse) Fourier transforming the Wigner distribution $W_{h_L}(t, \omega)$ in (7), whose oscillatory structure is described by the magnitude

$$\left|A_{h_L}(\theta, \tau)\right| = \frac{e^{-\mu|\tau|}}{\sqrt{4\mu^2 + \theta^2}}. \tag{43}$$

To illustrate our result, we show $W_{h_L}(t, \omega)$ in Fig. 1, and $\left|A_{h_L}(\theta, \tau)\right|$ in Fig. 2, for the case $\mu = 5$. From Fig. 1 we see that, at $t = 0$, the delta function at the

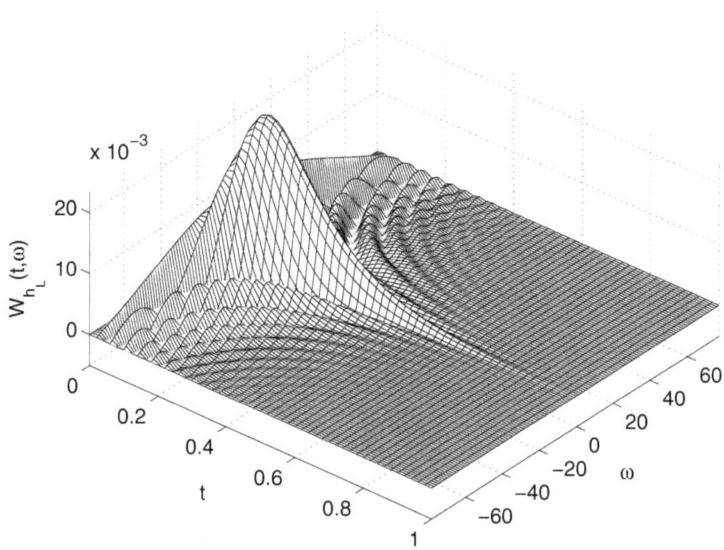

Fig. 1 Wigner distribution of the Green's function for the Langevin equation. The delta function at the input generates a time-frequency response made by an initial spread over all frequencies, which then concentrates on the zero frequency. This first-order equation can be interpreted as a system with a resonance at the zero frequency. The arc-shaped waves propagating from the origin are interference terms

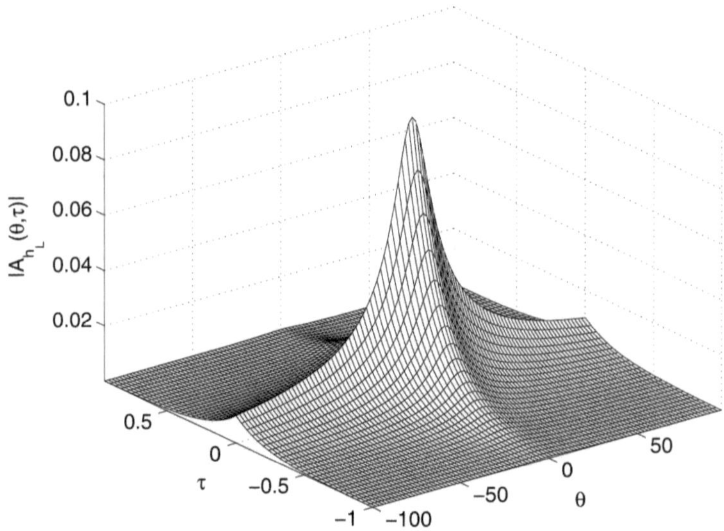

Fig. 2 Magnitude of the ambiguity function of the Green's function for the Langevin equation. This function has a peak at the origin of the ambiguity plane, and has tails on the θ and τ axes. These tails are mainly due to the interference terms of the Wigner distribution $W_{h_L}(t, \omega)$ in Fig. 1

input generates an initial spread over all frequencies, which then concentrates on the zero frequency. Therefore, this first-order equation can be interpreted as a resonant system whose resonance frequency is zero. The arc-shaped waves propagating from the origin of the ambiguity plane are interference terms. As Fig. 2 shows, the frequency spectrum of the Wigner distribution $W_{h_L}(t, \omega)$ is mainly concentrated on the origin, an expected result since $\left|A_{h_L}(\theta, \tau)\right|$ is made by the product of the Cauchy-like distribution $1/\sqrt{4\mu^2 + \theta^2}$ and the symmetric exponential function $e^{-\mu|\tau|}$. The tails of the ambiguity function are mainly due to the interference terms of $W_{h_L}(t, \omega)$, which oscillates more than the resonant component at $\omega = 0$. The component at $t = 0$ contributes also to the tails of the ambiguity function.

3.2 The Ambiguity Function for the Harmonic Oscillator

The Green's function for the harmonic oscillator can be written as [7]

$$h(t) = \frac{1}{\omega_c} h_L(t) \sin \omega_c t. \tag{44}$$

By using the properties (13) and (25) we immediately obtain

$$A_h(\theta, \tau) = \frac{1}{2\omega_c^2} A_{h_L}(\theta, \tau) \cos \omega_c \tau - \frac{1}{4\omega_c^2} A_{h_L}(\theta - 2\omega_c, \tau) - \frac{1}{4\omega_c^2} A_{h_L}(\theta + 2\omega_c, \tau). \tag{45}$$

An alternative way to derive this result is to apply the properties (27), (29), and (33) to $W_h(t, \omega)$ in (6), obtaining

$$A_h(\theta, \tau) = \frac{1}{4\omega_c^2} A_{h_L}(\theta, \tau) e^{i\omega_c \tau} + \frac{1}{4\omega_c^2} A_{h_L}(\theta, \tau) e^{-i\omega_c \tau} \tag{46}$$

$$- \frac{1}{4\omega_c^2} A_{h_L}(\theta + 2\omega_c, \tau) - \frac{1}{4\omega_c^2} A_{h_L}(\theta - 2\omega_c, \tau). \tag{47}$$

Combining the first two terms returns (45).

To illustrate our result, we show $W_h(t, \omega)$ in Fig. 3 and $|A_h(\theta, \tau)|$ in Fig. 4, for $\mu = 5$ and $\omega_c = 60$. From Fig. 3 we see that the input delta function at $t = 0$ generates an initial spread over all frequencies, which eventually concentrates on the resonance frequency ω_c, and on its symmetric counterpart at $-\omega_c$. The oscillating components between these two resonances are interference terms. From (6), aside from the constants, the resonance at frequency ω_c is described by the term $W_{h_L}(t, \omega - \omega_c)$ (its negative counterpart by $W_{h_L}(t, \omega + \omega_c)$), whereas the interference terms between the two resonances are described by the oscillating term $W_{h_L}(t, \omega) \cos 2\omega_c t$. Figure 4 shows that the ambiguity function is made by three components. The component centered about the origin represents the resonant components at ω_c and $-\omega_c$, which are merged in the single term $A_{h_L}(\theta, \tau) \cos \omega_c \tau$ in (45). In the ambiguity domain, the interference terms are instead split up in the

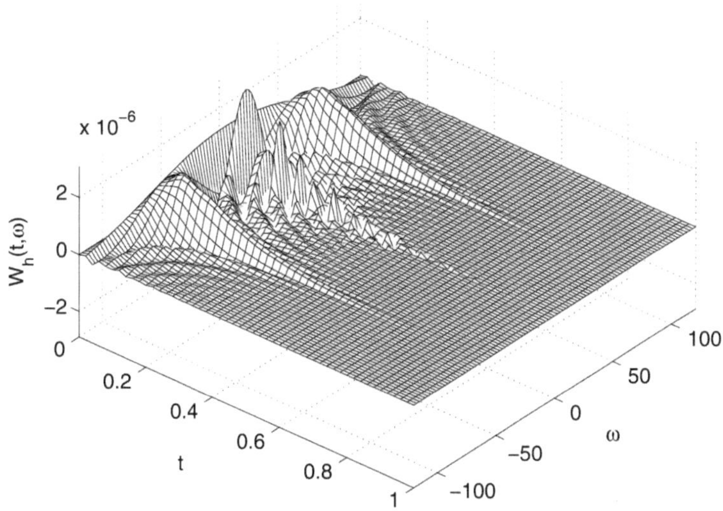

Fig. 3 Wigner distribution of the Green's function for the harmonic oscillator. The delta function at the input generates a time-frequency response made by an initial spread over all frequencies, which then concentrates on the resonant frequency ω_c, as well as on its symmetric counterpart at $-\omega_c$. The oscillating components centered about the time axis are interference terms

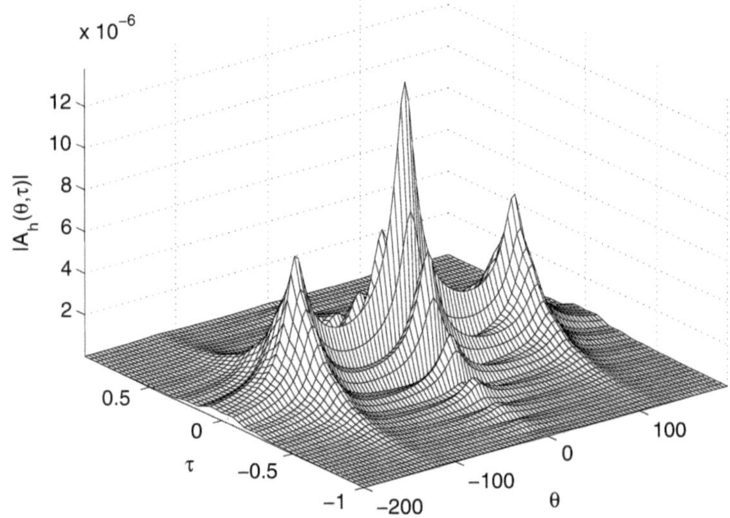

Fig. 4 Magnitude of the ambiguity function of the Green's function for the harmonic oscillator. This function is made by three components. The first has a peak at the origin, and it represents the resonances at ω_c and $-\omega_c$, merged together in the ambiguity domain. The other two components are located on the $\tau = 0$ axis, at $\theta = 2\omega_c$ and $\theta = -2\omega_c$, and they represent the interference terms of the Wigner distribution $W_h(t, \omega)$ in Fig. 3. These interference terms can be filtered out by a proper masking of the ambiguity function

two terms $A_{h_L}(\theta - 2\omega_c, \tau)$ and $A_{h_L}(\theta + 2\omega_c, \tau)$ in (45), which, in Fig. 4, correspond to the two components centered about $\theta = 2\omega_c$, $\tau = 0$, and $\theta = -2\omega_c$, $\tau = 0$.

The interference terms can be reduced by filtering the ambiguity function through the product

$$M_h(\theta, \tau) = G(\theta, \tau) A_h(\theta, \tau), \tag{48}$$

where $G(\theta, \tau)$ is the filter and $M_h(\theta, \tau)$ is the filtered ambiguity function. Since, as previously discussed, the interference terms are located on the $\tau = 0$ axis and centered about the frequencies $\pm 2\omega_c$, an effective choice for the cut-off frequency θ_c of the filter can be $\theta_c < \omega_c$. Therefore, the specifications for the lowpass filter are $|G(\theta, \tau)| = 1$ for $\theta \leq \theta_c$, and $|G(\theta, \tau)| = 0$ for $\theta > \theta_c$, whereas no filtering is needed on the τ axis. Because of (3), the parameters of the interference mitigation filter are linked to the coefficients μ and ω_0 of the differential equation governing the harmonic oscillator. We also note that, in the time-frequency domain, the filtering (48) corresponds to the smoothing [3]

$$C_h(t, \omega) = \int_{-\infty}^{+\infty} \int_{-\infty}^{+\infty} g(t - t', \omega - \omega') W_h(t', \omega') dt' d\omega', \tag{49}$$

where

$$g(t, \omega) = \frac{1}{4\pi^2} \int_{-\infty}^{+\infty} \int_{-\infty}^{+\infty} G(\theta, \tau) e^{-i\theta t - i\tau\omega} d\theta d\tau. \tag{50}$$

Note that, in general, the filtering (48) does not produce a proper Wigner distribution, because not every real function of time and frequency is a Wigner distribution. This fact is known as the representability problem [3]. Anyway, filtering is advantageous because the resulting smoothed Wigner distribution clearly highlights the time-frequency spectrum of systems modeled by differential equations, as shown in [14].

Furthermore, for an arbitrary input $f(t)$, the Wigner distribution $W_x(t, \omega)$ of the output of the harmonic oscillator is given by the convolution (5) between the Wigner distribution $W_h(t, \omega)$ of the impulse response and the Wigner distribution $W_f(t, \omega)$ of the input. Clearly, $W_f(t, \omega)$ is, in general, affected by interference terms, which can be strong, and, consequently, the resulting output $W_x(t, \omega)$ can also have strong interference terms. In general, the structure of such interference terms depends on the type of input signal. Nevertheless, they will have a highly oscillatory nature, therefore the common countermeasure of smoothing them can still be applied.

4 Summary of Results

The Langevin equation defined as

$$\frac{dx(t)}{dt} + \mu x(t) = f(t), \tag{51}$$

with damping coefficient $\mu > 0$ has a Green's function given by

$$h_L(t) = u(t) e^{-\mu t}, \tag{52}$$

whose corresponding ambiguity function is

$$A_{h_L}(\theta, \tau) = \frac{e^{(-\mu + i\theta/2)|\tau|}}{2\mu - i\theta}. \tag{53}$$

The harmonic oscillator defined as

$$\frac{d^2x(t)}{dt^2} + 2\mu \frac{dx(t)}{dt} + \omega_0^2 x(t) = f(t), \tag{54}$$

where $\mu < \omega_0$, has a Green's function given by

$$h(t) = \frac{1}{\omega_c} h_L(t) \sin \omega_c t, \tag{55}$$

where

$$\omega_c = \sqrt{\omega_0^2 - \mu^2}. \tag{56}$$

The corresponding ambiguity function is given by

$$A_h(\theta, \tau) = \frac{1}{2\omega_c^2} A_{h_L}(\theta, \tau) \cos \omega_c \tau - \frac{1}{4\omega_c^2} A_{h_L}(\theta - 2\omega_c, \tau) - \frac{1}{4\omega_c^2} A_{h_L}(\theta + 2\omega_c, \tau). \tag{57}$$

5 Conclusions

We have obtained the ambiguity function of the Green's function for the harmonic oscillator. The obtained result has a simple connection to the ambiguity function of the Green's function for the Langevin equation. The ambiguity function for the harmonic oscillator is made by three terms. The first, centered about the origin of the ambiguity domain, describes the resonant behavior of the harmonic oscillator. The second and third terms, located away from the origin of the ambiguity domain, represent the interference terms of the Wigner distribution of the Green's function. These interference terms can be filtered out by masking the ambiguity function, an operation corresponding to smoothing the Wigner distribution in the time-frequency domain.

Appendix

By using the property (10), the ambiguity function of the Green's function for the Langevin equation can be obtained from

$$A_{h_L}(\theta, \tau) = \int_{-\infty}^{+\infty} \int_{-\infty}^{+\infty} W_{h_L}(t, \omega) e^{i\theta t + i\tau\omega} dt d\omega. \tag{58}$$

Substituting $W_{h_L}(t,\omega)$ from (7) gives

$$A_{h_L}(\theta,\tau) = \int_{-\infty}^{+\infty}\int_{-\infty}^{+\infty} u(t)e^{-2\mu t}\frac{\sin 2\omega t}{\pi\omega}e^{i\theta t+i\tau\omega}dtd\omega, \tag{59}$$

$$= \int_{-\infty}^{+\infty}\frac{1}{\pi\omega}\left[\frac{1}{2i}\int_0^{+\infty}e^{(-2\mu+i(\theta+2\omega))t}dt - \frac{1}{2i}\int_0^{+\infty}e^{(-2\mu+i(\theta-2\omega))t}dt\right]e^{i\tau\omega}d\omega, \tag{60}$$

$$= \int_{-\infty}^{+\infty}\frac{1}{\pi\omega}\left[\frac{1}{2i}\frac{1}{2\mu-i(\theta+2\omega)} - \frac{1}{2i}\frac{1}{2\mu-i(\theta-2\omega)}\right]e^{i\tau\omega}d\omega, \tag{61}$$

$$= \frac{2}{\pi}\int_{-\infty}^{+\infty}\frac{1}{(2\mu-i\theta)^2+4\omega^2}e^{i\tau\omega}d\omega, \tag{62}$$

$$= \frac{e^{(-\mu+i\theta/2)|\tau|}}{2\mu-i\theta}. \tag{63}$$

which is (42).

References

1. G. Birkhoff, G.-C. Rota, *Ordinary Differential Equations* (Wiley, London, 1989)
2. C.J. Savant, *Fundamentals of the Laplace Transformation* (McGraw-Hill: New York, 1962)
3. L. Cohen, *Time-Frequency Analysis* (Prentice-Hall, Upper Saddle River, 1995)
4. B. Boashash (ed.), *Time-Frequency Signal Analysis and Processing: A Comprehensive Reference* (Academic Press, London, 2015)
5. T.A.C.M. Claasen, W.F.G. Mecklenbrauker, The Wigner distribution – a tool for time-frequency signal analysis. Part I: continuous time signals. Philips J. Res. **35**(3), 217–250 (1980)
6. E.P. Wigner, On the quantum correction for thermodynamic equilibrium. Phys. Rev. **40**, 749–759 (1932)
7. L. Galleani, A time-frequency relationship between the Langevin equation and the harmonic oscillator, in *Pseudo-Differential Operators: Groups, Geometry and Applications*, ed. by M.W. Wong, H. Zhu (Birkhäuser, Basel, 2007), pp. 119–131
8. A.H. Zemanian, *Distribution Theory and Transform Analysis* (Dover, New York, 1987)
9. L. Cohen, The history of noise. IEEE Signal Proc. Mag. **22**(6), 20–45 (2005)
10. F. Hlawatsch, Interference terms in the Wigner distribution. Digital Signal Process. **84**, 363–367 (1984)
11. H.I. Choi, W.J. Williams, Improved time-frequency representation of multicomponent signals using exponential kernels. IEEE Trans. Acoust. Speech Signal Process. **37**(6), 862–871 (1989)
12. J. Jeong, W.J. Williams, Kernel design for reduced interference distributions. IEEE Trans. Signal Process. **40**(2), 402–412 (1992)
13. L. Galleani, Time-frequency representation of MIMO dynamical systems. IEEE Trans. Signal Process. **61**(17), 4309–4317 (2013)

14. L. Galleani, Response of dynamical systems to nonstationary inputs. IEEE Trans. Signal Process. **60**(11), 5775–5786 (2012)
15. L. Galleani, The transient spectrum of a random system. IEEE Trans. Signal Process. **58**(10), 5106–5117 (2010)
16. L. Galleani, L. Cohen, Direct time-frequency characterization of linear systems governed by differential equations. IEEE Signal Process. Lett. **11**(9), 721–724, (2004)

On the Solvability in the Sense of Sequences for Some Non-Fredholm Operators Related to the Anomalous Diffusion

Vitali Vougalter and Vitaly Volpert

Abstract We study solvability of some linear nonhomogeneous elliptic problems and prove that under reasonable technical conditions the convergence in $L^2(\mathbb{R}^d)$ of their right sides implies the existence and the convergence in $H^{2s}(\mathbb{R}^d)$ of the solutions. The equations involve the second order non-Fredholm differential operators raised to certain fractional powers s and we use the methods of spectral and scattering theory for Schrödinger type operators developed in our preceding work (Volpert and Vougalter, Electron J Differ Equ 160:16 pp, 2013).

Keywords Solvability conditions · Non-Fredholm operators · Sobolev spaces

Mathematics Subject Classification (2000) 35J10, 35P10, 47F05

1 Introduction

Consider the equation

$$(-\Delta + V(x))u - au = f, \tag{1.1}$$

where $u \in E = H^2(\mathbb{R}^d)$ and $f \in F = L^2(\mathbb{R}^d)$, $d \in \mathbb{N}$, a is a constant and $V(x)$ is a function decaying to 0 at infinity. If $a \geq 0$, then the essential spectrum of the operator $A : E \to F$ corresponding to the left side of equation (1.1) contains the origin. As a consequence, such operator does not satisfy the Fredholm property. Its

V. Vougalter (✉)
Department of Mathematics, University of Toronto, Toronto, ON, Canada
e-mail: vitali@math.toronto.edu

V. Volpert
Institute Camille Jordan, CNRS, UMR 5208, University Lyon 1, Villeurbanne, France
e-mail: volpert@math.univ-lyon1.fr

S. Molahajloo, M. W. Wong (eds.), *Analysis of Pseudo-Differential Operators*, Trends in Mathematics, https://doi.org/10.1007/978-3-030-05168-6_10

image is not closed, for $d > 1$ the dimension of its kernel and the codimension of its image are not finite. The present article is devoted to the studies of some properties of the operators of this kind raised to a fractional power. We recall that elliptic problems with non-Fredholm operators were treated extensively in recent years (see [17, 19–25], also [3]) along with their potential applications to the theory of reaction-diffusion equations (see [7, 8]). In the particular case when $a = 0$ the operator A satisfies the Fredholm property in some properly chosen weighted spaces (see [1–5]). However, the case with $a \neq 0$ is significantly different and the method developed in these articles cannot be applied.

One of the important questions concerning problems with non-Fredholm operators is their solvability. We address it in the following setting. Let f_n be a sequence of functions in the image of the operator A, such that $f_n \to f$ in $L^2(\mathbb{R}^d)$ as $n \to \infty$. Denote by u_n a sequence of functions from $H^2(\mathbb{R}^d)$ such that

$$Au_n = f_n, \ n \in \mathbb{N}.$$

Because the operator A does not satisfy the Fredholm property, the sequence u_n may not be convergent. We call a sequence u_n such that $Au_n \to f$ a solution in the sense of sequences of equation $Au = f$ (see [16]). If such sequence converges to a function u_0 in the norm of the space E, then u_0 is a solution of this problem. Solution in the sense of sequences is equivalent in this sense to the usual solution. However, in the case of the non-Fredholm operators, this convergence may not hold or it can occur in some weaker sense. In this case, solution in the sense of sequences may not imply the existence of the usual solution. In the present article we will find sufficient conditions of equivalence of solutions in the sense of sequences and the usual solutions. In the other words, we will determine the conditions on sequences f_n under which the corresponding sequences u_n are strongly convergent. Solvability in the sense of sequences for the sums of non-Fredholm Schrödinger type operators was studied in [26]. In this work we deal with the situation when a second order differential operator without Fredholm property is raised to a certain fractional power. The resulting operator will be defined via the spectral calculus.

Let us consider the equation

$$(-\Delta)^s u - au = f(x), \ x \in \mathbb{R}^d, \ d \in \mathbb{N}, \tag{1.2}$$

where $s \in (0, 1)$, $a \geq 0$ is a constant and the right side is square integrable. The operator $(-\Delta)^s$ is actively used, for instance in the studies of the anomalous diffusion problems (see, e.g., [27] and the references therein). Anomalous diffusion can be described as a random process of particle motion characterized by the probability density distribution of jump length. The moments of this density distribution are finite in the case of normal diffusion, but this is not the case for the anomalous diffusion. Asymptotic behavior at infinity of the probability density function determines the value of the power of the Laplacian (see [13]). The problem analogous to (1.2) but with the standard Laplacian in the context of the solvability in the sense of sequences was studied in [18]. The case when the power of the negative Laplace operator $s = \frac{1}{2}$ was treated recently in [29]. Evidently, for the operator $(-\Delta)^s - a : H^{2s}(\mathbb{R}^d) \to L^2(\mathbb{R}^d)$ the essential spectrum fills the semi-axis $[-a, \infty)$ such that its inverse from $L^2(\mathbb{R}^d)$ to $H^{2s}(\mathbb{R}^d)$ is not bounded.

Let us write down the corresponding sequence of equations with $n \in \mathbb{N}$ as

$$(-\Delta)^s u_n - au_n = f_n(x), \ x \in \mathbb{R}^d, \ d \in \mathbb{N}, \tag{1.3}$$

where the right sides converge to the right side of (1.2) in $L^2(\mathbb{R}^d)$ as $n \to \infty$. The inner product of two functions

$$(f(x), g(x))_{L^2(\mathbb{R}^d)} := \int_{\mathbb{R}^d} f(x)\bar{g}(x)dx, \tag{1.4}$$

with a slight abuse of notations when these functions are not square integrable. Indeed, if $f(x) \in L^1(\mathbb{R}^d)$ and $g(x)$ is bounded, then clearly the integral in the right side of (1.4) makes sense, like in the case of functions involved in the orthogonality relations of Theorems 1.1 and 1.2 below. Let us use the space $H^{2s}(\mathbb{R}^d)$ equipped with the norm

$$\|u\|^2_{H^{2s}(\mathbb{R}^d)} := \|u\|^2_{L^2(\mathbb{R}^d)} + \|(-\Delta)^s u\|^2_{L^2(\mathbb{R}^d)}. \tag{1.5}$$

Throughout the article, the sphere of radius $r > 0$ in $\mathbb{R}^d$ centered at the origin will be designated by S^d_r. When $r = 1$, such unit sphere will be denoted by S^d and $|S^d|$ will stand for its Lebesgue measure. The unit ball in $\mathbb{R}^d$ centered at the origin will be designated by B^d and $|B^d|$ will denote its Lebesgue measure. Let us first formulate the solvability conditions for problem (1.2).

Theorem 1.1 *Let $f(x) \in L^2(\mathbb{R}^d)$, $d \in \mathbb{N}$, and $s \in (0, 1)$.*

a) Let $a = 0$, $d = 1$. If $s \in \left(0, \frac{1}{4}\right)$ and in addition $f(x) \in L^1(\mathbb{R})$, then Eq. (1.2) has a unique solution $u(x) \in H^{2s}(\mathbb{R})$.

Suppose that $s \in \left[\frac{1}{4}, \frac{3}{4}\right)$ and in addition $xf(x) \in L^1(\mathbb{R})$. Then problem (1.2) admits a unique solution $u(x) \in H^{2s}(\mathbb{R})$ if and only if the equality

$$(f(x), 1)_{L^2(\mathbb{R})} = 0 \tag{1.6}$$

holds.

Suppose that $s \in \left[\frac{3}{4}, 1\right)$ and in addition $x^2 f(x) \in L^1(\mathbb{R})$. Then Eq. (1.2) has a unique solution $u(x) \in H^{2s}(\mathbb{R})$ if and only if orthogonality conditions (1.6) along with

$$(f(x), x)_{L^2(\mathbb{R})} = 0 \tag{1.7}$$

hold.

b) Let $a = 0,\ d = 2$. *Then when* $s \in \left(0, \frac{1}{2}\right)$ *and additionally* $f(x) \in L^1(\mathbb{R}^2)$, *Eq.* (1.2) *admits a unique solution* $u(x) \in H^{2s}(\mathbb{R}^2)$.

Suppose that $s \in \left[\frac{1}{2}, 1\right)$ *and additionally* $xf(x) \in L^1(\mathbb{R}^2)$. *Then Eq.* (1.2) *has a unique solution* $u(x) \in H^{2s}(\mathbb{R}^2)$ *if and only if*

$$(f(x), 1)_{L^2(\mathbb{R}^2)} = 0 \tag{1.8}$$

holds.

c) Let $a = 0,\ d = 3$. *If* $s \in \left(0, \frac{3}{4}\right)$ *and in addition* $f(x) \in L^1(\mathbb{R}^3)$, *then problem* (1.2) *has a unique solution* $u(x) \in H^{2s}(\mathbb{R}^3)$.

Suppose that $s \in \left[\frac{3}{4}, 1\right)$ *and in addition* $xf(x) \in L^1(\mathbb{R}^3)$. *Then Eq.* (1.2) *admits a unique solution* $u(x) \in H^{2s}(\mathbb{R}^3)$ *if and only if*

$$(f(x), 1)_{L^2(\mathbb{R}^3)} = 0 \tag{1.9}$$

holds.

d) If $a = 0,\ d \geq 4$ *with* $s \in (0, 1)$ *and additionally* $f(x) \in L^1(\mathbb{R}^d)$, *then problem* (1.2) *possesses a unique solution* $u(x) \in H^{2s}(\mathbb{R}^d)$.

e) Suppose that $a > 0,\ d = 1$ *with* $s \in (0, 1)$ *and in addition* $xf(x) \in L^1(\mathbb{R})$. *Then Eq.* (1.2) *admits a unique solution* $u(x) \in H^{2s}(\mathbb{R})$ *if and only if*

$$\left(f(x), \frac{e^{\pm i a^{\frac{1}{2s}} x}}{\sqrt{2\pi}}\right)_{L^2(\mathbb{R})} = 0 \tag{1.10}$$

holds.

f) Suppose that $a > 0,\ d \geq 2$ *with* $s \in (0, 1)$ *and additionally* $xf(x) \in L^1(\mathbb{R}^d)$. *Then problem* (1.2) *has a unique solution* $u(x) \in H^{2s}(\mathbb{R}^d)$ *if and only if*

$$\left(f(x), \frac{e^{ipx}}{(2\pi)^{\frac{d}{2}}}\right)_{L^2(\mathbb{R}^d)} = 0, \quad p \in S^d_{a^{\frac{1}{2s}}} \tag{1.11}$$

holds.

Then we turn our attention to the issue of the solvability in the sense of sequences for our problem.

Theorem 1.2 *Let* $n \in \mathbb{N}$ *and* $f_n(x) \in L^2(\mathbb{R}^d)$, $d \in \mathbb{N}$, *such that* $f_n(x) \to f(x)$ *in* $L^2(\mathbb{R}^d)$ *as* $n \to \infty$.

a) Let $a = 0,\ d = 1$. *If* $s \in \left(0, \frac{1}{4}\right)$ *and additionally* $f_n(x) \in L^1(\mathbb{R})$, $n \in \mathbb{N}$, *such that* $f_n(x) \to f(x)$ *in* $L^1(\mathbb{R})$ *as* $n \to \infty$, *then Eqs.* (1.2) *and* (1.3) *admit unique solutions* $u(x) \in H^{2s}(\mathbb{R})$ *and* $u_n(x) \in H^{2s}(\mathbb{R})$, *respectively, such that* $u_n(x) \to u(x)$ *in* $H^{2s}(\mathbb{R})$ *as* $n \to \infty$.

Suppose that $s \in \left[\frac{1}{4}, \frac{3}{4}\right)$. *Let in addition* $xf_n(x) \in L^1(\mathbb{R}),\ n \in \mathbb{N}$, *such that* $xf_n(x) \to xf(x)$ *in* $L^1(\mathbb{R})$ *as* $n \to \infty$ *and the orthogonality conditions*

$$(f_n(x), 1)_{L^2(\mathbb{R})} = 0 \tag{1.12}$$

hold for all $n \in \mathbb{N}$. *Then Eqs.* (1.2) *and* (1.3) *admit unique solutions* $u(x) \in H^{2s}(\mathbb{R})$ *and* $u_n(x) \in H^{2s}(\mathbb{R})$, *respectively, such that* $u_n(x) \to u(x)$ *in* $H^{2s}(\mathbb{R})$ *as* $n \to \infty$.

Suppose that $s \in \left[\frac{3}{4}, 1\right)$. *Let in addition* $x^2 f_n(x) \in L^1(\mathbb{R}),\ n \in \mathbb{N}$, *such that* $x^2 f_n(x) \to x^2 f(x)$ *in* $L^1(\mathbb{R})$ *as* $n \to \infty$ *and the orthogonality conditions*

$$(f_n(x), 1)_{L^2(\mathbb{R})} = 0, \quad (f_n(x), x)_{L^2(\mathbb{R})} = 0 \tag{1.13}$$

hold for all $n \in \mathbb{N}$. *Then Eqs.* (1.2) *and* (1.3) *have unique solutions* $u(x) \in H^{2s}(\mathbb{R})$ *and* $u_n(x) \in H^{2s}(\mathbb{R})$, *respectively, such that* $u_n(x) \to u(x)$ *in* $H^{2s}(\mathbb{R})$ *as* $n \to \infty$.

b) *Let* $a = 0,\ d = 2$. *If* $s \in \left(0, \frac{1}{2}\right)$ *and additionally* $f_n(x) \in L^1(\mathbb{R}^2),\ n \in \mathbb{N}$, *such that* $f_n(x) \to f(x)$ *in* $L^1(\mathbb{R}^2)$ *as* $n \to \infty$, *then Eqs.* (1.2) *and* (1.3) *possess unique solutions* $u(x) \in H^{2s}(\mathbb{R}^2)$ *and* $u_n(x) \in H^{2s}(\mathbb{R}^2)$, *respectively, such that* $u_n(x) \to u(x)$ *in* $H^{2s}(\mathbb{R}^2)$ *as* $n \to \infty$.

Suppose that $s \in \left[\frac{1}{2}, 1\right)$. *Let in addition* $xf_n(x) \in L^1(\mathbb{R}^2),\ n \in \mathbb{N}$, *such that* $xf_n(x) \to xf(x)$ *in* $L^1(\mathbb{R}^2)$ *as* $n \to \infty$ *and the orthogonality relations*

$$(f_n(x), 1)_{L^2(\mathbb{R}^2)} = 0 \tag{1.14}$$

hold for all $n \in \mathbb{N}$. *Then Eqs.* (1.2) *and* (1.3) *admit unique solutions* $u(x) \in H^{2s}(\mathbb{R}^2)$ *and* $u_n(x) \in H^{2s}(\mathbb{R}^2)$, *respectively, such that* $u_n(x) \to u(x)$ *in* $H^{2s}(\mathbb{R}^2)$ *as* $n \to \infty$.

c) *Let* $a = 0,\ d = 3$. *Suppose that* $s \in \left(0, \frac{3}{4}\right)$ *and additionally* $f_n(x) \in L^1(\mathbb{R}^3),\ n \in \mathbb{N}$, *such that* $f_n(x) \to f(x)$ *in* $L^1(\mathbb{R}^3)$ *as* $n \to \infty$. *Then problems* (1.2) *and* (1.3) *possess unique solutions* $u(x) \in H^{2s}(\mathbb{R}^3)$ *and* $u_n(x) \in H^{2s}(\mathbb{R}^3)$, *respectively, such that* $u_n(x) \to u(x)$ *in* $H^{2s}(\mathbb{R}^3)$ *as* $n \to \infty$.

Suppose that $s \in \left[\frac{3}{4}, 1\right)$. *Let in addition* $xf_n(x) \in L^1(\mathbb{R}^3),\ n \in \mathbb{N}$, *such that* $xf_n(x) \to xf(x)$ *in* $L^1(\mathbb{R}^3)$ *as* $n \to \infty$ *and*

$$(f_n(x), 1)_{L^2(\mathbb{R}^3)} = 0 \tag{1.15}$$

holds for all $n \in \mathbb{N}$. *Then Eqs.* (1.2) *and* (1.3) *have unique solutions* $u(x) \in H^{2s}(\mathbb{R}^3)$ *and* $u_n(x) \in H^{2s}(\mathbb{R}^3)$, *respectively, such that* $u_n(x) \to u(x)$ *in* $H^{2s}(\mathbb{R}^3)$ *as* $n \to \infty$.

d) *Let* $a = 0,\ d \geq 4$ *with* $s \in (0, 1)$ *and additionally* $f_n(x) \in L^1(\mathbb{R}^d),\ n \in \mathbb{N}$, *such that* $f_n(x) \to f(x)$ *in* $L^1(\mathbb{R}^d)$ *as* $n \to \infty$. *Then problems* (1.2) *and* (1.3) *possess unique solutions* $u(x) \in H^{2s}(\mathbb{R}^d)$ *and* $u_n(x) \in H^{2s}(\mathbb{R}^d)$*, respectively, such that* $u_n(x) \to u(x)$ *in* $H^{2s}(\mathbb{R}^d)$ *as* $n \to \infty$.

e) Let $a > 0,\ d = 1$ *with* $s \in (0, 1)$ *and in addition* $xf_n(x) \in L^1(\mathbb{R}),\ n \in \mathbb{N}$*, such that* $xf_n(x) \to xf(x)$ *in* $L^1(\mathbb{R})$ *as* $n \to \infty$*. Let*

$$\left(f_n(x), \frac{e^{\pm i a^{\frac{1}{2s}} x}}{\sqrt{2\pi}}\right)_{L^2(\mathbb{R})} = 0 \tag{1.16}$$

hold for all $n \in \mathbb{N}$*. Then Eqs. (1.2) and (1.3) admit unique solutions* $u(x) \in H^{2s}(\mathbb{R})$ *and* $u_n(x) \in H^{2s}(\mathbb{R})$*, respectively, such that* $u_n(x) \to u(x)$ *in* $H^{2s}(\mathbb{R})$ *as* $n \to \infty$.

f) Let $a > 0,\ d \geq 2$ *with* $s \in (0, 1)$ *and additionally* $xf_n(x) \in L^1(\mathbb{R}^d),\ n \in \mathbb{N}$*, such that* $xf_n(x) \to xf(x)$ *in* $L^1(\mathbb{R}^d)$ *as* $n \to \infty$*. Let*

$$\left(f_n(x), \frac{e^{ipx}}{(2\pi)^{\frac{d}{2}}}\right)_{L^2(\mathbb{R}^d)} = 0, \quad p \in S^d_{a^{\frac{1}{2s}}} \tag{1.17}$$

hold for all $n \in \mathbb{N}$*. Then problems (1.2) and (1.3) have unique solutions* $u(x) \in H^{2s}(\mathbb{R}^d)$ *and* $u_n(x) \in H^{2s}(\mathbb{R}^d)$*, respectively, such that* $u_n(x) \to u(x)$ *in* $H^{2s}(\mathbb{R}^d)$ *as* $n \to \infty$.

Let us note that when $a = 0$ each of the cases $a) - d)$ above contains the situation when orthogonality conditions are not required.

We use the hat symbol to denote the standard Fourier transform

$$\widehat{f}(p) := \frac{1}{(2\pi)^{\frac{d}{2}}} \int_{\mathbb{R}^d} f(x) e^{-ipx} dx, \ p \in \mathbb{R}^d, \ d \in \mathbb{N}, \tag{1.18}$$

such that

$$\|\widehat{f}(p)\|_{L^\infty(\mathbb{R}^d)} \leq \frac{1}{(2\pi)^{\frac{d}{2}}} \|f(x)\|_{L^1(\mathbb{R}^d)}. \tag{1.19}$$

In the second part of the article we consider the equation

$$(-\Delta + V(x))^s u - au = f(x), \quad x \in \mathbb{R}^3, \quad a \geq 0, \quad s \in (0, 1), \tag{1.20}$$

with the square integrable right side. The corresponding sequence of equations for $n \in \mathbb{N}$ will be

$$(-\Delta + V(x))^s u_n - au_n = f_n(x), \quad x \in \mathbb{R}^3, \quad a \geq 0, \tag{1.21}$$

with $s \in (0, 1)$ and the right sides converging to the right side of (1.20) in $L^2(\mathbb{R}^3)$ as $n \to \infty$. Note that the situation when the power $s = \dfrac{1}{2}$ was studied in the recent work [29]. Let us make the following technical assumptions on the scalar potential involved in the problems above. Note that the conditions on $V(x)$, which is shallow and short-range will be analogous to those stated in Assumption 1.1 of [22] (see also [20, 23]). The essential spectrum of such a Schrödinger operator $-\Delta + V(x)$ fills the nonnegative semi-axis (see, e.g., [10]).

Assumption 1.3 *The potential function* $V(x) : \mathbb{R}^3 \to \mathbb{R}$ *satisfies the estimate*

$$|V(x)| \le \frac{C}{1 + |x|^{3.5+\delta}}$$

with some $\delta > 0$ *and* $x = (x_1, x_2, x_3) \in \mathbb{R}^3$ *a.e. such that*

$$4^{\frac{1}{9}}\frac{9}{8}(4\pi)^{-\frac{2}{3}}\|V\|^{\frac{1}{9}}_{L^\infty(\mathbb{R}^3)}\|V\|^{\frac{8}{9}}_{L^{\frac{4}{3}}(\mathbb{R}^3)} < 1 \quad \text{and} \quad \sqrt{c_{HLS}}\|V\|_{L^{\frac{3}{2}}(\mathbb{R}^3)} < 4\pi. \tag{1.22}$$

Here and further down C will stand for a finite positive constant and c_{HLS} given on p.98 of [12] is the constant in the Hardy-Littlewood-Sobolev inequality

$$\left|\int_{\mathbb{R}^3}\int_{\mathbb{R}^3}\frac{f_1(x)f_1(y)}{|x-y|^2}dxdy\right| \le c_{HLS}\|f_1\|^2_{L^{\frac{3}{2}}(\mathbb{R}^3)}, \quad f_1 \in L^{\frac{3}{2}}(\mathbb{R}^3).$$

By virtue of Lemma 2.3 of [22], under Assumption 1.3 above on the potential function, the operator $-\Delta + V(x)$ on $L^2(\mathbb{R}^3)$ is self-adjoint and unitarily equivalent to $-\Delta$ via the wave operators (see [11, 15])

$$\Omega^{\pm} := s - \lim_{t\to\mp\infty} e^{it(-\Delta+V)}e^{it\Delta},$$

where the limit is understood in the strong L^2 sense (see, e.g., [14] p.34, [6] p.90). Hence $(-\Delta + V(x))^s$ on $L^2(\mathbb{R}^3)$ defined via the spectral calculus has only the essential spectrum

$$\sigma_{ess}((-\Delta + V(x))^s - a) = [-a, \infty)$$

and no nontrivial $L^2(\mathbb{R}^3)$ eigenfunctions. By means of the spectral theorem, its functions of the continuous spectrum satisfy

$$(-\Delta + V(x))^s\varphi_k(x) = |k|^{2s}\varphi_k(x), \quad k \in \mathbb{R}^3, \tag{1.23}$$

in the integral formulation the Lippmann-Schwinger equation for the perturbed plane waves (see, e.g., [14] p.98)

$$\varphi_k(x) = \frac{e^{ikx}}{(2\pi)^{\frac{3}{2}}} - \frac{1}{4\pi}\int_{\mathbb{R}^3}\frac{e^{i|k||x-y|}}{|x-y|}(V\varphi_k)(y)dy \tag{1.24}$$

and the orthogonality relations

$$(\varphi_k(x), \varphi_q(x))_{L^2(\mathbb{R}^3)} = \delta(k-q), \quad k, q \in \mathbb{R}^3. \tag{1.25}$$

Particularly, when the vector $k = 0$, we have $\varphi_0(x)$. Let us denote the generalized Fourier transform with respect to these functions using the tilde symbol as

$$\tilde{f}(k) := (f(x), \varphi_k(x))_{L^2(\mathbb{R}^3)}, \ k \in \mathbb{R}^3. \tag{1.26}$$

(1.26) is a unitary transform on $L^2(\mathbb{R}^3)$. The integral operator involved in (1.24) is being denoted as

$$(Q\varphi)(x) := -\frac{1}{4\pi}\int_{\mathbb{R}^3} \frac{e^{i|k||x-y|}}{|x-y|}(V\varphi)(y)dy, \quad \varphi \in L^\infty(\mathbb{R}^3).$$

We consider $Q : L^\infty(\mathbb{R}^3) \to L^\infty(\mathbb{R}^3)$. Under Assumption 1.3, via Lemma 2.1 of [22] the operator norm $\|Q\|_\infty$ is bounded above by the quantity $I(V)$, which is the left side of the first inequality in (1.22), such that $I(V) < 1$. Corollary 2.2 of [22] under our assumptions gives us the bound

$$|\tilde{f}(k)| \le \frac{1}{(2\pi)^{\frac{3}{2}}}\frac{1}{1 - I(V)}\|f(x)\|_{L^1(\mathbb{R}^3)}. \tag{1.27}$$

We have the following result concerning the solvability of equation (1.20).

Theorem 1.4 *Let Assumption 1.3 hold and $f(x) \in L^2(\mathbb{R}^3)$.*

a) Let $a = 0$, $s \in \left(0, \frac{3}{4}\right)$ and additionally $f(x) \in L^1(\mathbb{R}^3)$. Then Eq. (1.20) possess a unique solution $u(x) \in L^2(\mathbb{R}^3)$.

Let $a = 0$, $s \in \left[\frac{3}{4}, 1\right)$ and in addition $xf(x) \in L^1(\mathbb{R}^3)$. Then problem (1.20) admits a unique solution $u(x) \in L^2(\mathbb{R}^3)$ if and only if

$$(f(x), \varphi_0(x))_{L^2(\mathbb{R}^3)} = 0 \tag{1.28}$$

holds.

b) Let $a > 0$, $s \in (0, 1)$ and in addition $xf(x) \in L^1(\mathbb{R}^3)$. Then Eq. (1.20) has a unique solution $u(x) \in L^2(\mathbb{R}^3)$ if and only if

$$(f(x), \varphi_k(x))_{L^2(\mathbb{R}^3)} = 0, \quad k \in S^3_{a^{\frac{1}{2s}}} \tag{1.29}$$

holds.

Our final main statement is devoted to the solvability in the sense of sequences of problem (1.20).

Theorem 1.5 *Let Assumption 1.3 hold, $n \in \mathbb{N}$ and $f_n(x) \in L^2(\mathbb{R}^3)$, such that $f_n(x) \to f(x)$ in $L^2(\mathbb{R}^3)$ as $n \to \infty$.*

a) Let $a = 0$. If $s \in \left(0, \frac{3}{4}\right)$ and additionally $f_n(x) \in L^1(\mathbb{R}^3)$, $n \in \mathbb{N}$, such that $f_n(x) \to f(x)$ in $L^1(\mathbb{R}^3)$ as $n \to \infty$, then Eqs. (1.20) and (1.21) possess unique solutions $u(x) \in L^2(\mathbb{R}^3)$ and $u_n(x) \in L^2(\mathbb{R}^3)$, respectively, such that $u_n(x) \to u(x)$ in $L^2(\mathbb{R}^3)$ as $n \to \infty$.

Suppose that $s \in \left[\frac{3}{4}, 1\right)$. Let in addition $xf_n(x) \in L^1(\mathbb{R}^3)$, $n \in \mathbb{N}$, such that $xf_n(x) \to xf(x)$ in $L^1(\mathbb{R}^3)$ as $n \to \infty$ and

$$(f_n(x), \varphi_0(x))_{L^2(\mathbb{R}^3)} = 0 \tag{1.30}$$

holds for all $n \in \mathbb{N}$. Then Eqs. (1.20) and (1.21) admit unique solutions $u(x) \in L^2(\mathbb{R}^3)$ and $u_n(x) \in L^2(\mathbb{R}^3)$, respectively, such that $u_n(x) \to u(x)$ in $L^2(\mathbb{R}^3)$ as $n \to \infty$.

b) Suppose that $a > 0$, $s \in (0, 1)$. Let in addition $xf_n(x) \in L^1(\mathbb{R}^3)$, $n \in \mathbb{N}$, such that $xf_n(x) \to xf(x)$ in $L^1(\mathbb{R}^3)$ as $n \to \infty$ and

$$(f_n(x), \varphi_k(x))_{L^2(\mathbb{R}^3)} = 0, \quad k \in S^3_{a^{\frac{1}{2s}}} \tag{1.31}$$

holds for all $n \in \mathbb{N}$. Then problems (1.20) and (1.21) have unique solutions $u(x) \in L^2(\mathbb{R}^3)$ and $u_n(x) \in L^2(\mathbb{R}^3)$, respectively, such that $u_n(x) \to u(x)$ in $L^2(\mathbb{R}^3)$ as $n \to \infty$.

Let us note that (1.28) and (1.29) are the orthogonality relations to the functions of the continuous spectrum of our Schrödinger operator, as distinct from the Limiting Absorption Principle in which one needs to orthogonalize to the standard Fourier harmonics (see, e.g., Lemma 2.3 and Proposition 2.4 of [9]).

2 Solvability in the Sense of Sequences in the Free Laplacian Case

Proof of Theorem 1.1 Let us note that the case a) of the theorem was stated in Lemma 4.1 of [28] and the case c) in Lemma 5 of [27].

Clearly, if $u(x) \in L^2(\mathbb{R}^d)$ is a solution of (1.2) with a square integrable right side, it belongs to $H^{2s}(\mathbb{R}^d)$ as well. Indeed, in this case from (1.2) we easily deduce $(-\Delta)^s u(x) \in L^2(\mathbb{R}^d)$, such that via norm definition (1.5) we have $u(x) \in H^{2s}(\mathbb{R}^d)$.

To prove the uniqueness of solutions for our equation, let us suppose that (1.2) has two square integrable solutions $u_1(x)$ and $u_2(x)$. Then their difference $w(x) := u_1(x) - u_2(x) \in L^2(\mathbb{R}^d)$ as well. Obviously, it is a solution of the equation

$$(-\Delta)^s w = aw.$$

Since the operator $(-\Delta)^s$ has no nontrivial square integrable eigenfunctions in the whole space, we have $w(x) = 0$ a.e. in $\mathbb{R}^d$.

We apply the standard Fourier transform (1.18) to both sides of problem (1.2) with $a = 0$. This gives us

$$\widehat{u}(p) = \frac{\widehat{f}(p)}{|p|^{2s}}\chi_{\{|p|\le 1\}} + \frac{\widehat{f}(p)}{|p|^{2s}}\chi_{\{|p|>1\}}. \tag{2.1}$$

Here and further down χ_A will denote the characteristic function of a set $A \subseteq \mathbb{R}^d$. Clearly, the second term in the right side of (2.1) can be bounded from above in the absolute value by $|\widehat{f}(p)| \in L^2(\mathbb{R}^d)$ due to one of our assumptions.

First we consider the case b) of the theorem when the dimension of the problem $d = 2$. Let us estimate the first term in the right side of (2.1) from above in the absolute value using (1.19) by $\frac{\|f(x)\|_{L^1(\mathbb{R}^2)}}{2\pi|p|^{2s}}\chi_{\{|p|\le 1\}}$. It can be easily verified that such expression is square integrable when $s \in \left(0, \frac{1}{2}\right)$.

To treat the case when $s \in \left(\frac{1}{2}, 1\right)$, we use the formula

$$\widehat{f}(p) = \widehat{f}(0) + \int_0^{|p|} \frac{\partial \widehat{f}(s,\sigma)}{\partial s} ds.$$

Here and throughout the article σ will denote the angle variables on the sphere. This enables us to express the first term in the right side of (2.1) as

$$\frac{\widehat{f}(0)}{|p|^{2s}}\chi_{\{|p|\le 1\}} + \frac{\int_0^{|p|} \frac{\partial \widehat{f}(s,\sigma)}{\partial s} ds}{|p|^{2s}}\chi_{\{|p|\le 1\}}. \tag{2.2}$$

Note that by means of the definition of the Fourier transform (1.18), we easily derive for the space of an arbitrary dimension

$$\left|\frac{\partial \widehat{f}(p)}{\partial |p|}\right| \le \frac{\|xf(x)\|_{L^1(\mathbb{R}^d)}}{(2\pi)^{\frac{d}{2}}}, \quad d \in \mathbb{N}. \tag{2.3}$$

Therefore, the second term in (2.2) can be bounded from above in the absolute value by

$$\frac{\|xf(x)\|_{L^1(\mathbb{R}^2)}}{2\pi}|p|^{1-2s}\chi_{\{|p|\le 1\}} \in L^2(\mathbb{R}^2).$$

It can be easily verified that the first term in (2.2) is square integrable if and only if $\widehat{f}(0)$ vanishes, which is equivalent to orthogonality relation (1.8).

Then we turn our attention to the case d) of the theorem. Let us estimate the first term in the right side of (2.1) from above in the absolute value via (1.19) by

$\frac{\|f(x)\|_{L^1(\mathbb{R}^d)}}{(2\pi)^{\frac{d}{2}}|p|^{2s}}\chi_{\{|p|\le 1\}}$, $d \ge 4$. It can be easily checked that this expression is square integrable for $s \in (0, 1)$.

Let us apply the standard Fourier transform (1.18) to both sides of Eq. (1.2) when $a > 0$. This yields

$$\widehat{u}(p) = \frac{\widehat{f}(p)}{|p|^{2s} - a}.$$

First of all we consider the case e) of the theorem, namely when the dimension of the problem $d = 1$. For $s \in (0, 1)$ we define the following sets on the real line

$$I_\delta^+ := [a^{\frac{1}{2s}} - \delta, a^{\frac{1}{2s}} + \delta], \quad I_\delta^- := [-a^{\frac{1}{2s}} - \delta, -a^{\frac{1}{2s}} + \delta], \quad 0 < \delta < a^{\frac{1}{2s}}, \tag{2.4}$$

such that

$$I_\delta := I_\delta^+ \cup I_\delta^-, \quad \mathbb{R} = I_\delta \cup I_\delta^c.$$

Here and further down $A^c \subseteq \mathbb{R}^d$ stands for the complement of the set $A \subseteq \mathbb{R}^d$. This allows us to express $\widehat{u}(p)$ as the sum

$$\frac{\widehat{f}(p)}{|p|^{2s} - a}\chi_{I_\delta^c} + \frac{\widehat{f}(p)}{p^{2s} - a}\chi_{I_\delta^+} + \frac{\widehat{f}(p)}{(-p)^{2s} - a}\chi_{I_\delta^-}. \tag{2.5}$$

Evidently, we have $I_\delta^c = I_\delta^{c+} \cup I_\delta^{c-}$, where

$$I_\delta^{c+} := I_\delta^c \cap \mathbb{R}^+, \quad I_\delta^{c-} := I_\delta^c \cap \mathbb{R}^-. \tag{2.6}$$

Here $\mathbb{R}^+$ and $\mathbb{R}^-$ are the nonnegative and the negative semi-axes of the real line, respectively. Clearly,

$$\left|\frac{\widehat{f}(p)}{p^{2s} - a}\chi_{I_\delta^{c+}}\right| \le C|\widehat{f}(p)| \in L^2(\mathbb{R})$$

due to one of our assumptions. Analogously,

$$\left|\frac{\widehat{f}(p)}{(-p)^{2s} - a}\chi_{I_\delta^{c-}}\right| \le C|\widehat{f}(p)| \in L^2(\mathbb{R}).$$

We express

$$\widehat{f}(p) = \widehat{f}(a^{\frac{1}{2s}}) + \int_{a^{\frac{1}{2s}}}^{p} \frac{d\widehat{f}(s)}{ds} ds.$$

(2.3) easily gives us the upper bound

$$\left|\frac{\int_{a^{\frac{1}{2s}}}^{p}\frac{d\widehat{f}(s)}{ds}ds}{p^{2s}-a}\chi_{I_\delta^+}\right| \le \frac{1}{\sqrt{2\pi}}\|xf(x)\|_{L^1(\mathbb{R})}\left|\frac{p-a^{\frac{1}{2s}}}{p^{2s}-a}\right|\chi_{I_\delta^+} \le$$

$$\le C\|xf(x)\|_{L^1(\mathbb{R})}\chi_{I_\delta^+} \in L^2(\mathbb{R}).$$

Apparently,

$$\frac{\widehat{f}(a^{\frac{1}{2s}})}{p^{2s}-a}\chi_{I_\delta^+} \in L^2(\mathbb{R})$$

if and only if $\widehat{f}(a^{\frac{1}{2s}})$ vanishes, which is equivalent to the orthogonality condition

$$\left(f(x), \frac{e^{ia^{\frac{1}{2s}}x}}{\sqrt{2\pi}}\right)_{L^2(\mathbb{R})} = 0, \quad s \in (0,1).$$

To study the singularity of the problem on the negative semi-axis, we apply the formula

$$\widehat{f}(p) = \widehat{f}(-a^{\frac{1}{2s}}) + \int_{-a^{\frac{1}{2s}}}^{p}\frac{d\widehat{f}(s)}{ds}ds.$$

Via (2.3) we have the upper bound

$$\left|\frac{\int_{-a^{\frac{1}{2s}}}^{p}\frac{d\widehat{f}(s)}{ds}ds}{(-p)^{2s}-a}\chi_{I_\delta^-}\right| \le \frac{1}{\sqrt{2\pi}}\|xf(x)\|_{L^1(\mathbb{R})}\left|\frac{p+a^{\frac{1}{2s}}}{(-p)^{2s}-a}\right|\chi_{I_\delta^-} \le$$

$$\le C\|xf(x)\|_{L^1(\mathbb{R})}\chi_{I_\delta^-} \in L^2(\mathbb{R}).$$

Evidently,

$$\frac{\widehat{f}(-a^{\frac{1}{2s}})}{(-p)^{2s}-a}\chi_{I_\delta^-} \in L^2(\mathbb{R})$$

if and only if $\widehat{f}(-a^{\frac{1}{2s}}) = 0$, which is equivalent to the orthogonality relation

$$\left(f(x), \frac{e^{-ia^{\frac{1}{2s}}x}}{\sqrt{2\pi}}\right)_{L^2(\mathbb{R})} = 0, \quad s \in (0,1).$$

We complete the proof of the theorem with establishing the part f). When the dimension $d \geq 2$, we define the set

$$A_\delta := \{p \in \mathbb{R}^d \mid a^{\frac{1}{2s}} - \delta \leq |p| \leq a^{\frac{1}{2s}} + \delta\}, \quad 0 < \delta < a^{\frac{1}{2s}} \tag{2.7}$$

and express

$$\widehat{u}(p) = \frac{\widehat{f}(p)}{|p|^{2s} - a} \chi_{A_\delta} + \frac{\widehat{f}(p)}{|p|^{2s} - a} \chi_{A_\delta^c}. \tag{2.8}$$

Obviously, we have the estimate from above

$$\left| \frac{\widehat{f}(p)}{|p|^{2s} - a} \chi_{A_\delta^c} \right| \leq C |\widehat{f}(p)| \in L^2(\mathbb{R}^d)$$

via one of our assumptions. To treat the first term in the right side of (2.8), we will use the representation formula

$$\widehat{f}(p) = \widehat{f}(a^{\frac{1}{2s}}, \sigma) + \int_{a^{\frac{1}{2s}}}^{|p|} \frac{\partial \widehat{f}(s, \sigma)}{\partial s} ds.$$

Inequality (2.3) enables us to estimate

$$\left| \frac{\int_{a^{\frac{1}{2s}}}^{|p|} \frac{\partial \widehat{f}(s,\sigma)}{\partial s} ds}{|p|^{2s} - a} \chi_{A_\delta} \right| \leq \frac{\|xf(x)\|_{L^1(\mathbb{R}^d)}}{(2\pi)^{\frac{d}{2}}} \left| \frac{|p| - a^{\frac{1}{2s}}}{|p|^{2s} - a} \right| \chi_{A_\delta} \leq$$

$$\leq C \|xf(x)\|_{L^1(\mathbb{R}^d)} \chi_{A_\delta} \in L^2(\mathbb{R}^d).$$

It can be easily verified that the remaining term

$$\frac{\widehat{f}(a^{\frac{1}{2s}}, \sigma)}{|p|^{2s} - a} \chi_{A_\delta} \in L^2(\mathbb{R}^d)$$

if and only if $\widehat{f}(a^{\frac{1}{2s}}, \sigma)$ vanishes, which is equivalent to orthogonality relation (1.11) for the dimensions $d \geq 2$. ∎

Then we proceed to establishing the solvability in the sense of sequences for our equation in the no potential case.

Proof of Theorem 1.2 Suppose $u(x)$ and $u_n(x)$, $n \in \mathbb{N}$ are the unique solutions of equations (1.2) and (1.3) in $H^{2s}(\mathbb{R}^d)$, $d \in \mathbb{N}$ with $a \geq 0$, respectively, $s \in (0, 1)$ and it is known that $u_n(x) \to u(x)$ in $L^2(\mathbb{R}^d)$ as $n \to \infty$. Then $u_n(x) \to u(x)$ in $H^{2s}(\mathbb{R}^d)$ as $n \to \infty$ as well. Indeed,

$$(-\Delta)^s (u_n(x) - u(x)) = a(u_n(x) - u(x)) + f_n(x) - f(x),$$

which clearly gives us

$$\|(-\Delta)^s(u_n(x)-u(x))\|_{L^2(\mathbb{R}^d)} \le a\|u_n(x)-u(x)\|_{L^2(\mathbb{R}^d)}+\|f_n(x)-f(x)\|_{L^2(\mathbb{R}^d)} \to 0$$

as $n \to \infty$ via our assumptions. Norm definition (1.5) yields $u_n(x) \to u(x)$ in $H^{2s}(\mathbb{R}^d)$ as $n \to \infty$.

If $u(x)$ and $u_n(x)$, $n \in \mathbb{N}$ are the unique solutions of equations (1.2) and (1.3) in $H^{2s}(\mathbb{R}^d)$, $d \in \mathbb{N}$, respectively with $a = 0$ as in the cases a)-d) of the theorem, by applying the standard Fourier transform (1.18) we easily derive

$$\widehat{u}_n(p) - \widehat{u}(p) = \frac{\widehat{f}_n(p) - \widehat{f}(p)}{|p|^{2s}}\chi_{\{|p|\le 1\}} + \frac{\widehat{f}_n(p) - \widehat{f}(p)}{|p|^{2s}}\chi_{\{|p|>1\}}. \tag{2.9}$$

Evidently, the second term in the right side of equality (2.9) can be estimated from above in the absolute value in the space of any dimension by $|\widehat{f}_n(p) - \widehat{f}(p)|$, such that

$$\left\|\frac{\widehat{f}_n(p) - \widehat{f}(p)}{|p|^{2s}}\chi_{\{|p|>1\}}\right\|_{L^2(\mathbb{R}^d)} \le \|f_n(x) - f(x)\|_{L^2(\mathbb{R}^d)} \to 0, \quad n \to \infty$$

due to one of our assumptions.

First we treat the case a) of the theorem when the dimension $d = 1$. Then, when $s \in \left(0, \frac{1}{4}\right)$ via the part a) of Theorem 1.1, Eq. (1.2) and each of equations (1.3) admit unique solutions $u(x) \in H^{2s}(\mathbb{R})$ and $u_n(x) \in H^{2s}(\mathbb{R})$, $n \in \mathbb{N}$, respectively. Clearly, the first term in the right side of equality (2.9) can be bounded from above in the absolute value via (1.19) by

$$\frac{1}{\sqrt{2\pi}}\|f_n(x) - f(x)\|_{L^1(\mathbb{R})}\frac{\chi_{\{|p|\le 1\}}}{|p|^{2s}},$$

such that its $L^2(\mathbb{R})$ norm can be estimated from above by

$$\frac{1}{\sqrt{\pi}}\|f_n(x) - f(x)\|_{L^1(\mathbb{R})}\frac{1}{\sqrt{1-4s}} \to 0, \quad n \to \infty$$

due to one of our assumptions and with $s \in \left(0, \frac{1}{4}\right)$. This shows that in this case $u_n(x) \to u(x)$ in $L^2(\mathbb{R})$ as $n \to \infty$.

Then we turn our attention to the case of $s \in \left[\frac{1}{4}, \frac{3}{4}\right)$. Note that by means of the parts a) and b) of Lemma 4.1 below, under our assumptions we have $f_n(x) \in L^1(\mathbb{R})$, $n \in \mathbb{N}$, such that $f_n(x) \to f(x)$ in $L^1(\mathbb{R})$ as $n \to \infty$. Then, via (1.12) we obtain

$$|(f(x), 1)_{L^2(\mathbb{R})}| = |(f(x) - f_n(x), 1)_{L^2(\mathbb{R})}| \le \|f_n(x) - f(x)\|_{L^1(\mathbb{R})} \to 0$$

as $n \to \infty$. Thus,

$$(f(x), 1)_{L^2(\mathbb{R})} = 0 \tag{2.10}$$

holds. By means of the part a) of Theorem 1.1, when $s \in \left[\frac{1}{4}, \frac{3}{4}\right)$, Eqs. (1.2) and (1.3) admit unique solutions $u(x), u_n(x) \in H^{2s}(\mathbb{R}),\ n \in \mathbb{N}$, respectively. Orthogonality relations (2.10) and (1.12) yield

$$\widehat{f}(0) = 0, \quad \widehat{f_n}(0) = 0, \quad n \in \mathbb{N}$$

in this case. This allows us to use the expressions

$$\widehat{f}(p) = \int_0^p \frac{d\widehat{f}(s)}{ds} ds, \quad \widehat{f_n}(p) = \int_0^p \frac{d\widehat{f_n}(s)}{ds} ds, \quad n \in \mathbb{N},$$

which enables us to write the first term in the right side of equality (2.9) as

$$\frac{\int_0^p \left(\frac{d\widehat{f_n}(s)}{ds} - \frac{d\widehat{f}(s)}{ds}\right) ds}{|p|^{2s}} \chi_{\{|p| \le 1\}}. \tag{2.11}$$

Using inequality (2.3), we easily estimate

$$\left|\frac{d\widehat{f_n}(p)}{dp} - \frac{d\widehat{f}(p)}{dp}\right| \le \frac{1}{\sqrt{2\pi}} \|x f_n(x) - x f(x)\|_{L^1(\mathbb{R})}, \tag{2.12}$$

such that expression (2.11) can be bounded from above in the absolute value by

$$\frac{1}{\sqrt{2\pi}} \|x f_n(x) - x f(x)\|_{L^1(\mathbb{R})} |p|^{1-2s} \chi_{\{|p| \le 1\}}.$$

Hence, we arrive at

$$\left\| \frac{\int_0^p \left(\frac{d\widehat{f_n}(s)}{ds} - \frac{d\widehat{f}(s)}{ds}\right) ds}{|p|^{2s}} \chi_{\{|p| \le 1\}} \right\|_{L^2(\mathbb{R})} \le \frac{1}{\sqrt{\pi}} \|x f_n(x) - x f(x)\|_{L^1(\mathbb{R})} \to 0$$

as $n \to \infty$ due to one of our assumptions. This implies that

$$u_n(x) \to u(x) \quad in \quad L^2(\mathbb{R}), \quad n \to \infty$$

when the dimension $d = 1$ and $a = 0$ with $s \in \left[\frac{1}{4}, \frac{3}{4}\right)$.

Then we proceed to the proof of the theorem when the power of the negative Laplacian $s \in \left[\frac{3}{4}, 1\right)$ in dimension $d = 1$ with $a = 0$. By means of the parts c) and

d) of Lemma 4.1 below under our assumptions we have $xf_n(x) \in L^1(\mathbb{R}),\ n \in \mathbb{N}$, such that $xf_n(x) \to xf(x)$ in $L^1(\mathbb{R})$ as $n \to \infty$. Then via the parts a) and b) of Lemma 4.1 we have $f_n(x) \in L^1(\mathbb{R}),\ n \in \mathbb{N}$, such that $f_n(x) \to f(x)$ in $L^1(\mathbb{R})$ as $n \to \infty$. Orthogonality condition (2.10) here can be easily obtained via the limiting argument as above. By means of the second orthogonality relation in (1.13), we derive

$$|(f(x), x)_{L^2(\mathbb{R})}| = |(f(x) - f_n(x), x)_{L^2(\mathbb{R})}| \le \|xf_n(x) - xf(x)\|_{L^1(\mathbb{R})} \to 0$$

as $n \to \infty$. Hence

$$(f(x), x)_{L^2(\mathbb{R})} = 0 \tag{2.13}$$

holds. By virtue of the part a) of Theorem 1.1, when $s \in \left[\frac{3}{4}, 1\right)$, Eqs. (1.2) and (1.3) possess unique solutions $u(x), u_n(x) \in H^{2s}(\mathbb{R}),\ n \in \mathbb{N}$, respectively. Via the definition of the standard Fourier transform (1.18), orthogonality relations (2.10), (1.13), and (2.13) give us for $n \in \mathbb{N}$

$$\widehat{f}(0) = 0, \quad \widehat{f_n}(0) = 0, \quad \frac{d\widehat{f}}{dp}(0) = 0, \quad \frac{d\widehat{f_n}}{dp}(0) = 0,$$

such that

$$\widehat{f}(p) = \int_0^p \left(\int_0^s \frac{d^2\widehat{f}(q)}{dq^2} dq \right) ds, \quad \widehat{f_n}(p) = \int_0^p \left(\int_0^s \frac{d^2\widehat{f_n}(q)}{dq^2} dq \right) ds, \quad n \in \mathbb{N}.$$

By means of definition (1.18), we easily estimate

$$\left| \frac{d^2\widehat{f}_n(p)}{dp^2} - \frac{d^2\widehat{f}(p)}{dp^2} \right| \le \frac{1}{\sqrt{2\pi}} \|x^2 f_n(x) - x^2 f(x)\|_{L^1(\mathbb{R})}.$$

This yields the inequality

$$|\widehat{f_n}(p) - \widehat{f}(p)| \le \frac{1}{\sqrt{2\pi}} \|x^2 f_n(x) - x^2 f(x)\|_{L^1(\mathbb{R})} \frac{p^2}{2},$$

which allows us to obtain the upper bound on the absolute value of the first term in the right side of identity (2.9) by

$$\frac{1}{2\sqrt{2\pi}} \|x^2 f_n(x) - x^2 f(x)\|_{L^1(\mathbb{R})} |p|^{2-2s} \chi_{\{|p| \le 1\}}.$$

Therefore,

$$\left\| \frac{\widehat{f_n}(p) - \widehat{f}(p)}{|p|^{2s}} \chi_{\{|p| \le 1\}} \right\|_{L^2(\mathbb{R})} \le \frac{1}{2\sqrt{\pi(5-4s)}} \|x^2 f_n(x) - x^2 f(x)\|_{L^1(\mathbb{R})} \to 0$$

when $n \to \infty$ as assumed. Thus

$$u_n(x) \to u(x) \quad in \quad L^2(\mathbb{R}), \quad n \to \infty$$

when the dimension $d = 1$ and $a = 0$ with $s \in \left[\frac{3}{4}, 1\right)$.

In the case of the dimension $d = 2$ and $a = 0$, let us first treat the situation when $s \in \left(0, \frac{1}{2}\right)$. Due to the part b) of Theorem 1.1, problem (1.2) and each of problems (1.3) have unique solutions $u(x) \in H^{2s}(\mathbb{R})$ and $u_n(x) \in H^{2s}(\mathbb{R}), \quad n \in \mathbb{N}$, respectively. Obviously, the first term in the right side of (2.9) can be estimated from above in the absolute value via (1.19) by

$$\frac{1}{2\pi} \|f_n(x) - f(x)\|_{L^1(\mathbb{R}^2)} \frac{\chi_{\{|p| \le 1\}}}{|p|^{2s}},$$

such that its $L^2(\mathbb{R}^2)$ norm can be bounded from above by

$$\frac{1}{2\sqrt{\pi(1-2s)}} \|f_n(x) - f(x)\|_{L^1(\mathbb{R}^2)} \to 0, \qquad n \to \infty$$

by means of one of our assumptions and with $s \in \left(0, \frac{1}{2}\right)$.

For the higher values of the power of the two-dimensional negative Laplacian $s \in \left(\frac{1}{2}, 1\right)$, the orthogonality relation

$$(f(x), 1)_{L^2(\mathbb{R}^2)} = 0 \tag{2.14}$$

can be derived via the easy limiting argument, analogously to (2.10). By virtue of the part b) of Theorem 1.1, problems (1.2) and (1.3) possess unique solutions $u(x) \in H^{2s}(\mathbb{R}^2)$ and $u_n(x) \in H^{2s}(\mathbb{R}^2)$, $n \in \mathbb{N}$, respectively. Orthogonality relations (2.14) and (1.12) imply

$$\widehat{f}(0) = 0, \quad \widehat{f_n}(0) = 0, \quad n \in \mathbb{N}$$

when the dimension $d = 2$ and $a = 0$ with $s \in \left(\frac{1}{2}, 1\right)$. This enables us to express

$$\widehat{f}(p) = \int_0^{|p|} \frac{\partial \widehat{f}(s, \sigma)}{\partial s} ds, \quad \widehat{f_n}(p) = \int_0^{|p|} \frac{\partial \widehat{f_n}(s, \sigma)}{\partial s} ds, \quad n \in \mathbb{N} \tag{2.15}$$

and to write the first term in the right side of identity (2.9) as

$$\frac{\int_0^{|p|}\left(\frac{\partial \widehat{f_n}(s,\sigma)}{\partial s}-\frac{\partial \widehat{f}(s,\sigma)}{\partial s}\right)ds}{|p|^{2s}}\chi_{\{|p|\leq 1\}}. \tag{2.16}$$

Inequality (2.3) gives us

$$\left|\frac{\partial \widehat{f_n}(p)}{\partial |p|}-\frac{\partial \widehat{f}(p)}{\partial |p|}\right|\leq \frac{1}{2\pi}\|xf_n(x)-xf(x)\|_{L^1(\mathbb{R}^2)}. \tag{2.17}$$

Thus, expression (2.16) can be bounded from above in the absolute value by

$$\frac{1}{2\pi}\|xf_n(x)-xf(x)\|_{L^1(\mathbb{R}^2)}|p|^{1-2s}\chi_{\{|p|\leq 1\}}.$$

Hence

$$\left\|\frac{\int_0^{|p|}\left(\frac{\partial \widehat{f_n}(s,\sigma)}{\partial s}-\frac{\partial \widehat{f}(s,\sigma)}{\partial s}\right)ds}{|p|^{2s}}\chi_{\{|p|\leq 1\}}\right\|_{L^2(\mathbb{R}^2)}\leq \frac{\|xf_n(x)-xf(x)\|_{L^1(\mathbb{R}^2)}}{2\sqrt{2\pi(1-s)}}\to 0$$

as $n\to\infty$ via one of our assumptions. Therefore,

$$u_n(x)\to u(x)\quad in\quad L^2(\mathbb{R}^2),\quad n\to\infty$$

when the dimension $d=2$ and $a=0$ with $s\in\left(\frac{1}{2},1\right)$.

Let us proceed to the proof of the part c) of the theorem, when the dimension $d=3$ and $a=0$ with $s\in\left(0,\frac{3}{4}\right)$. In such case, by virtue of the part c) of Theorem 1.1, problems (1.2) and (1.3) admit unique solutions $u(x)$ and $u_n(x)$, $n\in\mathbb{N}$, respectively, belonging to $H^{2s}(\mathbb{R}^3)$. Using (1.19), we obtain the upper bound on the first term in the right side of (2.9) in the absolute value by

$$\frac{\|f_n(x)-f(x)\|_{L^1(\mathbb{R}^3)}}{(2\pi)^{\frac{3}{2}}|p|^{2s}}\chi_{\{|p|\leq 1\}},$$

such that its $L^2(\mathbb{R}^3)$ norm can be estimated from above by

$$\frac{1}{\pi\sqrt{2(3-4s)}}\|f_n(x)-f(x)\|_{L^1(\mathbb{R}^3)}\to 0,\quad n\to\infty$$

via one of our assumptions. Thus,

$$u_n(x)\to u(x)\quad in\quad L^2(\mathbb{R}^3),\quad n\to\infty$$

in the case of the dimension $d=3$ and $a=0$ with $s\in\left(0,\frac{3}{4}\right)$.

For the higher values of the power of the three-dimensional negative Laplacian $s \in \left[\frac{3}{4}, 1\right)$, the orthogonality condition

$$(f(x), 1)_{L^2(\mathbb{R}^3)} = 0 \tag{2.18}$$

can be obtained via the trivial limiting argument, similarly to (2.10). By means of the part c) of Theorem 1.1, Eqs. (1.2) and (1.3) have unique solutions $u(x) \in H^{2s}(\mathbb{R}^3)$ and $u_n(x) \in H^{2s}(\mathbb{R}^3)$, $n \in \mathbb{N}$, respectively. Orthogonality conditions (2.18) and (1.15) yield

$$\widehat{f}(0) = 0, \quad \widehat{f_n}(0) = 0, \quad n \in \mathbb{N}$$

when the dimension $d = 3$ and $a = 0$ with $s \in \left[\frac{3}{4}, 1\right)$. This allows us to obtain here the expressions analogous to (2.15). Let us use the three-dimensional analog of inequality (2.17) to derive the upper bound on the first term in the right side of (2.9) in the absolute value by

$$\frac{\|xf_n(x) - xf(x)\|_{L^1(\mathbb{R}^3)}}{(2\pi)^{\frac{3}{2}}} |p|^{1-2s} \chi_{\{|p| \le 1\}},$$

such that its $L^2(\mathbb{R}^3)$ norm can be estimated from above by

$$\frac{1}{\pi\sqrt{2(5-4s)}} \|xf_n(x) - xf(x)\|_{L^1(\mathbb{R}^3)} \to 0, \quad n \to \infty$$

due to one of our assumptions. Therefore,

$$u_n(x) \to u(x) \quad in \quad L^2(\mathbb{R}^3), \quad n \to \infty$$

in the case of the dimension $d = 3$ and $a = 0$ with $s \in \left[\frac{3}{4}, 1\right)$.

Then we turn our attention to the case d) of the theorem. By virtue of the part d) of Theorem 1.1 Eqs. (1.2) and (1.3) admit unique solutions $u(x) \in H^{2s}(\mathbb{R}^3)$ and $u_n(x) \in H^{2s}(\mathbb{R}^3)$, $n \in \mathbb{N}$, respectively. Using inequality (1.19), we estimate the first term in the right side of (2.9) in the absolute value by

$$\frac{\|f_n(x) - f(x)\|_{L^1(\mathbb{R}^d)}}{(2\pi)^{\frac{d}{2}} |p|^{2s}} \chi_{\{|p| \le 1\}}, \quad d \ge 4,$$

such that its $L^2(\mathbb{R}^d)$ norm can be bounded from above by

$$\frac{1}{(2\pi)^{\frac{d}{2}}} \sqrt{\frac{|S^d|}{d-4s}} \|f_n(x) - f(x)\|_{L^1(\mathbb{R}^d)} \to 0, \quad n \to \infty$$

by virtue of the one of our assumptions. Hence,

$$u_n(x) \to u(x) \quad in \quad L^2(\mathbb{R}^d), \quad d \geq 4, \quad n \to \infty$$

when $a = 0$ and $s \in (0, 1)$.

If $u(x)$ and $u_n(x),\ n \in \mathbb{N}$ are the unique solutions of equations (1.2) and (1.3) in $H^{2s}(\mathbb{R}^d),\ d \in \mathbb{N}$, respectively with $a > 0$ as in the cases e) and f) of the theorem, by applying the standard Fourier transform (1.18) we easily obtain

$$\widehat{u}(p) = \frac{\widehat{f}(p)}{|p|^{2s} - a}, \quad \widehat{u}_n(p) = \frac{\widehat{f}_n(p)}{|p|^{2s} - a}, \quad n \in \mathbb{N}. \tag{2.19}$$

First of all, we consider the case e) of the theorem, when the dimension $d = 1$ and $a > 0$. Thus, due to the result of the part e) of Theorem 1.1, Eq. (1.3) has a unique solution $u_n(x) \in H^{2s}(\mathbb{R}),\ n \in \mathbb{N}$. Clearly, $f_n(x) \in L^1(\mathbb{R}),\ n \in \mathbb{N}$, such that $f_n(x) \to f(x)$ in $L^1(\mathbb{R})$ as $n \to \infty$ via the parts a) and b) of Lemma 4.1 below. By means of the limiting argument, analogously to the proof of (2.10) we obtain the orthogonality relations

$$\left(f(x), \frac{e^{\pm i a^{\frac{1}{2s}} x}}{\sqrt{2\pi}}\right)_{L^2(\mathbb{R})} = 0, \quad s \in (0, 1). \tag{2.20}$$

Then by virtue of the result of the part e) of Theorem 1.1, problem (1.2) admits a unique solution $u(x) \in H^{2s}(\mathbb{R})$. Using (2.19), we express $\widehat{u}_n(p) - \widehat{u}(p)$ as

$$\frac{\widehat{f}_n(p) - \widehat{f}(p)}{p^{2s} - a}\chi_{I_\delta^+} + \frac{\widehat{f}_n(p) - \widehat{f}(p)}{p^{2s} - a}\chi_{I_\delta^{c+}} +$$

$$+ \frac{\widehat{f}_n(p) - \widehat{f}(p)}{(-p)^{2s} - a}\chi_{I_\delta^-} + \frac{\widehat{f}_n(p) - \widehat{f}(p)}{(-p)^{2s} - a}\chi_{I_\delta^{c-}}, \tag{2.21}$$

with $I_\delta^+,\ I_\delta^-$ are given by (2.4) and $I_\delta^{c+},\ I_\delta^{c-}$ are defined in (2.6). Evidently, the second term in (2.21) can be estimated from above in the absolute value by $C|\widehat{f}_n(p) - \widehat{f}(p)|$, such that

$$\left\| \frac{\widehat{f}_n(p) - \widehat{f}(p)}{p^{2s} - a}\chi_{I_\delta^{c+}} \right\|_{L^2(\mathbb{R})} \leq C\|f_n(x) - f(x)\|_{L^2(\mathbb{R})} \to 0, \quad n \to \infty$$

as assumed. Analogously, the last term in (2.21) can be bounded from above in the absolute value by $C|\widehat{f}_n(p) - \widehat{f}(p)|$. Thus

$$\left\| \frac{\widehat{f}_n(p) - \widehat{f}(p)}{(-p)^{2s} - a}\chi_{I_\delta^{c-}} \right\|_{L^2(\mathbb{R})} \leq C\|f_n(x) - f(x)\|_{L^2(\mathbb{R})} \to 0, \quad n \to \infty$$

due to one of our assumptions. Orthogonality relations (2.20) and (1.16) give us

$$\widehat{f}(a^{\frac{1}{2s}}) = 0, \quad \widehat{f_n}(a^{\frac{1}{2s}}) = 0, \quad n \in \mathbb{N},$$

such that

$$\widehat{f}(p) = \int_{a^{\frac{1}{2s}}}^{p} \frac{d\widehat{f}(s)}{ds} ds, \quad \widehat{f_n}(p) = \int_{a^{\frac{1}{2s}}}^{p} \frac{d\widehat{f_n}(s)}{ds} ds, \quad n \in \mathbb{N},$$

which enables us to express the first term in (2.21) as

$$\frac{\int_{a^{\frac{1}{2s}}}^{p} \left[\frac{d\widehat{f_n}(s)}{ds} - \frac{d\widehat{f}(s)}{ds}\right] ds}{p^{2s} - a} \chi_{I_\delta^+}. \tag{2.22}$$

By means of (2.12), we obtain the upper bound on (2.22) in the absolute value by

$$\frac{1}{\sqrt{2\pi}} \|xf_n(x) - xf(x)\|_{L^1(\mathbb{R})} \left| \frac{p - a^{\frac{1}{2s}}}{p^{2s} - a} \right| \chi_{I_\delta^+} \leq C \|xf_n(x) - xf(x)\|_{L^1(\mathbb{R})} \chi_{I_\delta^+}.$$

Thus, the $L^2(\mathbb{R})$ norm of (2.22) can be estimated from above by

$$C \|xf_n(x) - xf(x)\|_{L^1(\mathbb{R})} \to 0, \quad n \to \infty$$

due to one of our assumptions. Orthogonality conditions (2.20) and (1.16) yield

$$\widehat{f}(-a^{\frac{1}{2s}}) = 0, \quad \widehat{f_n}(-a^{\frac{1}{2s}}) = 0, \quad n \in \mathbb{N}$$

with $s \in (0, 1)$. Therefore, at the negative singularity

$$\widehat{f}(p) = \int_{-a^{\frac{1}{2s}}}^{p} \frac{d\widehat{f}(s)}{ds} ds, \quad \widehat{f_n}(p) = \int_{-a^{\frac{1}{2s}}}^{p} \frac{d\widehat{f_n}(s)}{ds} ds, \quad n \in \mathbb{N}.$$

This gives us the upper bound on the third term in (2.21) in the absolute value by

$$\frac{1}{\sqrt{2\pi}} \|xf_n(x) - xf(x)\|_{L^1(\mathbb{R})} \left| \frac{p + a^{\frac{1}{2s}}}{(-p)^{2s} - a} \right| \chi_{I_\delta^-} \leq C \|xf_n(x) - xf(x)\|_{L^1(\mathbb{R})} \chi_{I_\delta^-}.$$

Thus, its $L^2(\mathbb{R})$ norm can be estimated from above by

$$C \|xf_n(x) - xf(x)\|_{L^1(\mathbb{R})} \to 0, \quad n \to \infty$$

as assumed. This shows that in dimension $d = 1$, when $a > 0$ and $s \in (0, 1)$ we have

$$u_n(x) \to u(x) \quad in \quad L^2(\mathbb{R}), \quad n \to \infty.$$

We conclude the proof of the theorem with treating the case f) when the dimension $d \geq 2$ and $a > 0$ with $s \in (0, 1)$. Then under our assumptions, by virtue of the part f) of Theorem 1.1, problem (1.3) has a unique solution $u_n(x) \in H^{2s}(\mathbb{R}^d)$, $n \in \mathbb{N}$. A trivial limiting argument analogous to the proof of (2.10) gives us

$$\left(f(x), \frac{e^{ipx}}{(2\pi)^{\frac{d}{2}}}\right)_{L^2(\mathbb{R}^d)} = 0, \quad p \in S^d_{a^{\frac{1}{2s}}}. \tag{2.23}$$

Then by means of the part f) of Theorem 1.1, problem (1.2) admits a unique solution $u(x) \in H^{2s}(\mathbb{R}^d)$. Using (2.19), we easily arrive at

$$\widehat{u}_n(p) - \widehat{u}(p) = \frac{\widehat{f}_n(p) - \widehat{f}(p)}{|p|^{2s} - a}\chi_{A_\delta} + \frac{\widehat{f}_n(p) - \widehat{f}(p)}{|p|^{2s} - a}\chi_{A_\delta^c}, \tag{2.24}$$

with the set A_δ defined in (2.7). Evidently, the second term in the right side of (2.24) can be estimated from above in the absolute value by $C|\widehat{f}_n(p) - \widehat{f}(p)|$. Thus,

$$\left\| \frac{\widehat{f}_n(p) - \widehat{f}(p)}{|p|^{2s} - a}\chi_{A_\delta^c} \right\|_{L^2(\mathbb{R}^d)} \leq C\|f_n(x) - f(x)\|_{L^2(\mathbb{R}^d)} \to 0$$

as $n \to \infty$ via one of our assumptions. Orthogonality relations (2.23) and (1.17) imply that

$$\widehat{f}(a^{\frac{1}{2s}}, \sigma) = 0, \quad \widehat{f}_n(a^{\frac{1}{2s}}, \sigma) = 0, \quad n \in \mathbb{N},$$

such that

$$\widehat{f}(p) = \int_{a^{\frac{1}{2s}}}^{|p|} \frac{\partial \widehat{f}(s, \sigma)}{\partial s} ds, \quad \widehat{f}_n(p) = \int_{a^{\frac{1}{2s}}}^{|p|} \frac{\partial \widehat{f}_n(s, \sigma)}{\partial s} ds, \quad n \in \mathbb{N}.$$

By means of the definition of the Fourier transform (1.18), analogously to inequalities (2.12) and (2.17) in lower dimensions, we easily obtain

$$\left| \frac{\partial \widehat{f}_n(p)}{\partial |p|} - \frac{\partial \widehat{f}(p)}{\partial |p|} \right| \leq \frac{1}{(2\pi)^{\frac{d}{2}}} \|xf_n(x) - xf(x)\|_{L^1(\mathbb{R}^d)}.$$

We derive the upper bound in the absolute value on the first term in the right side of (2.24) by

$$\frac{1}{(2\pi)^{\frac{d}{2}}}\|xf_n(x)-xf(x)\|_{L^1(\mathbb{R}^d)}\left|\frac{|p|-a^{\frac{1}{2s}}}{|p|^{2s}-a}\right|\chi_{A_\delta}\leq C\|xf_n(x)-xf(x)\|_{L^1(\mathbb{R}^d)}\chi_{A_\delta}.$$

This implies that

$$\left\|\frac{\widehat{f_n}(p)-\widehat{f}(p)}{|p|^{2s}-a}\chi_{A_\delta}\right\|_{L^2(\mathbb{R}^d)}\leq C\|xf_n(x)-xf(x)\|_{L^1(\mathbb{R}^d)}\to 0,\quad n\to\infty$$

as assumed. Therefore, in dimensions $d\geq 2$, when $a>0$ and $s\in(0,1)$, we have

$$u_n(x)\to u(x)\quad in\quad L^2(\mathbb{R}^d)$$

as $n\to\infty$. ■

3 Solvability in the Sense of Sequences with a Scalar Potential

Proof of Theorem 1.4 Note that the case a) of the theorem is the result of Lemma 7 of [27]. Then we proceed to proving the case of $a>0$.

To prove the uniqueness of solutions of our equation, let us suppose that there exist both $u_1(x)$ and $u_2(x)$ which are square integrable in $\mathbb{R}^3$ and solve (1.20). Then their difference $w(x):=u_1(x)-u_2(x)\in L^2(\mathbb{R}^3)$ is a solution of the problem

$$(-\Delta+V(x))^s w=aw,\quad s\in(0,1).$$

The fact that the operator $(-\Delta+V(x))^s$ has no nontrivial $L^2(\mathbb{R}^3)$ eigenfunctions as discussed above yields that $w(x)$ vanishes a.e. in $\mathbb{R}^3$.

Let us apply the generalized Fourier transform (1.26) with the functions of the continuous spectrum of the Schrödinger operator to both sides of problem (1.20), which yields

$$\tilde{u}(k)=\frac{\tilde{f}(k)}{|k|^{2s}-a},\quad s\in(0,1).$$

We introduce the spherical layer in the space of three dimensions as

$$B_\delta:=\{k\in\mathbb{R}^3\mid a^{\frac{1}{2s}}-\delta\leq|k|\leq a^{\frac{1}{2s}}+\delta\},\ 0<\delta<a^{\frac{1}{2s}}. \tag{3.1}$$

This allows us to express

$$\tilde{u}(k) = \frac{\tilde{f}(k)}{|k|^{2s} - a}\chi_{B_\delta} + \frac{\tilde{f}(k)}{|k|^{2s} - a}\chi_{B_\delta^c}. \tag{3.2}$$

The second term in the right side of (3.2) can be trivially bounded from above in the absolute value by

$$C|\tilde{f}(k)| \in L^2(\mathbb{R}^3),$$

because $f(x)$ is square integrable as assumed. We express

$$\tilde{f}(k) = \tilde{f}(a^{\frac{1}{2s}}, \sigma) + \int_{a^{\frac{1}{2s}}}^{|k|} \frac{\partial \tilde{f}(q,\sigma)}{\partial q} dq.$$

Therefore, the first term in the right side of (3.2) can be written as

$$\frac{\tilde{f}(a^{\frac{1}{2s}}, \sigma)}{|k|^{2s} - a}\chi_{B_\delta} + \frac{\int_{a^{\frac{1}{2s}}}^{|k|} \frac{\partial \tilde{f}(q,\sigma)}{\partial q} dq}{|k|^{2s} - a}\chi_{B_\delta}. \tag{3.3}$$

The second term in sum (3.3) can be easily bounded above in the absolute value by

$$\|\nabla_k \tilde{f}(k)\|_{L^\infty(\mathbb{R}^3)} \left| \frac{|k| - a^{\frac{1}{2s}}}{|k|^{2s} - a} \right| \chi_{B_\delta} \le C\|\nabla_k \tilde{f}(k)\|_{L^\infty(\mathbb{R}^3)}\chi_{B_\delta} \in L^2(\mathbb{R}^3).$$

Note that under the stated assumptions $\nabla_k \tilde{f}(k) \in L^\infty(\mathbb{R}^3)$ due to Lemma 2.4 of [22]. Apparently, the first term in (3.3) is square integrable if and only if $\tilde{f}(a^{\frac{1}{2s}}, \sigma)$ vanishes, which yields orthogonality relation (1.29). ■

Then we proceed to the establishing of our last main statement dealing with the solvability in the sense of sequences.

Proof of Theorem 1.5 In the case a) when $s \in \left(0, \frac{3}{4}\right)$ problems (1.20) and (1.21) have unique solutions $u(x),\ u_n(x) \in L^2(\mathbb{R}^3),\ n \in \mathbb{N}$, respectively due to the part a) of Theorem 1.4 above. Let us apply the generalized Fourier transform (1.26) to both sides of equations (1.20) and (1.21). We obtain

$$\tilde{u}(k) = \frac{\tilde{f}(k)}{|k|^{2s}}, \quad \tilde{u}_n(k) = \frac{\tilde{f}_n(k)}{|k|^{2s}}, \quad n \in \mathbb{N}.$$

Therefore

$$\tilde{u}_n(k) - \tilde{u}(k) = \frac{\tilde{f}_n(k) - \tilde{f}(k)}{|k|^{2s}}\chi_{\{|k|\le 1\}} + \frac{\tilde{f}_n(k) - \tilde{f}(k)}{|k|^{2s}}\chi_{\{|k|>1\}}. \tag{3.4}$$

Obviously, the second term in the right side of (3.4) can be easily estimated from above in the absolute value by $|\tilde{f}_n(k) - \tilde{f}(k)|$. Hence

$$\left\| \frac{\tilde{f}_n(k) - \tilde{f}(k)}{|k|^{2s}} \chi_{\{|k|>1\}} \right\|_{L^2(\mathbb{R}^3)} \le \|f_n(x) - f(x)\|_{L^2(\mathbb{R}^3)} \to 0, \quad n \to \infty$$

via one of our assumptions. Using (1.27) we obtain the upper bound for the first term in the right side of (3.4) in the absolute value by

$$\frac{1}{(2\pi)^{\frac{3}{2}}} \frac{1}{1 - I(V)} \|f_n(x) - f(x)\|_{L^1(\mathbb{R}^3)} \frac{\chi_{\{|k|\le 1\}}}{|k|^{2s}}.$$

Apparently, this yields

$$\left\| \frac{\tilde{f}_n(k) - \tilde{f}(k)}{|k|^{2s}} \chi_{\{|k|\le 1\}} \right\|_{L^2(\mathbb{R}^3)} \le$$

$$\le \frac{1}{\sqrt{2(3-4s)}\pi} \frac{1}{1 - I(V)} \|f_n(x) - f(x)\|_{L^1(\mathbb{R}^3)} \to 0, \quad n \to \infty$$

due to one of our assumptions. Therefore, $u_n(x) \to u(x)$ in $L^2(\mathbb{R}^3)$ as $n \to \infty$ in the case when the parameter $a = 0$ and $s \in \left(0, \frac{3}{4}\right)$.

Then we turn our attention to the situation when $a = 0$ and $s \in \left[\frac{3}{4}, 1\right)$. By means of orthogonality relation (1.30) along with the Corollary 2.2 of [22] and the part b) of Lemma 4.1 below we obtain

$$|(f(x), \varphi_0(x))_{L^2(\mathbb{R}^3)}| = |(f(x) - f_n(x), \varphi_0(x))_{L^2(\mathbb{R}^3)}| \le$$

$$\le \frac{1}{(2\pi)^{\frac{3}{2}}} \frac{1}{1 - I(V)} \|f_n(x) - f(x)\|_{L^1(\mathbb{R}^3)} \to 0, \quad n \to \infty.$$

Therefore,

$$(f(x), \varphi_0(x))_{L^2(\mathbb{R}^3)} = 0 \tag{3.5}$$

holds. Hence Eqs. (1.20) and (1.21) admit unique solutions $u(x)$, $u_n(x) \in L^2(\mathbb{R}^3)$, $n \in \mathbb{N}$, respectively, via the part a) of Theorem 1.4. As discussed above, it is sufficient to consider the first term in the right side of (3.4). Orthogonality relations (3.5) and (1.30) yield

$$\tilde{f}(0) = 0, \quad \tilde{f}_n(0) = 0, \quad n \in \mathbb{N},$$

such that

$$\tilde{f}(k) = \int_0^{|k|} \frac{\partial \tilde{f}(s,\sigma)}{\partial s} ds, \quad \tilde{f}_n(k) = \int_0^{|k|} \frac{\partial \tilde{f}_n(s,\sigma)}{\partial s} ds, \quad n \in \mathbb{N}.$$

This enables us to estimate the first term in the right side of (3.4) from above in the absolute value by $\|\nabla_k[\tilde{f}_n(k) - \tilde{f}(k)]\|_{L^\infty(\mathbb{R}^3)}|k|^{1-2s}\chi_{\{|k|\le 1\}}$. Therefore

$$\left\| \frac{\tilde{f}_n(k) - \tilde{f}(k)}{|k|^{2s}} \chi_{\{|k|\le 1\}} \right\|_{L^2(\mathbb{R}^3)} \le C\|\nabla_k[\tilde{f}_n(k) - \tilde{f}(k)]\|_{L^\infty(\mathbb{R}^3)} \to 0, \quad n \to \infty$$

due to Lemma 3.4 of [18] under the given assumptions. This shows that $u_n(x) \to u(x)$ in $L^2(\mathbb{R}^3)$ as $n \to \infty$ when $a = 0$ and $s \in \left[\frac{3}{4}, 1\right)$.

We complete the proof of the theorem by establishing the result of the part b). By virtue of the limiting argument similar to the proof of relation (3.5), we have

$$(f(x), \varphi_k(x))_{L^2(\mathbb{R}^3)} = 0, \quad k \in S^3_{a^{\frac{1}{2s}}}, \quad s \in (0,1). \tag{3.6}$$

Thus by means of the result the part b) of Theorem 1.4, Eqs. (1.20) and (1.21) possesses unique solutions $u(x), u_n(x) \in L^2(\mathbb{R}^3)$. We apply the generalized Fourier transform (1.26) to both sides of problems (1.20) and (1.21). Hence, we obtain

$$\tilde{u}_n(k) - \tilde{u}(k) = \frac{\tilde{f}_n(k) - \tilde{f}(k)}{|k|^{2s} - a}\chi_{B_\delta} + \frac{\tilde{f}_n(k) - \tilde{f}(k)}{|k|^{2s} - a}\chi_{B_\delta^c} \tag{3.7}$$

with B_δ defined in (3.1). Obviously, the second term in the right side of (3.7) can be estimated from above in the absolute value by $C|\tilde{f}_n(k) - \tilde{f}(k)|$, such that

$$\left\| \frac{\tilde{f}_n(k) - \tilde{f}(k)}{|k|^{2s} - a} \chi_{B_\delta^c} \right\|_{L^2(\mathbb{R}^3)} \le C\|f_n(x) - f(x)\|_{L^2(\mathbb{R}^3)} \to 0, \quad n \to \infty$$

due to one of our assumptions. By means of orthogonality conditions (3.6) and (1.31), we have

$$\tilde{f}(a^{\frac{1}{2s}}, \sigma) = 0, \quad \tilde{f}_n(a^{\frac{1}{2s}}, \sigma) = 0, \quad n \in \mathbb{N}.$$

This gives us the representations

$$\tilde{f}(k) = \int_{a^{\frac{1}{2s}}}^{|k|} \frac{\partial \tilde{f}(s,\sigma)}{\partial s} ds, \quad \tilde{f}_n(k) = \int_{a^{\frac{1}{2s}}}^{|k|} \frac{\partial \tilde{f}_n(s,\sigma)}{\partial s} ds, \quad n \in \mathbb{N},$$

such that the first term in the right side of (3.7) can be expressed as

$$\frac{\int_{a^{\frac{1}{2s}}}^{|k|}\left[\frac{\partial \tilde{f}_n(s,\sigma)}{\partial s} - \frac{\partial \tilde{f}(s,\sigma)}{\partial s}\right]ds}{|k|^{2s} - a}\chi_{B_\delta}. \tag{3.8}$$

Clearly, (3.8) can be trivially estimated from above in the absolute value by

$$\|\nabla_k[\tilde{f}_n(k) - \tilde{f}(k)]\|_{L^\infty(\mathbb{R}^3)}\left|\frac{|k| - a^{\frac{1}{2s}}}{|k|^{2s} - a}\right|\chi_{B_\delta} \le C\|\nabla_k[\tilde{f}_n(k) - \tilde{f}(k)]\|_{L^\infty(\mathbb{R}^3)}\chi_{B_\delta}.$$

Therefore, the $L^2(\mathbb{R}^3)$ norm of (3.8) can be bounded from above by

$$C\|\nabla_k[\tilde{f}_n(k) - \tilde{f}(k)]\|_{L^\infty(\mathbb{R}^3)} \to 0, \quad n \to \infty$$

by virtue of Lemma 3.4 of [18] under the given assumptions. This yields that $u_n(x) \to u(x)$ in $L^2(\mathbb{R}^3)$ as $n \to \infty$ when $a > 0$ with $s \in (0, 1)$. ■

4 Auxiliary Results

The following technical lemma is useful for proving the solvability in the sense of sequences in our theorems. Note that its parts a) and b) were established in Lemma 6 of [29].

Lemma 4.1

a) Let $f(x) \in L^2(\mathbb{R}^d)$, $d \in \mathbb{N}$ and $xf(x) \in L^1(\mathbb{R}^d)$. Then $f(x) \in L^1(\mathbb{R}^d)$.
b) Let $n \in \mathbb{N}$, $f_n(x) \in L^2(\mathbb{R}^d)$, $d \in \mathbb{N}$, such that $f_n(x) \to f(x)$ in $L^2(\mathbb{R}^d)$ as $n \to \infty$. Let $xf_n(x) \in L^1(\mathbb{R}^d)$, such that $xf_n(x) \to xf(x)$ in $L^1(\mathbb{R}^d)$ as $n \to \infty$. Then $f_n(x) \to f(x)$ in $L^1(\mathbb{R}^d)$ as $n \to \infty$.
c) Let $f(x) \in L^2(\mathbb{R}^d)$, $d \in \mathbb{N}$ and $x^2 f(x) \in L^1(\mathbb{R}^d)$. Then $xf(x) \in L^1(\mathbb{R}^d)$.
d) Let $n \in \mathbb{N}$, $f_n(x) \in L^2(\mathbb{R}^d)$, $d \in \mathbb{N}$, such that $f_n(x) \to f(x)$ in $L^2(\mathbb{R}^d)$ as $n \to \infty$. Let $x^2 f_n(x) \in L^1(\mathbb{R}^d)$, such that $x^2 f_n(x) \to x^2 f(x)$ in $L^1(\mathbb{R}^d)$ as $n \to \infty$. Then $xf_n(x) \to xf(x)$ in $L^1(\mathbb{R}^d)$ as $n \to \infty$.

Proof To prove the part c) of the lemma, we express the norm $\|xf(x)\|_{L^1(\mathbb{R}^d)}$ as

$$\int_{|x|\le 1}|x||f(x)|dx + \int_{|x|>1}|x||f(x)|dx \le \int_{|x|\le 1}|f(x)|dx + \int_{|x|>1}|x|^2|f(x)|dx.$$

This sum can be easily bounded from above via the Schwarz inequality by

$$\|f(x)\|_{L^2(\mathbb{R}^d)}\sqrt{|B^d|} + \|x^2 f(x)\|_{L^1(\mathbb{R}^d)} < \infty$$

as assumed. Let us complete the proof of the lemma with establishing its part d). Clearly, the norm $\|xf_n(x) - xf(x)\|_{L^1(\mathbb{R}^d)}$ can be written as

$$\int_{|x|\le 1} |x||f_n(x) - f(x)|dx + \int_{|x|>1} |x||f_n(x) - f(x)|dx \le$$

$$\le \int_{|x|\le 1} |f_n(x) - f(x)|dx + \int_{|x|>1} |x|^2|f_n(x) - f(x)|dx.$$

By means of the Schwarz inequality this sum can be trivially estimated from above by

$$\|f_n(x) - f(x)\|_{L^2(\mathbb{R}^d)}\sqrt{|B^d|} + \|x^2 f_n(x) - x^2 f(x)\|_{L^1(\mathbb{R}^d)} \to 0, \quad n \to \infty$$

due to our assumptions. ■

References

1. C. Amrouche, F. Bonzom, Mixed exterior Laplace's problem. J. Math. Anal. Appl. **338**(1), 124–140 (2008)
2. C. Amrouche, V. Girault, J. Giroire, Dirichlet and Neumann exterior problems for the *n*-dimensional Laplace operator: an approach in weighted Sobolev spaces. J. Math. Pures Appl. (9) **76**(1), 55–81 (1997)
3. N. Benkirane, Propriétés d'indice en théorie höldérienne pour des opérateurs elliptiques dans R^n. CRAS, Série I **307**(11), 577–580 (1988)
4. P. Bolley, T.L. Pham, Propriétés d'indice en théorie höldérienne pour des opérateurs différentiels elliptiques dans R^n. J. Math. Pures Appl. (9) **72**(1), 105–119 (1993)
5. P. Bolley, T.L. Pham, Propriété d'indice en théorie Höldérienne pour le problème extérieur de Dirichlet. Commun. Partial Differ. Equ. **26**(1–2), 315–334 (2001)
6. H.L. Cycon, R.G. Froese, W. Kirsch, B. Simon, *Schrödinger Operators with Application to Quantum Mechanics and Global Geometry* (Springer, Berlin, 1987), 319 pp.
7. A. Ducrot, M. Marion, V. Volpert, Systémes de réaction-diffusion sans propriété de Fredholm. CRAS **340**(9), 659–664 (2005)
8. A. Ducrot, M. Marion, V. Volpert, Reaction-diffusion problems with non-Fredholm operators. Adv. Differ. Equ. **13**(11–12), 1151–1192 (2008)
9. M. Goldberg, W. Schlag, A limiting absorption principle for the three-dimensional Schrödinger equation with L^p potentials. Int. Math. Res. Not. **75**, 4049–4071 (2004)
10. B.L.G. Jonsson, M. Merkli, I.M. Sigal, F. Ting, *Applied Analysis*, In preparation.
11. T. Kato, Wave operators and similarity for some non-selfadjoint operators. Math. Ann. **162**, 258–279 (1965/1966)
12. E.H. Lieb, M. Loss, *Analysis*. Graduate Studies in Mathematics, vol. 14 (American Mathematical Society, Providence, 1997), 278 pp.
13. R. Metzler, J. Klafter, The random walk's guide to anomalous diffusion: a fractional dynamics approach. Phys. Rep. **339**, 1–77 (2000)
14. M. Reed, B. Simon, *Methods of Modern Mathematical Physics. III. Scattering Theory* (Academic Press, New York, 1979), 463 pp.

15. I. Rodnianski, W. Schlag, Time decay for solutions of Schrödinger equations with rough and time-dependent potentials. Invent. Math. **155**(3), 451–513 (2004)
16. V. Volpert, *Elliptic Partial Differential Equations. Volume 1: Fredholm Theory of Elliptic Problems in Unbounded Domains* (Birkhäuser, Basel, 2011), 639 pp.
17. V. Volpert, V. Vougalter, On the solvability conditions for a linearized Cahn-Hilliard equation. Rend. Istit. Mat. Univ. Trieste **43**, 1–9 (2011)
18. V. Volpert, V. Vougalter, Solvability in the sense of sequences to some non-Fredholm operators. Electron. J. Differ. Equ. **2013**(160), 1–16 (2013)
19. V. Volpert, B. Kazmierczak, M. Massot, Z.Peradzynski, Solvability conditions for elliptic problems with non-Fredholm operators. Appl. Math. **29**(2), 219–238 (2002)
20. V. Vougalter, V. Volpert, On the solvability conditions for some non Fredholm operators. Int. J. Pure Appl. Math. **60**(2), 169–191 (2010)
21. V. Vougalter, V. Volpert, Solvability relations for some non Fredholm operators. Int. Electron. J. Pure Appl. Math. **2**(1), 75–83 (2010)
22. V. Vougalter, V. Volpert, Solvability conditions for some non-Fredholm operators. Proc. Edinb. Math. Soc. (2) **54**(1), 249–271 (2011)
23. V. Vougalter, V. Volpert, On the solvability conditions for the diffusion equation with convection terms. Commun. Pure Appl. Anal. **11**(1), 365–373 (2012)
24. V. Vougalter, V. Volpert, Solvability conditions for a linearized Cahn-Hilliard equation of sixth order. Math. Model. Nat. Phenom. **7**(2), 146–154 (2012)
25. V. Vougalter, V. Volpert, Solvability conditions for some linear and nonlinear non-Fredholm elliptic problems. Anal. Math. Phys. **2**(4), 473–496 (2012)
26. V. Vougalter, V. Volpert, On the solvability in the sense of sequences for some non-Fredholm operators. Dyn. Partial Differ. Equ. **11**(2), 109–124 (2014)
27. V. Vougalter, V. Volpert, Existence of stationary solutions for some integro-differential equations with anomalous diffusion. J. Pseudo-Differ. Oper. Appl. **6**(4), 487–501 (2015)
28. V. Vougalter, V. Volpert, On the solvability of some systems of integro-differential equations with anomalous diffusion, in *Pseudo-Differential Operators: Groups, Geometry and Applications*. Trends in Mathematics (Birkhäuser/Springer, Cham, 2017), pp. 161–179
29. V. Vougalter, V. Volpert, Solvability in the sense of sequences for some non Fredholm operators related to the superdiffusion. J. Pseudo-Differ. Oper. Appl. **9**(1), 25–46 (2018)

MIX
Papier aus verantwortungsvollen Quellen
Paper from responsible sources
FSC® C105338

If you have any concerns about our products,
you can contact us on
ProductSafety@springernature.com

In case Publisher is established outside the EU,
the EU authorized representative is:
Springer Nature Customer Service Center GmbH
Europaplatz 3, 69115 Heidelberg, Germany

Printed by Libri Plureos GmbH
in Hamburg, Germany